AF333876

ALS

Advances in Life Sciences

The Peptidergic Neuron

Edited by
B. Krisch
R. Mentlein

Birkhäuser Verlag
Basel · Boston · Berlin

Editors

Prof. Dr. Brigitte Krisch
Prof. Dr. Rolf Mentlein
Anatomisches Institut der Universität zu Kiel
Olshausenstrasse 40
D-24098 Kiel
Germany

A CIP catalogue record for this book is available from the Library of Congress, Washington D.C., USA

Deutsche Bibliothek Cataloging-in-Publication Data
The peptidergic neuron / ed. by B. Krisch : R. Mentlein. -
Basel ; Boston ; Berlin : Birkhäuser, 1996
   (Advances in life sciences)
   ISBN 3-7643-5314-7 (Basel ...)
   ISBN 0-8176-5314-7 (Boston)
NE: Krisch, Brigitte [Hrsg.]

© 1996 Birkhäuser Verlag, P.O. Box 133, CH-4010 Basel, Switzerland
Camera-ready copy prepared by the editors
Printed on acid-free paper produced from chlorine-free pulp. TCF ∞
Printed in Germany
ISBN 3-7643-5314-7
ISBN 0-8176-5314-7

9 8 7 6 5 4 3 2 1

Contents

III. Neuropeptide receptors

IV. Comparative aspects

This volume is dedicated to Berta Scharrer

Berta Scharrer

In memoriam Berta Scharrer

A. Oksche

Justus Liebig University of Giessen, Department of Anatomy and Cell Biology, Aulweg 123, D-35392 Giessen, Germany

We are very saddened by the loss of Professor Berta Scharrer, a founder of neuroendocrinology and the spiritual leader of our group. Berta Scharrer died in New York on the 23rd of July 1995 at the age of 88. There are very few scientists whose discoveries have marked the advent of a new discipline; Berta Scharrer was one of those pioneers. We have always been impressed by the broad spectrum of her knowledge and her keen judgment of the quality of scientific communications. We recognize that this judgment reflects an unusual perception into the basis of natural science and is evidence for an exceptional personality.

The scientific career of Berta Scharrer has been crowned with great success. The concept of neurosecretion developed by Ernst and Berta Scharrer between 1928 and 1937, and later extended by Wolfgang Bargmann, forms the foundation for contemporary neuroendocrinology, particularly the concept of peptidergic neurons in vertebrates and invertebrates. Today, we know that secretory nerve cells are widely distributed over the central and the peripheral nervous systems, including the autonomic nervous system. The neuropeptides, whose chemical nature has largely been deciphered, serve in both vertebrates and invertebrates for the maintenance of the organism and for the preservation of the species.

Berta Vogel Scharrer was born on the 1st of December, 1904 in Munich. Her early scientific work in the laboratory of Karl von Frisch, Nobel Laureate 1973, was concerned with chemoreception in bees. She received her Ph. D. from the University of Munich in

1930. Soon thereafter she became research associate in the Research Institute for Psychiatry in Munich (1931 - 1934) where her scientific activities were mainly focussed on the microbiology of spirochetes and their occurence in the central nervous system.

In 1934, Berta Vogel married Ernst Scharrer, a disciple of Karl von Frisch and her fellow in the institute. In 1928, Ernst Scharrer had discovered colloid-like inclusions in the magnocellular preoptic nucleus of a teleost, the European minnow. He interpreted this phenomenon as a manifestation of endocrine activity, which was further supported by the rich vascularization of this area and an obvious correlation with the hypophysis. After their marriage, Ernst and Berta Scharrer became an congenial team, dividing the animal kingdom such that Ernst studied vertebrates and Berta invertebrates.

During 1934, Ernst and Berta Scharrer moved from Munich to Frankfurt (Main) where they became associated with the Neurological Institute (Edinger Institute). For the development of the concept of neurosecretion the years in Frankfurt (1934 - 1937) were of extraordinary importance.

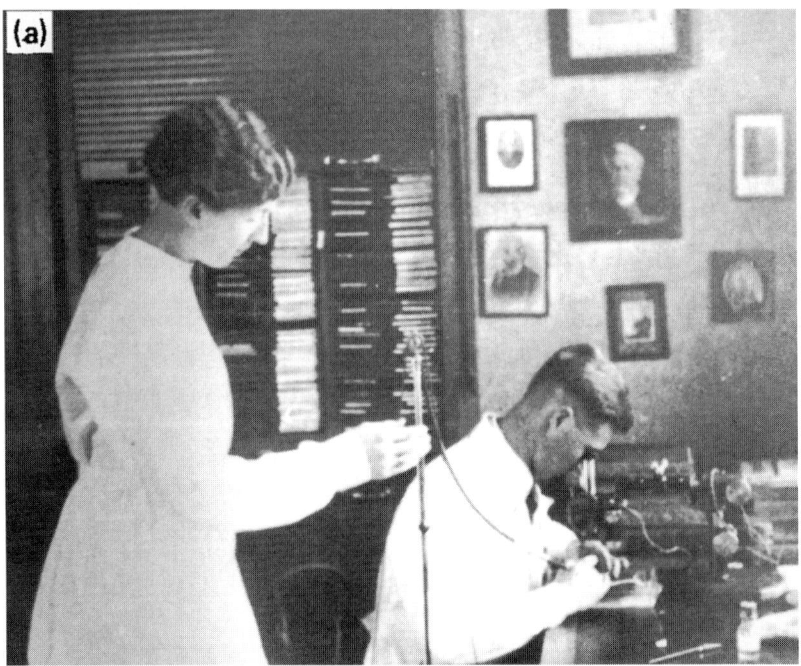

Berta and Ernst Scharrer in their laboratory at the Edinger Institute in Frankfurt / Main.

In 1937, Berta Scharrer published a representive paper on neurosecretion in the central nervous system of invertebrates. This paper was followed by a comprehensive review by Ernst and Berta Scharrer on gland-like nerve cells and neurosecretory organs both in vertebrates and invertebrates. These two papers (Naturwissenschaften; Biological Review) can be regarded as the conceptual basis of neuroendocrinology.

After Ernst and Berta Scharrer had left Germany in 1937, Chicago, New York, Cleveland and Denver became their further academic stations. Following a year (1937 - 1938) as Research Associate in the Department of Anatomy at the University of Chicago and two years (1938 - 1940) at the Rockefeller Institute in New York, Berta Scharrer became instructor (1940 - 1946) in the Department of Anatomy at the Western Reserve University in Cleveland, Ohio. In 1941, she published a paper on the neurosecretory cells in the central nervous sysem of cockroaches. Since that time, *Leucophaea madrae* became her favorite experimental animal. Berta Scharrer emphasized that certain principles found in an insect may deliver the key for understanding basic phenomena in mammals, including man.

The idea of a secretory activity of nerve cells was so revolutionary, that it originally met with considerable opposition. The breakthrough for Ernst and Berta Scharrer occured during their Denver period (1947 - 1955), in continual exchange of thoughts with Wolfgang Bargmann of Kiel, a close friend since the early Frankfurt days. In 1954, Ernst and Berta Scharrer published several important surveys on neurosecretion and hormones produced by neurosecretory cells; these reviews belong to the classics in neuroendocrinology.

In 1955, Ernst and Berta Scharrer founded the Department of Anatomy at the Albert Einstein College of Medicine in New York. In 1963, they published a comprehensive monograph on „Neuroendocrinology", a classic document of the state-of-the-art in the field. After the sudden death of Ernst Scharrer in 1965, Berta Scharrer continued to interpret the role of neurosecretory cells in the central nervous system. She extended the concept of neurosecretion to an overall concept of peptidergic neurons. Furthermore, Berta Scharrer was one of the early defenders of the unifying concept of a diffuse neuroendocrine system. It is amazing to see to what extent new molecular evidence has been integrated into the original framework of the concept of neurosecretion.

During the last decade of her life Berta Scharrer was fascinated by the interrelationships between the neuroendocrine system and the immune apparatus. Again, the comparative approach proved to be very fruitful; remarkable parallelisms between invertebrates and vertebrates could be established. Berta Scharrer continued her research work long past retirement age; her last paper (with Stefano and collaborators) dealing with neuroimmunological questions was submitted for publication two days before her death.

For 22 years (1955 - 1977) Berta Scharrer was Professor of Anatomy at the Albert Einstein College of Medicine in New York. For two periods she was the acting head of the department. Her former students remember her as a highly devoted and respected academic teacher. In 1978, she became Distinguished Professor Emerita of Anatomy and Structural Biology and of Neuroscience. The extensive editorial work of Berta Scharrer has been outstanding, and her dedication has been greatly admired by the scientific community all around the world.

The international recognition for the scientific achievements of Berta Scharrer has been mirrored by the honours she has received. Eleven universities have bestowed honory doctores upon her, including Harvard University. Among her numerous medals and prizes only a few will be mentioned here: the Kraepelin Gold Medal of the Max Planck Society (1978), the F.C. Koch Award of the Endocrine Society (1980), the Henry Gray Award of the American Association of Anatomists (1982), the Schleiden Medal of the German Academy of Sciences Leopoldina (1983), and the National Medal of Science of the United States of America (1985). Berta Scharrer was a member of the National Academy of Sciences in the USA, the American Academy of Arts and Sciences, the German Academy of Sciences Leopoldina, and several other European academies and scientific societies.

Berta Scharrer was admired for her integrity, modesty, warmth and her willingness to help. Her devotion to humanitarian goals has earned our special respect and esteem. We are indeed most privileged to have known her. She will long be remembered by her friends and colleagues.

PREFACE

In the twenties and thirties of our century, Berta and Ernst Scharrer discovered morphological evidence that the hypothalamus exhibits secretory and hormonal functions which have been called neurosecretion. At first fiercely debated, the phenomenon was demonstrated beyond any doubt by Wolfgang Bargmann (1949) who applied Gomori's staining method to the hyopthalamo-neurohyopophyseal system. A few decades later, a great number of neuropeptides were identified as neurosecretory signal substances. In the sixties, the concept of neurosecretion was extended and became an overall concept of peptidergic neurons. Neuropeptides rank among the phylogenetically oldest interneuronal signal substances, mostly coexisting with other transmitters. From the viewpoint of molecular biology and physiology their biosynthesis, transport within and release from the neuron is quite different from other transmitters (biogenic amines, acetylcholine, amino acids). The neurohormonal function is not unique to peptididergic neurons but rather depends on the position of the neuron in relation to the particular midsagittal organization of the brain. Neuropeptides - including the "classical" hypothalamic and hypophyseal hormones - were not only detected in other parts of the brain and in peripheral organs; they are not only synthesized by neurons, but also by several types of non-neuronal cells. Moreover, in recent years, neuropeptides have been identified as intercellular messengers between neurons and the immune system. Thus, it appears that neuropeptides constitute one of the most important intercellular signal substances acting not only as interneuronal messenger molecules, but also as hormones, paracrine and autocrine signal substances in various organs. Neuropeptide research is therefore still one of the most rapidly expanding fields in biomedicine.

Since 1953 Symposia on Neurosecretion have been taking place every four years in different parts of the world (see list on page XVII) gathering scientists from various fields of research. The limited number of participants is the tremendous advantage of these meetings, allowing stimulating contacts and discussions among scientists who otherwise, due to their diverse background would have little chance to meet one another. The proceedings of the Symposia of Neurosecretion always mirror the development of techniques and concepts and can be regarded as vivid and condensed chapters of scientific history.

This volume contains contributions from the "12th International Symposium on Neurosecretion - The Peptidergic Neuron" which was held in Kiel/Germany from September 20-22, 1995. The invited lectures about the molecular aspects of processing, release and degradation of neuropeptides, receptors and signal transduction, comparative aspects of neurosecretion, behavioural and immunoregulatory effects of neuropeptides and their involvement in pathology review the state of the art in these fields of topical interest; several selected contributions will illustrate these topics from special viewpoints.

The symposium and the proceedings were generously supported by the Deutsche Forschungsgemeinschaft (Bonn), by the state government of Schleswig- Holstein (Kiel) and the Christian-Albrechts-Universität zu Kiel, by a colleague who wishes to remain anonymous and the following companies (in alphabetical order): BACHEM Biochemica (Heidelberg), Biotrend Chemikalien (Cologne), Brunswiker Universitätsbuchhandlung (Kiel), Dako Immunochemicals (Hamburg), Ferring (Malmö/Kiel), Hugo Hamann (Kiel), E. Merck (Darmstadt), Nunc (Wiesbaden), Saxon Biochemicals (Hannover), Serva (Heidelberg) and Zeiss (Oberkochen/Hamburg).

The editors are grateful to the members of the scientific commitee for reviewing the manuscripts, to the students Ulf Prien and Philipp Roeben for computer and editorial assistance and numerous colleagues of the Department of Anatomy, University of Kiel, for their tireless help in the preparation and realization of the symposium.

Brigitte Krisch

Rolf Mentlein

INTERNATIONAL SYMPOSIA ON NEUROSECRETION

1^{rst} International Symposium on Neurosecretion. Naples 1953
Publ. Staz. Zool. Napoli 24, Suppl. 1-98 (1954)
2nd International Symposium on Neurosecretion. Lund 1957
II. International Symposium on Neurosecretion. II. Internationales Symposium über Neurosekretion (W. Bargmann, B. Hanström, B. und E. Scharrer, Eds.), Springer, Berlin Göttingen Heidelberg 1958
3rd International Symposium on Neurosecretion. Bristol 1961
Neurosecretion (H. Heller, R.B. Clark, Eds.), Mem. Soc. Endocrinol. 12, Academic Press London New York 1962
4th International Symposium on Neurosecretion. Strassbourg 1966
Neurosecretion (F. Stutinsky, Ed.), Springer, Berlin Göttingen   Heidelberg 1967
5th International Symposium on Neurosecretion. Kiel 1969
Aspects of Neuroendocrinology (W. Bargmann, B. Scharrer, Eds.), Springer, Berlin Göttingen Heidelberg 1970
6th International Symposium on Neurosecretion. London 1973
Neurosecretion - The Final Neuroendocrine Pathway (F. Knowles, L. Vollrath, Eds.) Springer, Berlin Göttingen Heidelberg 1974
7th International Symposium on Neurosecretion. Leningrad 1976
8th International Symposium on Neurosecretion. Friday Harbor 1980
Neurosecretion: Molecules, Cells, Systems (D.S. Farner, K. Lederis, Eds.), Plenum Press New York 1981
9th International Symposium on Neurosecretion. Tokyo, Susono-shi 1984
Neurosecretion and the Biology of Neuropeptides (H. Kobayashi, H.A. Bern, A. Urano, Eds.), Japan Scientific Societies Press, Tokyo, and Springer, Berlin Heidelberg New York Tokyo 1985
10th International Symposium on Neurosecretion. Bristol 1987
Neurosecretion : Cellular Aspects of the Production and Release of Neuropeptides (B.T. Pickering, J.B.Wakerley, A.J.S. Summerlee, Eds.), Plenum Press, New York London 1988
11th International Symposium on Neurosecretion. Amsterdam 1991
The Peptidergic Neuron, Progr. Brain Res. 92 , 1992
12th International Symposium on Neurosecretion. Kiel 1995
Birkhäuser Verlag, Basel (1996)

CORRESPONDING AUTHORS

Dr. Bridget Baker
School Of Biology and Biochemistry
Bath University
Bath Avon BA2 7AY
U.K.

Prof. Dr. Karl Bauer
Abt. Experimentelle Endokrinologie
Max-Planck-Institut
Postfach 630 309, Feodor Lynen-Str. 7
30625 Hannover
Germany

Prof. Dr. Bela Bohus
Dept. of Animal Physiology
University Groningen
P.O. 14
9750 AA Haren
Netherlands

Dr. Karl H. Bohuslavizki
Klinik für Nuklearmedizin
Christian - Albrechts - Universität Kiel
Arnhold-Heller-Str. 9
24105 Kiel
Germany

Prof. Dr. Michel Chrétien
Clinical Research Institute
University of Montreal
Quebec City, G1V 4G2
Canada

Prof. Dr. Shigeo Daikoku
Tokushima Research Institute
Otsuka Pharmaceutical Co., Ltd.
Tokushima
Japan

Dr. Camilla Davoli
1. St. Medicina Sperimentale C.N.R.
Viale Marx 15 / 43
00137 Roma
Italy

Dr. Jacques H.B. Diederen
Department Of Experimental Zoology
Padualaan 8
3584 CH Utrecht
Netherlands

Dr. Hanne Duve
School Of Biological Sciences
University Of London
Mile End Road
London E1 4NS
U.K.

Prof. Dr. Yasuhisa Endo
Department of Applied Biology
Kyoto Institute of Technology
Matsugasaki, Sakyo
606 Kyoto
Japan

Prof. Dr. T. Fujita
Department of Oral Anatomy
The Nippon Dental University
Harnaura-cho 1-8
951 Niigata
Japan

Dr. D.W. Golding
Department of Marine Sciences
Newcastle University
Ridley Building
Newcastle upon Tyne NE1 7RU
U.K.

Prof. Dr. C. Grimmelikhuijzen
Cellebiologisk Anatomisk Laboratorium
Kopenhagen University
Universitetsparken 15
2100 Kopenhagen
Denmark

Dr. Susumu Hyodo
Department of Biology
University of Tokyo
3-8-1 Komaba, Meguro-ku
Tokyo 153
Japan

Prof. Dr. A.J. Kenny
Le Presbytère
St. Etienne d'Albagan
34390 Hérault
France

Prof. Dr. Kunio Koshimura
First Division, Department of Medicine
Shimane Medical University
89 - 1 Emya-cho
693 Izumo
Japan

Prof. Dr. Brigitte Krisch
Anatomisches Institut
Christian - Albrechts - Universität Kiel
Otto-Hahn-Platz 8
24118 Kiel
Germany

Dr. Yoke Peng Loh
Bldg 49, Rm 5A38
NIH, 9000 Rockville Pike
Bethesda, MD 20892
USA

Prof. Dr. J. Meldolesi
Dept. Biol. & Technol. Research
Hospitale San Raffaele
Via Olgettina
Milano 20132
Italy

Prof. Dr. Rolf Mentlein
Anatomisches Institut
Christian - Albrechts - Universität Kiel
Otto-Hahn-Platz 8
24118 Kiel
Germany

Ralf Nehring
Institut für Zellbiochemie und klinische
Neurobiologie
Universitäts-Klinikum Eppendorf
Martinistr. 52
20246 Hamburg
Germany

Prof. Dr. Dr. h.c. A. Oksche
Institut für Anatomie und Zytobiologie
Justus-Liebig-Universität
Aulweg 123
35385 Gießen
Germany

Dr. James Olcese
Institute for Hormone Research
Grandweg 64
22529 Hamburg
Germany

Prof. Dr. Jonathan Pevsner
Neurology Department
The Kennedy Krieger Institute
707 N. Broadway
Baltimore, MD 21224
USA

Prof. Dr. H.L. Fehm
Klinische Forschergruppe
Universität Lübeck
Haus 23 a, Ratzeburger Allee 160
23538 Lübeck
Germany

Dr. Ada Rafaeli
The Volcani Center, Aro
Department of Stored Product
P.O.Box 6, Bet Dagan
50250
Israel

Prof. Dr. Terry Reisine
Department of Pharmacology
University of Pennsylvania
36th and Hamilton Walk
Philadelphia, PA 19104
USA

XXII

Prof. Dr. D. Richter
Institut für Zellbiochemie
Universität Hamburg
Martinistraße 52
20246 Hamburg
Germany

Prof. Dr. Dr. h.c. Y. Sano
Sakyoku
Jyodji
Kamibabacho 67
606 Kyoto
Japan

Prof. Dr. Helmut Schwarzberg
Abteilung Neurophysiologie (Haus 24)
Otto-von-Guericke-Universität
Leipziger Str. 44
239120 Magdeburg
Germany

Dr. Gisela Segond von Banchet
Physiologisches Institut
Universität Würzburg
Röntgenring 9
97070 Würzburg
Germany

Prof. Dr. H. Imboden
Department of Neurobiology
University of Berne
Erlachstr. 9a
3012 Bern
Switzerland

Prof. Dr. G.B. Stefano
Dept. Biological Sciences
SUNY / Old Westbury
Science Building, Rm S231, P.O.Box 210
Old Westbury, New York
USA

Dr. Masakazu Suzuki
School Of Biology
Bath University
Claverton Down
Bath Avon BAZ 7AY
U.K.

Prof. Dr. D.F. Swaab
Netherlands Institute for Brain Research
Meibergdreef 33
1105 AZ Amsterdam Zuidost
Netherlands

Prof. Dr. Alan Thorpe
School of Biological Sciences
University of London
Mile End Road
London E1 4NS
U.K.

Prof. Dr. J.W. Truman
Department of Zoology
University of Washington
NJ-15
Seattle, WA 98195
USA

Dr. Andreas Turzynski
Institut für Pathologie der Charité
Humboldt Universität
Schumannstraße 20/21
10177 Berlin
Germany

Prof. Dr. A. Urano
Division of Biological Sciences
Hokkaido University
Sapporo, Hokkaido 060
Japan

Dr. Peter Verhaert
Zoological Institute
Naansestraat 59
3000 Leuven
Belgium

Dr. Rainer Wegerhoff
Biochemisches Institut
Christian - Albrechts - Universität Kiel
Olshausenstr. 40
24098 Kiel
Germany

Dr. Heike Wolf
Klinik für Nuklearmedizin
Christian - Albrechts - Universität Kiel
Arnold-Heller-Str. 9
24105 Kiel
Germany

I. General considerations on peptidergic neurons

The Peptidergic Neuron
B. Krisch and R. Mentlein (eds)
© 1996 Birkhäuser Verlag Basel/Switzerland

Retrospective and prospectives for research on neurosecretion

Y. Sano

*Department of Anatomy and Neurobiology, Kyoto Prefectural University of Medicine
Kawaramachi-Hirokoji, Kamigyo-ku, Kyoto 602, Japan*

The 12th International Symposium on Neurosecretion (Sept. 20-22, 1995) was held at the University of Kiel, where Wolfgang Bargmann (1906-1978) chaired the Department of Anatomy. Along with Ernst Scharrer and Berta Scharrer, Bargmann demonstrated the existence of hormone-producing neurons and established the study of neurosecretion. This monograph presents the proceedings of the symposium.

In 1928, Ernst Scharrer (1905-1965) discovered neurons with the cytological correlates of secretory activities in the preoptic nucleus of *Phoxinus laevis L.* From 1931, at the Edinger Institute of Neurology in Frankfurt, he vigourously pursued his research and established the existence of similar secretory neurons in a variety of vertebrates. In 1933, Bargmann became a research assistant in the Department of Anatomy in Frankfurt/Main, and befriended Scharrer; thus he came to see the exquisite histological specimens of the neurons containing stainable granules prepared by Scharrer. Shortly afterwards, in 1934, Ernst and Berta were married; they moved to the United States in 1937.

Soon after the end of the Second World War, in February 1946, Bargmann was invited to chair the Department of Anatomy in Kiel. At that time, most of Kiel had been devasted by the war and the town was in ruins. The first letter Bargmann received from abroad after the war was from Scharrer, who wrote that, while his work on neurosecretion had shown progress, the relationship between the secretory activity of the hypothalamic cell bodies and the posterior lobe of the pituitary gland, where the processes of these cells terminate, was still not clear. This led Bargmann, who was interested in research on the hypothalamo-hypophyseal system, to think of studying neurosecretion himself.

In the spring of 1948, when research again became possible, two excellent students, Werner Creutzfeldt (later Professor of Medicine at Göttingen University) and Walter Hild (later Professor of Anatomy at University of Texas at Galveston) enrolled in Bargmann's department for doctoral study. Bargmann assigned Creutzfeldt to confirm the results of a study on the degeneration of pancreatic islets in alloxan diabetes reported by Dunn and his colleagues of England during the war. Hild was assigned to study the neurosecretory cells of the hypothalamus in lower vertebrates.

Bargmann suggested to Creutzfeldt that he use chrome hematoxylin-phloxine staining, introduced by Gomori in 1941, to identify the α- and β-cells of the pancreatic islet. One day, Bargmann conceived the using of the Gomori method to stain sections of the pituitary for the student's histological drill. His technician, K. Jacob, used a specimen from the hypophysis of a dog, which, by chance, also included the hypothalamus. When Jacob saw the entire posterior lobe of the hypophysis stained intensely blue, as if dipped in ink, she thought that she must have made a mistake in the procedure. However, Bargmann instinctively realized that the continuous, blue-stained tract running from the supraoptic nucleus to the posterior lobe of the pituitary which had troubled Scharrer. Bargmann told Hild to recommense his research using the Gomori stain and wrote a paper titled „Über die neurosekretorische Verknüpfung von Hypothalamus und Hypophyse" in *Zeitschrift für Zellforschung und mikroskopische Anatomie*, of which journal he was an editor.

Bargmann, with his knowledge of the endocrine system, proposed that the hematoxylin-stainable material was a carrier substance for a hormone which was produced in the cell bodies, transported intra-axonally, and released into the blood stream from the nerve terminals. P. Weiss and H.B. Hiscoe had published their historical paper on axonal transport in 1948, but their work had yet to gain wide acceptance and Bargmann was not acquainted with it at the time. The neuroscience community had not at this stage accepted the idea that neurons could produce hormones, and Bargmann's hypothesis was rejected by most as being too radical. The main objections were: (1) why should the hormone of the posterior hypophysis be produced in the hypothalamus, at a distance from the site of its release; and (2) why should the axon, whose role is impulse conduction, also transport materials?

Fortunately, neurosecretory materials in insects and crustaceans have similar staining affinities to those of the mammalian hypothalamus. Many workers on invertebrates embraced Bargmann's hypothesis and developed their research without hindrance from classical neuroscientists. In 1950, Bargmann was invited to the United States and was reunited with the Scharrers. He and Ernst Scharrer published a paper together (Bargmann and Scharrer, 1951) and made plans for an International Symposium on Neurosecretion. Their plan was realized in 1953 when the first Symposium was held in the Stazione Zoologica in Naples with the help of Drs. Reinhard and Peter Dohrn. Since then, the Symposium has been held approximately every 4 years in different parts of the world, and this year's symposium marks the twelvth of the series.

Neurosecretion was initially studied as a phenomenon only of specialized, hormone-producing neurons. However, other studies have revealed that virtually all neurons produce and release biologically active substances. Hormones had been considered unique because they act at a distance from their site of release, being carried to their targets via the blood stream. However, hormones, neurotransmitters and immunological mediators are all intercellular messenger substances and may have similar mechanisms of action.

Some researchers use the term neurosecretion to describe the mechanism of discharge of synaptic vesicles. We have learned that neuropeptides are co-stored with other neurotransmitters in the same axon terminals. In this way, neuropeptides - first identified as the products of neurosecretory cells - have been integrated into the spectrum of substances involved in interneuronal message transfer and their role as neurohormones seems to be a special, phylogenetically old property related to the particular bauplan of the brain. Thus, along with the advance of biomedical research, the nature of the Neurosecretion Symposium has changed greatly over the years.

Molecular biology has revealed many details of the synthesis and release of neuropeptides - however, shifting the horizon means opening a new one. So the machinery involved in the event of exocytosis, the molecular cascade initiated by the binding of a neuropeptide to one of its numerous receptors, and the regulation of these events in health and disease - these are the challenges for today's researchers on neuropeptides.

In this Symposium, the activity of peptide - secreting neurons have been discussed from three viewpoints: 1) biosynthesis of neuropeptides, 2) comparative aspects and targets for neuropeptides, and 3) integrative processes and neuropeptide receptors. In each neuron, not only neurotransmitters but various other substances, such as the proteins which constitute the membranes and cytoskeleton, various adhesion molecules, neurotrophic substances and others, are also produced. Neurons support their own life, exchange information with each other and regulate the whole body with its miriads of cells, throughout its natural life. In neuroscience research, which studies the function of this system, the role of the Symposium is not a small one. I sincerely hope that the seed planted by Ernst Scharrer, Berta Scharrer and Wolfgang Bargmann will continue its growth as a large tree, extending ist branches in all areas of neuroscience.

References

Bargmann, W. (1949) Über die neurosekretorische Verknüpfung von Hypothalamus und Neurohypophyse. *Z. Zellforsch.* 34: 610-634.
Bargmann, W. and Scharrer, E.A. (1951) The site of origin of the hormones of the posterior pituitary. *Am. Scientist* 39: 255-259.

The Peptidergic Neuron
B. Krisch and R. Mentlein (eds)
© 1996 Birkhäuser Verlag Basel/Switzerland

Patterns of peptide discharge - implications for Dale's principle

D.W. Golding

Biomedical EM Unit and Department of Marine Sciences and Coastal Management, University of Newcastle, Newcastle upon Tyne, NE1 7RU, UK

Summary. According to Dale's principle, a neuron uses the same transmitter(s) at all its branches, reflecting its metabolic unity. However, there are reasons to doubt the universal validity of this hypothesis. Peptide secretion involves a fundamentally different metabolic pathway from that followed by classical transmitters. Furthermore, peptides are stored in secretory granules and released by exocytosis from mainly nonsynaptic regions of the plasmalemma, whereas transmitters are typically discharged at specialised synaptic thickenings. Thus different regions of the neuronal surface are responsible for the discharge of different classes of chemical mediators - a finding quite at variance with Dale's principle. The principle retains valuable predictive powers if transmitters and peptides are considered in isolation from each other, although it will not be vindicated invariably even on this basis.

Introduction

Dale's principle was formulated by J.C. Eccles (Eccles et al., 1954) on the basis of a suggestion made by H.H. Dale (1935), that the identity of a chemical transmitter released by one branch of a neuron "would furnish a hint" as to the identity of the transmitter employed by other branches. The principle is commonly assumed to deny the possibility of the production of more than one transmitter by a neuron, but this is mistaken. Dale's words should more reasonably be taken as reflecting his ignorance of the phenomenon of contransmission, rather than any opposition to the idea - and Eccles accordingly interpreted the principle in terms of the operation of "the same transmission mechanism", i.e. the use of "the same transmitter substance or substances", by all the branches of a nerve cell, stemming from its metabolic unity (see reviews by Eccles, 1986; Lichtigfeld and Gillman, 1991).

The predictions of the principle have, of course, been confirmed with respect to many neuronal systems, the release of acetylcholine by vertebrate motoneurons at both their peripheral and central terminals being the classic example. Similarly, the peripheral (neurohaemal) and central (synaptic) terminals of endocrine neurons often show a remarkable identity of ultrastructure (review by Golding et al., 1988). However, although the principle is not inconsistent with cotransmission *per se*, the ability of a neuron to vary greatly the proportions in which it releases its secretory products, whether in relation to impulse frequency (e.g. see Agoston and Lisziewicz, 1989) or other physiological modulation (Swanson, 1991) stretches the principle to (and perhaps beyond) its limits.

Results and discussion

There is no *prima facie* reason to expect that Dale's principle will be upheld routinely. Epithelial cells have apical and basolateral domains and direct constitutive vesicular components of diverse chemical character to each. It seems inherently unlikely that neurons, with domains (somal/dendritic and axonal/terminal, respectively) corresponding to the above (Dotti et al., 1991) will not exploit this capability in regulated secretion. Indeed, there is evidence that they do so with respect to both peptides and non-peptide transmitters. For example, in the bag cells of *Aplysia*, different peptides processed from the same prohormone are segregated in the Golgi apparatus, packaged into different granules and released from different sets of terminals (Sossin et al., 1990). Nigrostriatal neurons apparently use different transmitters at polarized and symmetrical synapses, respectively (Hattori et al., 1991).

The discovery of the virtual ubiquity of peptide secretion in the nervous system has important implications for any assessment of Dale's principle. It has revealed that the secretory biology of the neuron is typically characterised by a fundamental metabolic dichotomy, rather than by unity, since peptide synthesis involves a different pathway from that followed by classical transmitters. Transmitters such as acetylcholine are fairly simple substances which are synthesised by enzymic actions. These processes take place mainly in the cytosol of the terminals, following which the transmitters are sequestered within synaptic vesicles by the action of proton-driven pumps. In contrast, peptides are elaborated in the cell body by ribosomes, passed into the lumen of the rough endoplasmatic reticulum and packaged into secretory granules (large dense core vesicles) by the Golgi apparatus (reviewed by Bean et al., 1994). Initially, large protein precursors are formed, and these are then cleaved to form smaller peptides. Furthermore, whereas synaptic vesicles are recycled locally after discharge, taking up fresh stores of transmitter from the cytosol, secretory granules cannot be replenished in this way and their membranes are probably degraded.

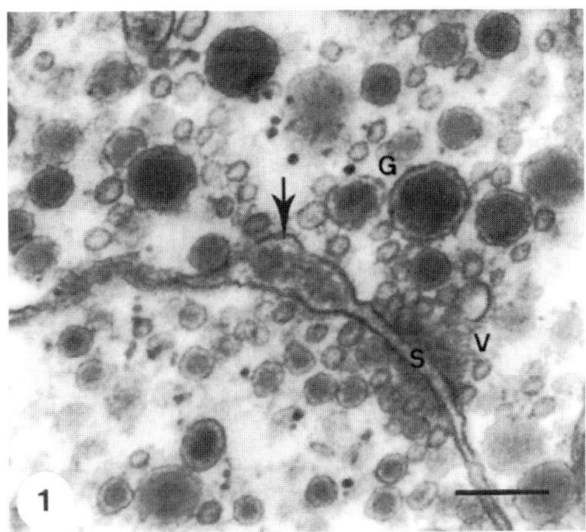

Fig. 1. Cerebral ganglion of *Lumbricus terrestris*, showing synaptic contact (S) with associated vesicle cluster (V) and nonsynaptic exocytosis (arrow) of secretory granules (G). x 52,000; bar = 200 nm.

Ultrastructural studies have shown that the sites of release of transmitters and peptides are often morphologically distinct. Transmitters are commonly discharged from synaptic vesicles by exocytosis from within highly localized 'active zones' marked by presynaptic thickenings or bars. In contrast, we observed that secretory granule exocytosis (Figs. 1 and 2) is mainly associated with the more expansive, morphologically undifferentiated regions of synaptic terminals (Golding and May, 1982) and this phenomenon of 'nonsynaptic' release has now been detected in a wide variety of systems, including the archetypal cholinergic terminals innervating the adrenal chromaffin glands of vertebrates (reviewed by Golding, 1994).

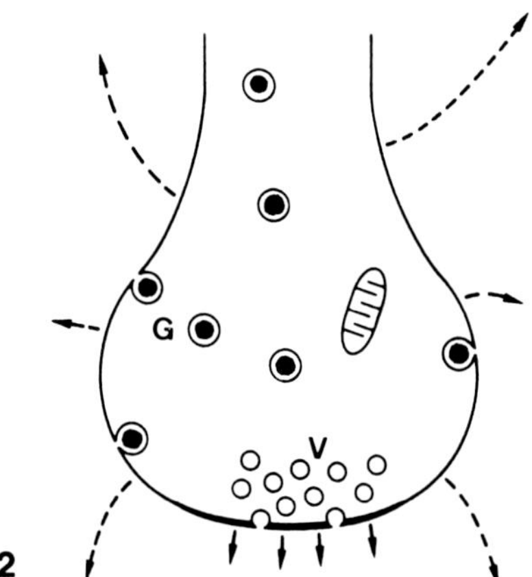

Fig. 2. Pattern of secretory discharge by a typical synaptic terminal, with a combination of synaptic exocytosis by vesicles (V) and nonsynaptic exocytosis by granules (G).

In some systems, e.g. the adrenal chromaffin gland of the goldfish, granule exocytosis is quite closely associated with synaptic sites in that it is largely restricted to areas of the plasmalemma lying adjacent to the postsynaptic cells (i.e. the 'parasynaptic' pattern). In other systems, e.g. the secretomotor synapses of the locust corpus cardiacum, granule discharge is more widely spread across the terminal surface. A similar lack of apparent targeting is shown by the neurosecretory fibres of the neural lobe of the rat pituitary, where the density of granule exocytosis from within the preterminal 'swellings' equals that of the 'endings' which make contact with the vascular lamella (Morris and Pow, 1991). Peptide release has even been encountered (albeit more rarely) in relation to preterminal axons and cell bodies (for references, see Golding, 1994).

More striking still, a cell may apparently bear one type of neurite engaging in discharge of both peptide and nonpeptide products, and another type engaged in only peptide release. For example, both secretory granules and small, clear ('synaptoid') vesicles are present in the neurohypophysis and there is a widening consensus that such vesicles are identical to synaptic vesicles (see reviews by Golding et al., 1988; Thomas-Reetz and De Camilli, 1994) - in the neural lobe, they may store and discharge glutamate (Meeker et al., 1991). In contrast, Pow and Morris (1989) have shown that the hypothalamic dendrites of the neurons release peptides by granule exocytosis, clusters of synaptoid vesicles being apparently absent from these fibres. Peptide discharge by dendrites is also thought to be a feature of dentate granular cells (Drake et al., 1994).

Conclusions

Different regions of the neuronal surface are responsible for the discharge of different classes of neurochemical mediators - a finding quite at variance with Dale's principle. However, the principle will doubtless continue to furnish valuable "hints" (i.e. have significant predictive powers) for transmitter identification.

Acknowledgements
The author gratefully acknowledges the assistance of Mr. R.M. Hewit.

References

Agoston, D.V. and Lisziewicz, J. (1989) Calcium uptake and protein phosphorylation in myenteric neurons, like the release of vasactive intestinal peptide and acetylcholine, are frequency dependent. *J. Neurochem.* 52: 1637-1640.

Bean, A.J., Zhang, X. and Hökfelt, T. (1994) Peptide secretion: what do we know? *FASEB J.* 8: 630-638.

Dale, H.H. (1935) Pharmacology at nerve endings. *Proc. R. Soc. Med.* 28: 319-332.

Dotti, C.G., Parton, R.G. and Simmons, K. (1991) Polarized sorting of glypiated proteins in hippocampal neurons. *Nature (London)* 349: 158-161.

Drake, C.T., Terman, G.W., Simmons M.L., Milner, T.A., Kunkel, D.D., Schwartzkroin, P.A. and Chavkin, C. (1994) Dynorphin opioids present in dentate granule cells may function as retrograde inhibitory neurotransmitters. *J. Neurosci.* 14: 3736-3750.

Eccles, J.C. (1986) Chemical transmission and Dale's principle. *Prog. Brain Res.* 68: 3-13.

Eccles, J.C., Fatt, P. and Koketsu, K. (1954) Cholinergic and inhibitory synapses in a pathway from motor axon collaterals to motoneurons. *J. Physiol.* 126: 524-562.

Golding, D.W. (1994) A pattern confirmed and refined - synaptic, nonsynaptic and parasynaptic exocytosis. *BioEssays* 16: 503-508.

Golding, D.W. and May, B.A. (1982) Duality of secretory inclusions in neurons - ultrastructure of the corresponding sites of release in invertebrate nervous systems. *Acta Zool. (Stockholm)* 63: 229-238.

Golding, D.W., Pow, D.V., Bayraktaroglu, E., May, B.A. and Hewit, R.M. (1988) Emerging identity in cytophysiology between synaptic and neurohaemal terminals. *In:* B.T. Pickering, A.J.S., Summerlee and J.B. Wakerlee (eds.): *Neurosecretion: Cellular Aspects of the Production and Release of Neuropeptides,* Plenum Press, New York, pp. 137-146.

Hattori, T., Takada, M., Moriizumi, T. and Van der Kooy, D. (1991) Single dopaminergic nigrostriatal neurons for two chemically distinct synaptic types: possible transmitter segregation within neurons. *J .Comp. Neurol.* 309: 391-401.

Lichtigfeld, F.J. and Gillman, M.A. (1991) In spite of its validity, has Dale's principle served its purpose - a scientific paradox. *Persp. Biol. Med.* 34: 239-253.

Meeker, R.B., Swanson, D.J., Greenwood, R.S. and Haywood, J.N. (1991) Ultrastructural distribution of glutamate immunoreactivity within neurosecretory endings and pituicytes of the rat neurohypophysis. *Brain Res.* 564: 181-193.

Morris, J.F. and Pow, D.V. (1991) Widespread release of neuropeptides in the central nervous system - quantitation of tannic acid-captured exocytoses. *Anat. Rec.* 231: 437-445.

Pow, D.V. and Morris, J.F. (1989) Dendrites of hypothalamic magnocellular neurons release neurohypophysial peptides by exocytosis. *Neuroscience* 32: 435-439.

Sossin, W.S., Sweet, C.A. and Scheller, R.H. (1990) Dale's hypothesis revisited: different neuropeptides derived from a common prohormone are targeted to different processes. *Proc. Natl. Acad. Sci. USA* 87: 4845-4848.

Swanson, L.W. (1991) Biochemical switching in hypothalamic circuits mediating response to stress. *Prog. Brain. Res.* 87: 181-200.

Thomas-Reetz, A.C. and De Camilli, P. (1994) A role for synaptic vesicles in non-neuronal cells: clues from pancreatic β cells and from chromaffin cells. *FASEB J.* 8: 209-216.

The Peptidergic Neuron
B. Krisch and R. Mentlein (eds)
© 1996 Birkhäuser Verlag Basel/Switzerland

The paraneuron revisited

T. Fujita

Department of Histology, Nippon Dental University School of Dentistry, Hamaura 1-8, Niigata, 951 Japan

Summary. The pancreatic endocrine cells reveal their paraneuronal nature by forming, with intrapancreatic neurons, a neuro-insular complex. Studies on this peculiar structure opened up a new field of neurosecretion in the pancreas. Recent studies on enterochromaffin cells, which constitute one of the most typical categories of paraneurons in the gut, are reviewed with special reference to their involvement in the pathogenesis of diarrhea and to the possibility of their partially exocrine activity. This review further deals with "Segi's cap", an endocrine cell aggregation at the villus tip in the small intestine in the human and bovine fetus. Recent studies suggest that the paraneurons forming Segi's cap migrate, during the perinatal period, into the interstitium, where they develop more neuronal features and are incorporated in the intramural nerve plexus.

Neuro-insular complex and neurosecretion

In 1958, I encountered in microscopic sections of the dog pancreas a peculiar structure called the neuro-insular complex. It consists of neurons (either somata or nerve fibre bundles) and of islet endocrine cells closely juxtaposed with each other. I studied the morphological features and the developmental aspect of the complex in several mammals and sent a paper reporting my findings to the late Professor Bargmann. This paper, suggesting a particular affinity between nervous and endocrine cells, was favoured by him and accepted for publication in the "Zeitschrift für Zellforschung" (Fujita, 1959). As a result of this connection, I was able to pursue my studies for two years in Bargmann's Anatomy Department in Kiel. Here, in the mecca of neurosecretion, I learned how similar in structure and function neurons and endocrine cells truly are.

Studies of mine and of co-workers subsequently extended the investigation into the occurrence of neuron-like endocrine/paracrine cells and the cell-biological features which they share with neurons. Our findings on the gut endocrine cells, which revealed a wide variety of neuronal features, prompted us in 1975 to propose the term and concept of the paraneuron (Fujita, 1976).

Meanwhile, we continued studies of the neuro-insular complex and reached the following conclusions (Fujita and Kobayashi, 1979):

1) Close juxtaposition of nervous and endocrine elements occurs both in the neuro-insular complexes and in ordinary islets of Langerhans which may receive numerous nerve fibres of intrapancreatic neurons (especially in the dog) or which may contain ganglion cells (as in the mink).

2) In the islets as well as in the neuro-insular complexes, either nerve fibres or short neuronal processes containing numerous vesicles and peptidergic (mostly immunoreactive for vasoactive intestinal peptide) granules terminate on the capillary wall - a feature typical of neurosecretory cells, being the morphological correlate of endocrine discharge into the blood.

3) The neuronal secretory products, together with islets cell hormones (paraneuronal products) are believed to regulate the exocrine functions of the pancreas, being conveyed via insulo-acinar portal vessels. The islets could be called the neuro-paraneuronal control centre of the pancreas (Fujita et al., 1981).

Neurosecretion in its original sense, i.e., the hemocrine release of peptidic neurohormones has been suggested by our research group to occur also in other digestive organs. Vasoactive intestinal peptide-(VIP-)containing nerve fibres are closely associated with blood capillaries in the small intestine in some mammals (Iwanaga et al., 1987), as are gastrin-releasing peptide-containing fibres in the gastric fundus of the rat (Iwanaga et al., 1989; Iwanaga, 1994).

Some features of paraneurons

Paraneurons comprise peptide-producing endocrine and sensory cells, sharing structural, functional and metabolic features with neurons (Fujita, 1976). A set of heterologous substances are known to constitute their secretory products: 1) fragments of prepropeptides, partly representing bioactive messengers; 2) amines or other "classical" messengers, including acetylcholine and GABA; 3) ATP and other adenine nucleotides; 4) chromogranins and other acidic carrier proteins. Chromogranins, especially chromogranin A, have been useful as markers of paraneurons (Fujita et al., 1988).

Since the 1980's, various neuron-specific proteins have been found to occur in the cytoplasmic matrix of paraneurons, although the origin of these cells is not invariably neurectodermal. Besides neuron-specific enolase, neurofilament protein and brain tubulin, we found Purkinje cell-specific protein or spot-35 protein, which was later identified as calbindin, to be one of the most extensively distributed markers for paraneurons (Iwanaga et al., 1985). More recent studies have shown that synaptophysin, synapsins and other membrane proteins of synaptic vesicles may represent useful markers for neurons and paraneurons (Gratzl and Langley, 1991).

Problems concerning enterochromaffin cells

Among gut endocrine cells, enterochromaffin cells have the widest distribution, but have been subject to the most limited investigation. Occurrence of subtypes was suggested in the 1970's, but has received little attention. Peptide secretion has been postulated and studied by several authors, but has not been established unequivocally.

Early studies by our group demonstrated that enterochromaffin cells in the small intestine of rabbits selectively undergo degranulation by exocytosis after intraluminal administration of cholera toxin (Fujita et al., 1974; Osaka et al., 1975). We conceived at that time that serotonin (5HT) released from the cells might cause diarrhea and recent studies suggest that this effect of serotonin is mediated by stimulation of VIP releas efrom nerve networks closely associated with the enterochromaffin cells (Cassuto et al., 1980; Iwanaga et al., 1993). Enterochromaffin cells in the gut are known to release serotonin in response not only to bacterial toxins, but to hyperosmolarity in the lumen and to mechanical stimulation of the mucosal surface (Neya et al., 1993).

Recently, researchers in Hannover identified guanylin as the long sought peptide hormone of the enterochromaffin cell (Cetin et al., 1994). Guanylin is known to act as an endogenous ligand of guanylate cyclase C and thus regulate intestinal fluid secretion. As guanylate cyclase C is located, however, on the intestial microvilli, Cetin and his associates (1994) postulated an exocrine release of guanylin from the apices of enterochromaffin cells; the action and fate of peptide released basally remain to be determined. An exocrine release of part of enterochromaffin cell secretory materials has been supported by electron microscopic observations of the apical accumulation of a small proportion of the secretory granules present in the cell (Nilsson et al., 1987; Cetin et al., 1994).

The problem of endocrine and exocrine partition (Fujita, 1983), i.e., whether endocrine cells might release part of their secretory materials in an exocrine fashion, is now a major problem in the field of endocrinology. Such a phenomenon would correspond to the bipolar (and multipolar) secretion in neurons (Fujita, 1983; Fujita et al., 1988) which is now being more and more widely accepted in the field of neurocytology.

Migration of paraneurons

A large aggregation of endocrine cells occurs in a concavity at the top of every intestinal villus in the human fetus (from the 5th month to the full term). This peculiar structure - which was accurately described by Segi (1935, 1936) - was rediscovered by us 40 years later and designated "Segi's cap" (Kobayashi et al., 1980). Cattle and pigs were also found to possess Segi's caps during fetal life, and we recently investigated the fate of the endocrine cells (which are paraneurons) of the cap in fetal and newborn cattle (Kasuya and Fujita, 1995).

In the bovine fetus, the endocrine cells of Segi's cap are apparently quite active and have a distinct polarity, partly reaching the lumen with their apical process; no sign of cell degeneration or apoptosis could be seen. At this stage, a part or the whole mass of the endocrine cells has already shifted to the subepithelial layer or deeper into the lamina propria mucosae. After birth this translocation is even more conspicious; Segi's cap itself disappears in a few days after birth. Some of the endocrine cells apparently migrate along nerve fibres and a few become associated with neurons of Meissner's plexuses. It was noted that the endocrine cells of the cap at its original, intraepithelial site are strongly immunoreactive for chromogranin A, whereas those located subepithelially and deeper show a progressive decline in the clarity of such immunostaining. In contrast, these latter cells become more and more intensely immunoreactive for PGP-9.5, a neuronal marker. This finding seems to support the view that paraneurons comprising Segi's cap obtain more neuron-like immunoaffinity while migrating deeper in association with nerves (Kasuya and Fujita, 1995).

More conclusive findings concerning the migration of paraneurons have been obtained by the study of olfactory cells. In fetal rats, some of these typical paraneurons originate from the nasal epithelium, but migrate into the brain. Assuming a neuronal shape and immunopositivity for LH-RH, they settle in the hypothalamus as bona fide luteinizing hormone-releasing hormone-(LH-RH-) producing neurons (Schwanzel-Fukuda and Pfaff, 1989; Daikoku-Ishido et al., 1990).

Migration of "enterochromaffin cells" and their conversion into intramural neurons in the gut was advocated by the French pathologist Masson (1924) as the "neurentoderm theory". This view has long been neglected or forgotten, but may now be worthy of re-evaluation, even though it could never be accepted in its original form.

References

Cassuto, J., Fahrenkrug, J., Jodal, M., Tuttle, R. and Lundgren, O. (1980) The role of 5-hydrotryptamine and vasoactive intestinal polypeptide in pathogenesis of choleraic secretion. *Acta Physiol. Scand.* 109: 37A (Abstract).

Cetin, Y., Kuhn, M., Kulaksiz, H., Adermann, K., Bargsten, G., Grube, D. and Forssmann, W.-G. (1994) Enterochromaffin cells of the digestive system: Cellular source of guanylin, a guanylate cyclase-activating peptide. *Proc. Natl. Acad. Sci. USA* 91: 2935-2939.

Daikoku-Ishido, H., Okamura, Y., Yanaihara, N. and Daikoku, S. (1990) Development of the hypothalamic luteinizing hormone-releasing hormone-containig system in the rat: in vivo and in transplantation studies. *Devel. Biol.* 140: 374-387.

Fujita, T. (1959) Histological studies on the neuro-insular complex in the pancreas of some mammals. *Z. Zellforsch.* 50: 94-109.

Fujita, T. (1976) The gastro-enteric endocrine cell and its paraneuronic nature. *In*: T. Fujita and R.E. Coupland (eds.): *Chromaffin, enterochromaffin and related cells,* Elsevier, Amsterdam-Oxford-New York, pp. 191-208.

Fujita, T. (1983) Messenger substances of neurons and paraneurons: Their chemical nature and the routes and ranges of their transport to targets. *Biomed. Res.* 4: 239-256.

Fujita, T. (1985) Neurosecretion and new aspects of neuroendocrinology. *In*: H. Kobayashi, H.A. Bern and A. Urano (eds.): *Neurosecretion and the biology of neuropeptides,* Japan Sientific Societies Press, Tokyo; Springer, Berlin-Heidelberg-New York-Tokyo, pp. 521-528.

Fujita, T., Osaka, M. and Yanatori, Y. (1974) Granule release of enterochromaffin cells by cholera enterotoxin in the rabbit. *Arch. Histol. Jap.* 36: 367-378.

Fujita, T. and Kobayashi, S. (1979) Proposal of a neurosecretory system in the pancreas. An electron microscope study in the dog. *Arch. Histol. Jap.* 42: 277-295.

Fujita, T., Kobayashi, S., Fujii, S., Iwanaga, T. and Serizawa, Y. (1981) Langerhans islets as the neuro-paraneuronal control center of the exocrine pancreas. *In*: M.I. Grossman, M.A.B. Brazier and J. Lechago (eds.): *Cellular basis of chemical messengers in the digestive system,* Academic Press, New York-London-Toronto-Sydney-San Francisco, pp. 231-242.

Fujita, T., Kanno, T. and Kobayashi, S. (1988) *The Paraneuron.* Springer Verlag, Tokyo-Berlin-Heidelberg.

Gratzl, M. and Langley, K. (1991) *Markers for neural and endocrine cells. Molecular and cell biology, diagnostic applications.* Weinheim, New York-Basel-Cambridge.

Iwanaga, T. (1994) Distribution, ultrastructure, and morphological change of neurons in the gastric mucosa with special reference to gastrin-releasing peptide-containing nerve fibers. *Gastrointestinal function: regulation and disturbances* (Excerpta Medica, Tokyo) 12: 43-57.

Iwanaga, T., Takahashi-Iwanaga, H., Fujita, T., Yamakuni, T. and Takahashi, Y. (1985) Immuno-histochemical demonstration of a cerebellar protein (spot 35 protein) in some sensory cells of guinea pig. *Biomed. Res.* 6: 329-334.

Iwanaga, T., Yui, R., Kuramoto, H. and Fujita, T. (1987) The paraneuron concept and its implications in neurobiology. *In*: B. Scharrer, H.-W. Korf and H.-G. Hartwig (eds.): *Functional morphology of neuroendocrine systems. Evolutionary and environmental aspects,* Springer Verlag, Berlin, pp. 139-149.

Iwanaga, T., Takahashi-Iwanaga, H. and Fujita, T. (1989) Possible mechanisms of GRP-containing nerves regulating the mucosal microcirculation in the gastric body. *Acta Med. Biol.* 37: 35-43.

Iwanaga, T., Ohtsuka, H. and Fujita, T. (1993) Enterochromaffin cells in the gut with special reference to their role in diarrhea. *Biomed. Res.* 14, Supplement 3: 1-11.

Kasuya, K. and Fujita, T. (1995) Perinatal changes in bovine Segi's caps with special reference to their endocrine cells migrating into the lamina propria. *Arch. Histol. Cytol.*, in press.

Kobayashi, S., Iwanaga, T. and Fujita, T. (1980) Segi's cap: Huge aggregation of basal-ganulated cells discovered by Segi (1935) on the intestinal villi of the human fetus. *Arch. Histol. Jap.* 43:-79-83.

Masson, P. (1924) Appendicite neurogéne et carcinoides. *Ann. Anat. Pathol.* 1: 1-59.

Neya, T., Mizutani, M. and Yamasato, T. (1993) Role of 5-HT$_3$ receptors in peristaltic reflex elicited by stroking the mucosa in the canine jejunum. *J. Physiol.* 471: 159-173.

Nilsson, O., Ahlmann, H., Geffard, M., Dahlström, A. and Ericsson, L.E. (1987) Bipolarity of duodenal enterochromaffin cells in the rat. *Cell Tissue Res.* 248: 49-54.

Osaka, M., Fujita, T. and Yanatori, Y. (1975) On the possible role of intestinal hormones as the diarrhetic messenger in cholera. *Virchows Arch. B (Cell Pathol.)* 18: 287-296.

Schwanzel-Fukuda, M. and Pfaff, W.D. (1989) Origin of luteinizing hormone-releasing hormone neurons. *Nature* 338: 161-164.

Segi, M. (1935) Über eine aus chromaffinen Darmzellen bestehende Struktur auf den Zotten vom Menschen-embryo (In Japanese with German summary). *Acta Anat. Nippon.* 8: 276-280.

Segi, M. (1936) Über die Entwicklung der verschiedenen Granulazellen im Darmepithel des Menschenembryo (In Japanese with German summary). *Acta Anat. Nippon.* 9: 850-937.

II. Biosynthesis, release and degradation of neuropeptides

Role of convertases in the processing of neuropeptides and neurotrophins

M. Chrétien, S. Benjannet, M. Marcinkiewicz, R. Day[1] and N.G. Seidah[1]

Clinical Research Institute of Montreal, Laboratories of Molecular ([1] Biochemical) Neuroendocrinology, 110 Pine Avenue West, Montreal, QC H2W 1R7, CANADA

Summary. The hypothesis that endoproteolytic cleavage of large precursor proteins produces hormonal peptides has been extended to neuropeptides and neurotrophins. The wide distribution of 7 known proprotein convertases in the brain, along with the presence of a large number of substrates, indicate that these enzymes participate in numerous brain functions. Co-localization of convertases with their putative substrates along with *in vivo* co-expression studies are key experiments to define the relationship between each individual substrate with its appropriate convertase. These developments bring new lights to the understanding of brain functions and development.

List of abbrevations:

ACTH	adrenocorticotropic hormone
ANF	atrial natruretic factor
BDNF	brain derived-neurotrophic factor
BPN'	bacterial subtilisin
CLIP	corticotropin-like intermediate lobe peptide
EGF	epidermal growth factor
END	endorphin
GDNF	glia-derived neurotrophic factor
HIV_{gp}	human immunodeficiency virus glycoprotein
IGF	insulin-like growth factor
JP	joining peptide
LHRH	luteinizing hormone-releasing hormone
LPH	lipotropic hormone
MSH	melanocyte-stimulating hormone
NGF	nerve growth factor
NT	neurotrophin
PA	pseudomonas antrax toxin
PACE 4	paired base amino acids-converting enzyme 4
PC	proprotein convertases
PDGF	platelet-derived growth factor
POMC	pro-opiomelanocortin
PTH	parathormone
PTP	phosphotyrosine phosphatase receptor
TGF	transforming growth factor

Introduction

(1) *The prohormone theory.* In 1967, Chrétien and Li, and Steiner et al. advanced the hypothesis that prohormones exist and need to be cleaved post-translationally in order to release bioactive end-products. The initial endoproteolytic cleavage occurs C-terminally to pairs of basic amino acids followed by the removal of the basic residues by exopeptidases. Further modifications can occur in the form of N-terminal acetylation, pyroglutamate formation or C-terminal amidation which are essential for the bioactivity of many peptides (Eipper et al., 1987). Since then, it became apparent that the limited proteolysis of precursors at specific pairs of basic residues and/or at single basic amino acids is a widespread mechanism by which the cell produces a repertoire of biologically active proteins and peptides (Seidah and Chrétien, 1994).

(2) *Proneuropeptides.* The participation of neuropeptides in the modulation of a variety of functions in the central nervous system is well established. Most neuropeptides are initially synthesized as inactive precursor proteins, which undergo an enzymatic cascade of posttranslational modification during their intracellular transport. Pro-opiomelanocortin (POMC) has been the key precursor studied by interest to β-endorphin as an important end-product of this precursor (Chrétien et al., 1979; Lazure et al., 1983). Soon after followed the precursors of enkephalin, dynorphin and other hypothalamic factors. The recent discovery of the amnesiac peptide by Feany and Quinn (1995) is the latest important addition to a long list of proneuropeptides.

(3) *Proneurotrophins.* Neurotrophins represent a family of chemically related proteins that promote the survival, growth and maintenance of neurons in the central and peripheral nervous systems. Over forty years ago, Levi-Montalcini and co-workers (1987) discovered nerve growth factor (NGF), the first member of the family. Since, four other members of the family have been identified: brain-derived neurotrophic factor (BDNF), neurotrophin-3 (NT3), neurotrophin-4/5 (NT-4/5), and neurotrophin-6 (NT-6). Neurotrophins are also synthesized as long inactive precursors and need intracellular cleavage to produce active

growth factors. The cleavage takes place following pairs of basic amino acids of the type I precursor motif (Fig. 1A) Arg-Xaa-(Lys/Arg)-Arg↓, where Xaa = Ser, Val or Arg for proNGF/proNT-4/5, proBDNF/proNT-6 and proNT3, respectively (Maisonpierre et al., 1990). This general motif is also found in the recently described glial-derived neurotrophic factor (GDNF, Xaa = Val). Thus the activation of neurotrophins requires the cleavage of the precursor at the typical recognition sequence for the convertases (Seidah et al., 1996).

A

Precursor protein	**Cleavage site sequence**
	P6 P5 P4 P3 P2 P1 P1' P2'
Type I precursors	[X- X- **R**- X- **K/R**- **R** ↓ X- X]
Neurotropines	
-(h)Pro-βNGF	Thr-His-**Arg**-Ser-**Lys**-**Arg** ↓ Ser-Ser
-(h)Pro-BDNF	Ser-Met-**Arg**-Val-**Arg**-**Arg** ↓ His-Ser
-(h)Pro-NT-3	Thr-Ser-**Arg**-Arg-**Lys**-**Arg** ↓ Tyr-Ala
-(r)Pro-NT-4	Ala-Asn-**Arg**-Ser-**Arg**-**Arg** ↓ Gly-Val
-(h)Pro-GDNF	Ile-Lys-**Arg**-Leu-**Lys**-**Arg** ↓ Ser-Pro
(h)Pro-PDGF-A	Pro-Ile-**Arg**-Arg-**Lys**-**Arg** ↓ Ser-Ile
(h)Pro-PDGF-B	Leu-Ala-**Arg**-Gly-**Arg**-**Arg** ↓ Ser-Leu
(h)Pro-TGF-β 1	Ser-Ser-**Arg**-His-**Arg**-**Arg** ↓ Ala-Leu
(h)Pro-von Willebrand factor	Ser-His-**Arg**-Ser-**Lys**-**Arg** ↓ Ser-Leu
(h)Insulin Pro-receptor	Pro-Ser-**Arg**-Lys-**Arg**-**Arg** ↓ Ser-Leu
(h)Pro-Factor IX	Leu-Asn-**Arg**-Pro-**Lys**-**Arg** ↓ Tyr-Asn
(h)Pro-Factor X	Leu-Glu-**Arg**-Arg-**Lys**-**Arg** ↓ Ser-Val
(h)Pro-7B2	Glu-Arg-**Arg**-Lys-**Arg**-**Arg** ↓ Ser-Val
PA from *B. anthracis*	Asn-Ser-**Arg**-Lys-**Lys**-**Arg** ↓ Ser-Thr
Diphteria toxin (DT)	Gly-Asn-**Arg**-Val-**Arg**-**Arg** ↓ Ser-Val
Hemagglutinin (HA)	Lys-Lys-**Arg** Glu **Lys**-**Arg** ↓ Gly-Leu
HIV-1 gp160	Val-Gln-**Arg**-Glu-**Lys**-**Arg** ↓ Ala-Val

Fig. 1A. Proposed classification of precursor proteins according to the basic amino acids motifs surrounding the cleavage site. The four types of precursors I, II, III and IV cover most of the known cleavage sites recognized by the subtilisin/kexin-like pro-protein convertases.

(4) *Proprotein convertases (PC).* Following a long search to identify the convertases, their molecular characterization has recently been achieved (for review see Seidah and Chrétien, 1992) and have provided the necessary tools to investigate the regulation of neuropeptide and neurotrophin expression at the level of the enzymes involved in their processing. With the knowledge accumulated over the last decade on the structural characteristics of numerous substrates, we can describe the sequence arrangements around the cleavage sites.

B

Precursor protein	**Cleavage site sequence**
	P6 P5 P4 P3 P2 P1 P1' P2'
Type II precursors	[X- X- X- X- **K/R- R** ⬇ X- X]
(m)POMC (JP/ACTH)	Pro-Arg-Glu-Gly-**Lys-Arg** ⬇ Ser-Tyr
(ACTH/β LPH)	Pro-Leu-Glu-Phe-**Lys-Arg** ⬇ Glu-Leu
(αMSH/CLIP)	Pro-Val-Gly-Lys-**Lys-Arg** ⬇ Arg-Pro
(γLPH/β End)	Pro-Pro-**Lys**-Asp-**Lys-Arg** ⬇ Tyr-Gly
(h)Pro-Insulin (B/C chain)	Thr-Pro-**Lys**-Thr-**Arg-Arg** ⬇ Glu-Ala
(C/A chain)	Gly-Ser-Leu-Gln-**Lys-Arg** ⬇ Gly-Ile
(r)Pro-LHRH	**Arg**-Pro-Gly-Gly-**Lys-Arg** ⬇ Asp-Ala
(h)Pro-Enkephalin	Gly-Gly-Phe-Met-**Lys-Lys** ⬇ Asp-Ala
	Met-Asp-Tyr-Gln-**Lys-Arg** ⬇ Tyr-Gly
	Gly-Gly-Phe-Leu-**Lys-Arg** ⬇ Phe-Ala
(r)Pro-Dynorphin	**Arg-Lys**-Gln-Ala-**Lys-Arg** ⬇ Tyr-Gly
	Glu-Asp-Leu-Tyr-**Lys-Arg** ⬇ Tyr-Gly
	Arg-Lys-Tyr-Pro-**Lys-Arg** ⬇ Ser-Ser
(h)Pro-PTH	**Lys**-Ser-Val-**Lys-Lys-Arg** ⬇ Ser-Val
(h)Pro-Renin	Ser-Gln-Pro-Met-**Lys-Arg** ⬇ Leu-Thr
(h)Pro-Albumin	**Arg**-Gly-Val-Phe-**Arg-Arg** ⬇ Asp-Ala
(h)Pro-Protein C	**Arg**-Ser-**His**-Leu-**Lys-Arg** ⬇ Asp-Thr
(h)α -fibronectin Pro-receptor	**His**-His-Gln-Gln-**Lys-Arg** ⬇ Glu-Ala

Fig. 1B. Proposed classification of precursor proteins according to the basic amino acids motifs surrounding the cleavage site. The four types of precursors I, II, III and IV cover most of the known cleavage sites recognized by the subtilisin/kexin-like pro-protein convertases.

As shown in Fig. 1A,B,C, substrates can be divided in 4 categories. The comparative architectural features of the 7 mammalian subtilisin-like pro-protein convertases known so far, and those of the subtilisin BPN' and the yeast kexin are presented in Fig. 2, showing that the sizes of the eukaryotic members vary from 637 to 969 amino acids. It is seen that all the family members contain a signal peptide enabling them to enter the secretory pathways. Intriguingly, PC2 is the only convertase having an Asp residue in place of the catalytically important Asn which is found in all other subtilisin-like proteinases. Finally, it is worth noting that only furin and PC7 have transmembrane domains (Seidah et al, 1996).

C

Precursor protein   **Cleavage site sequence**

P8 P7 P6 P5 P4 P3 P2 P1   P1' P2'

Type III precursors(Single R)   **(B)- X- (B)- X- (B)- X- X   R ⬇ X - X**

(r)Pro-Dynorphin (C-peptide)   **Arg**-Gln-Phe-Lys-Val-Val-Thr-**Arg** ⬇ Ser-Gln
(h)Pro-ANF   **Arg**-Ala-Leu-Leu-Thr-Ala-Pro-**Arg** ⬇ Ser-Leu
(h)Pro-Somatostatin (SS-28)   Glu-Met-**Arg**-Leu-Glu-Leu-Gln-**Arg** ⬇ Ser-Ala
(m)Pro-SMR1   Gly-Glu-Gly-Val-**Arg**-Gly-Pro-**Arg** ⬇ Arg-Gln
                                                          Pro-Arg-**Arg**-Gln-**His**-Asn-Pro-**Arg** ⬇ Arg-Gln
(m)Pro-IGF-I   Leu-Lys-Pro-Thr-**Lys**-Ala-Ala-**Arg** ⬇ Ser-Ile
(h)Pro-IGF-II   Ala-Thr-Pro-Ala-**Lys**-Ser-Glu-**Arg** ⬇ Asp-Val
(m)Pro-EGF (N-terminal)   Glu-Asp-Gly-His-**His**-Leu-Asp-**Arg** ⬇ Asn-Ser
(m)Pro-EGF (C-terminal)   Asp-Leu-**Arg**-Trp-Trp-Glu-Leu-**Arg** ⬇ His-Ala

P8 P7 P6 P5 P4 P3 P2 P1   P1' P2'

Type IV precursors(P2'= R/K)   **(B)- X- (B)- X- (B)- X- (B)-R/K ⬇ X -R/K**

(r)Pro-Mullerian Inhibiting Substance   Glu-Gly-**Arg**-Gly-**Arg**-Ala-Gly-**Arg** ⬇ Ser-**Lys**
(h)Pro-Pancreatic Polypeptide   Tyr-Arg-Pro-Arg-Tyr-Gly-**Lys**-**Arg** ⬇ His-**Lys**
(b)Pro-Substance K (pro-SK)   **His**-Gly-Gln-Leu-Ser-His-**Lys**-**Arg** ⬇ His-**Lys**
(r)Pro-Gastrin-I   Phe-Ile-Ala-Asp-Leu-Ser-**Lys**-**Lys** ⬇ Gln-**Arg**
(r)Pro-Glucagon   Gln-Trp-Leu-Met-Asn-Thr-**Lys**-**Arg** ⬇ Asn-**Arg**
(h)Pro-IGF-II   Glu-Ala-Phe-**Arg**-Glu-Ala-**Lys**-**Arg** ⬇ His-**Arg**
(r)Pro-PTP-μ receptor   Val-Glu-Glu-Glu-**Arg**-Pro-**Arg**-**Arg** ⬇ Thr-**Lys**

Fig. 1C. Proposed classification of precursor proteins according to the basic amino acids motifs surrounding the cleavage site. The four types of precursors I, II, III and IV cover most of the known cleavage sites recognized by the subtilisin/kexin-like pro-protein convertases.

M. Chrétien et al.

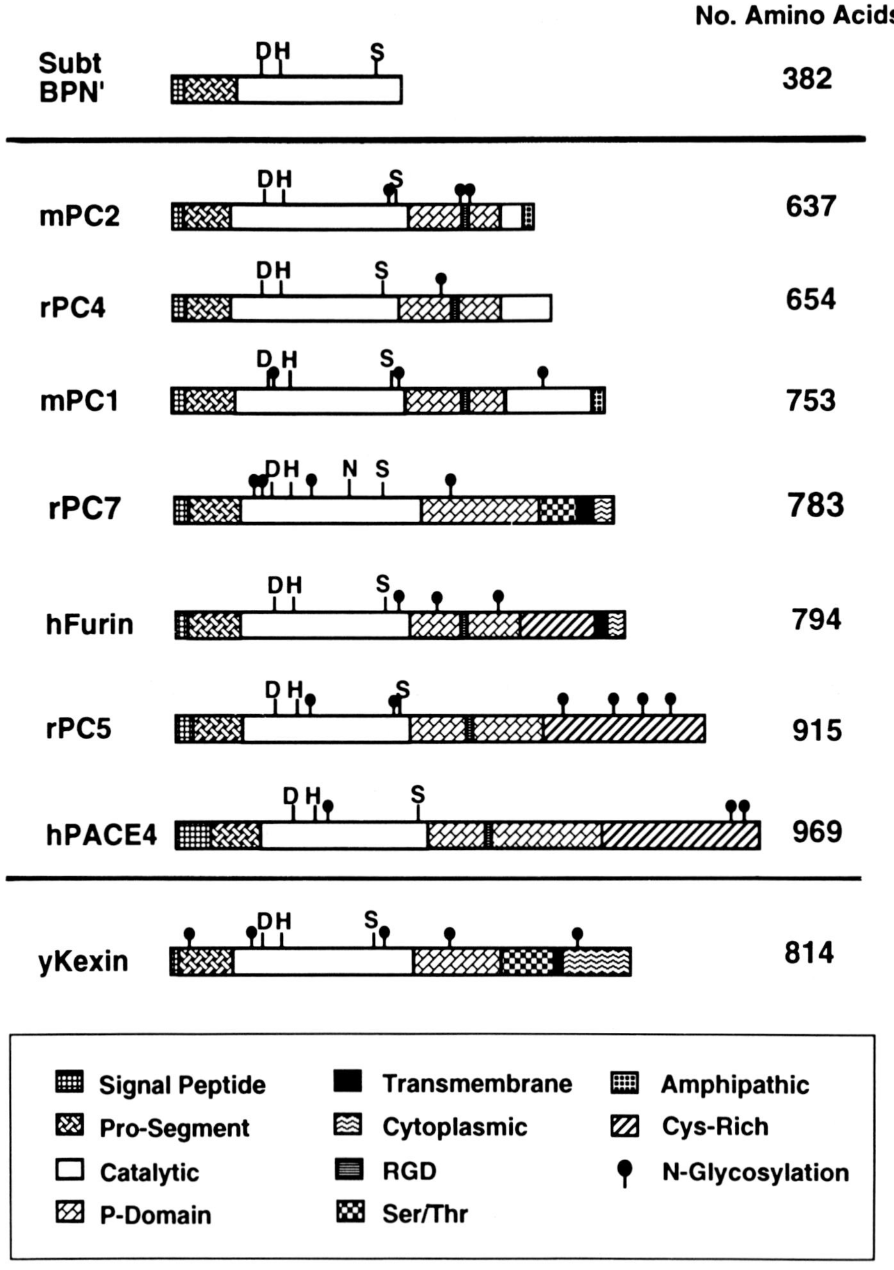

Fig. 2. Schematic representation of the family of mammalian prohormone convertases comparing structural and domain features of each. Also shown are the subtilisin BPN' (subt BPN') and yeast Kexin (yKexin).

Distribution of convertases in the nervous system

Using Northern blot analysis and *in situ* hybridization, our group and a number of collaborators (Marcinkiewicz et al., 1993a; Day et al., 1993; Dong et al., 1995; Schäfer et al., 1993) have examined the mRNA distribution of each convertase in the neuronal cells. The mRNAs coding for the prohormone processing enzymes, PC1, PC2, PC5, furin and PACE4 were localized throughout the rat and mouse brain. Although regional overlaps occurred, each enzyme exhibited an unique expression pattern. The genes coding for these prohormone convertases are widely expressed, often in the same brain nuclei, but usually with a distinct cellular distribution pattern. The results are summarized in Table I.

(1) *PC1, PC2 and furin.* High levels of expression of PC1 and PC2 mRNA were observed in neuropeptide rich regions such as the hypothalamus, hippocampus and cerebral cortex. These areas are also characterized by the relatively high expression of carboxypeptidase E and peptidylglycine α-amidating monoxygenase (Eipper et al., 1987). In general, PC1 mRNA distribution is more restricted, compared to PC2. An impressive example is the thalamus, where almost all the thalamic nuclei express PC2 mRNA, but PC1 mRNA is found at high levels only in the anterior subdivisions. These unique distribution patterns of PC1 and PC2 mRNAs and the observed differences in their relative ratios of expression at the cellular level suggests distinct roles in the activation of brain proproteins. This variation in cellular mRNA levels may have significance for region-specific post-translational processing. In other words, the same substrate proneuropeptide may well result in different biologically active end-products depending on the differential expression of PC1 and PC2. This notion has already been demonstrated to be valid in understanding the differential post-translational processing of pituitary pro-opiomelanocortionin (POMC), which occurs in the anterior lobe corticotrophs and in the intermediate lobe melanotrophs (Benjannet et al., 1991; Day et al., 1992; Marcinkiewicz et al., 1993b). In the hypothalamus and the nucleus tractus solitarius neurons, precise colocalization studies will be needed in order to establish the exact relationship of POMC with PC1 or PC2.

M. Chrétien et al.

Table I. Regional distribution of mRNAs coding for processing enzymes in the rat brain.

Region	PC1	PC2	PC5	Furin	PACE4
Telencephalon					
Olfactory bulb					
External Plexiform Layer	+++	++	++*	++++	++
Mitral cells	+	++	++++	++	0
Ependymal cells	0	0	0	++	+*
Granule cells	0/+	+	0	++	+*
Periglomular cells	+	++	++	++	++
Anterior olfactory nucl.	+++	++++	++++	+++	+++
Olfactory cortex	+++	++++	+++	+++	+++
Tenia tecta	++++	+++	0	+++	0
Cerebral cortex					
Layer I	0	0	0	++	0
Layer II	++	+++	0	++	0
Layer III	++	+++	+++	++	0
Layer IV	+	+	0	+++	0
Layer V	+++	+++	++++	++	+
Layer VI	+++	+++	++++	++	+
Hippocampal formation					
CA1	+++	+++++	++	++++	0
CA2	++	++++	+	++++	0
CA3	++++	+++++	+++++	++++	0
Scattered hilar cells	++++	+++	++++	++++	0
Dentate gyrus					
Molecular layer	0	0	0	++	0
Granular layer	+++++	+++	+	++++	0 to +
Islands of Calleja	0	0/+	0	+++++	0
Nucl. accumbens	+	++	0	++	+*
Caudate putamen	+++++*	+++	0	++	++*
Globus pallidus	0/+	+	++++*	++	0
Ventral pallidum	+	++	++++*	++	0
Amygdala					
Medial nucl.	+++	++	+++*	+	+
Central nucl.	0/+	0/+	+	+	++++
Anterior cortical nucl.	++	+	+++	+	++
Basolateral nucl.	+++	++++	+++	+++	0
Septum	++	++++	++	++	+
Nucl. of diagonal band	+++	++	+++	+	+
Subfornical organ	++++	++	++++	+++	+
Bed nucl. stria terminalis	+++	++	+	++	+
Medial preoptic area	+++	+	++	++	+*
Choroid plexus	0	0	0	++++	0
Ependyma of lat.ventricle	0	0	0	+++	0

Diencephalon

Thalamus					
Anteroventral nucl.	+++	++	0	++	+++
Anteromedial nucl.	++	++++	0/+	++	0/+
Anterodorsal nucl.	++++	++++	0/+	++	0/+
Centrolateral nucl.	++	++++	0/+	++	0/+
Centromedial nucl.	+++	++++	++++	++	0
Mediodorsal nucl.	+++	++++	0	++	+++
Paraventricular nucl.	+++	++++	+++++	++	++
Reuniens nucleus	0	+++	0/+	+	0/+
Rhomboid nucl.	+++	++++	+++	++	+*
Ventrolateral nucl.	0/+	++++	0	++	++++
Ventromedial nucl.	0/+	+++	0	++	0
Dorsal lat. geniculate nucl.	0	++++	0	++	+++
Medial geniculate nucl.					
Dorsal	0/+	++++	0	+	++
Ventral	0/+	+++	0	+	++
Medial	0/+	++++	0	+	++
Marginal zone	+++	+	0	+	++
Habenula					
Medial					
Lateral part	++	+	0	+++	+++++
Medial part	+	++++	0	+++	+++++
Lateral	++	+	0	+++	+
Hypothalamus					
Suprachiasmatic nucleus	+++	++	++*	++	+*
Periventricular nucleus	++++	+	+++*	+++	0
Supraoptic nucleus	+++++	+++	+++++	+++	0
Paraventricular nucleus					
Magnocellular part	+++++	++	+++++	+++	0
Parvocellular part	++	+++	+++*	++	++*
Arcuate nucl.	+++	++	++*	++	++++
Lateral hypothal.area	+++++*	++++*	++++	+++	+
Ventromedial nucleus	+	++	+	++	+
Dorsomedial nucleus	+++	+++	+++*	++	0
Medial mammillary nucl.	+	++++	+++	++	+
Lateral mammillary nucl.	+	+++	+	++	++
Mesencephalon					
Zona incerta	+++	++	++	+	0
Substantia nigra					
Pars compacta	++++	+++	+++*	+++	+*
Pars reticulata	+	+++	+++*	+	0
Ventral tegmental area	++	+++	+*	++	++*
Central Grey	++	+++	++*	+	+*
Oculomotor nucleus	+++	++++	+++	+	+
Red nucleus	++	++++	++++	++	+
Dorsal raphe nucl.	+++	+++	+++++	+	+*
Median raphe nucl.	+++	++	+++	+	+
Superior colliculus					
Superficial grey layer	+++	+	+++	+	+
Intermediate grey layer	++	+++	+++	+	+
Deep grey layer	+	+	+++	+/++	+
Inferior colliculus	+++	+++	++	+/++	0
Interpeduncular nucl.	+++	++/+++	++to++++	++	+

Pons and Medulla

Pontine nucl.	+++	++++	+++++	+	++++
Dorsal tegmental nucl.	++	+++	0	+	++
Dorsolateral	++++	++	+++	+	+
Locus coeruleus	++	++++	+++	++	0
Dorsal parabrachial nucl.	+++	++	++*	+	++
Ventral parabrachial nucl.	+++	+++	++*	+	++
Kölliker-Fuse nucl.	++++	++	++*	+	++
Ventral tegmental nucl.	+	+++	0	++	++
Dorsal cochlear nucl.	++++	+++	++	++	++
Lateral vestibular nucl.	++	+++	0	+	++
trapezoid body	++	++	0	+++	+++
Nucleus raphe magnus	+++	+	0	+	0
Nucl. raphe pallidus	+++	+	0	++	+++
Facial motor nucleus	+++	++++	++++	+++	0
Inferior olive	++	++++	++++	+	0
Prepositus hypoglossal nucl.	+++	+	++++*	+	0
Nucl. of the solitary tract					
Medial	++++	++	++++	+	0
Lateral	++	+++	++++	+	0
Cerebellar Cortex					
Molecular layer	0	0/+	0	++	+
Purkinje cell layer	++	++++	0	+++	+++++
Granular layer	0	++	+	+	+

This table presents a subjective evaluation of mRNA levels in different brain areas from three animals. Density: +++++, maximum density; ++++, very dense; +++ dense: ++, moderate; +, light; 0, below detection limit. An asterix indicates scattered cells with high labeling density.

Furthermore, similarities in distribution patterns of neuropeptides and convertases can point to potential precursor selectivity for each convertase. For example, the expression of cholecystokinin (CCK) in the neocortex and the thalamus closely resembles the distribution pattern of PC2. Other possible precursor substrates for PC2 could be proenkephalin, which shows a similar distribution pattern in the caudate putamen, and calcitonin gene-related peptide (CGRP), which is synthesized in the motor neurons of spinal cord, where PC2 is highly expressed. However, the situation in the brain is further complicated by the complex and multiple co-existence of several neuropeptides in the same neuron which is the rule rather than the exception (Hökfelt et al., 1980). For example, neurons of the magnocellular hypothalamic paraventricular and synaoptic nuclei produce almost all of the different neuropeptides present in hypothalamus. Thus, the relationship between one specific neuropeptide precursor as substrate for a particular convertase is possible when (i) their co-existence has been demonstrated in the same neurons and (ii) the co-expression studies confirm biochemically their relationship.

(2) *PC5 and PACE4*. From the detailed mapping of PC5 and PACE4 (Dong et al., 1995) it is noted that they exhibit distinct expression patterns which also differ from those of other convertases. In the central nervous system, PC5 gene expression was unique and more restricted than that of PACE4. For example, while PC5 expression could only be detected in neurons, PACE4 mRNA is found in subpopulations of both neuronal and non-neuronal cells.

PACE4 expression was very high in some restricted central nervous system areas, but many other regions expressed low to moderate levels of PACE4. Within the central nervous system, aside from certain expression "hot spots", PACE4 mRNA is distributed in a widespread manner at a low cellular level. Similar contrasting patterns of expression were also shown for furin, with high expression in the subsets of ependymal cells and the choroid plexus cells and much lower levels elsewhere (Day et al., 1993).

The expression of each convertase in specific brain regions raises interesting questions as to their physiological roles, the nature of their putative substrates and whether these proteinases have overlapping and/or distinct functions. For example, the hypothalamic paraventricular nucleus is rich in convertase expression and is known to synthesize many neuropeptides. Our data demonstrated that the expression of PC5 is predominant in magnocellular neurons in the anterior magnocellular and posterior magnocellular divisions of the paraventricular nucleus. The distribution pattern of PC5 positive cells is similar to that of the oxytocin producing cells. We have previously shown that PC1 and PC2 are also expressed in the paraventricular nucleus (Schäfer et al., 1993). In this nucleus, PC1 is abundantly expressed in the posterior magnocellular part but also in scattered magnocellular cells in its mediate parvocellular division of paraventricular nucleus. PC2 is more abundantly expressed in the parvocellular division of the paraventricular nucleus, but significant expression levels were also observed in the magnocellular cells. Taken together, these data suggest that vasopressin-containing magnocellular cells in the paraventricular nucleus are principally expressing PC1, while oxytocin containing magnocellular cells in this nucleus are principally expressing PC5. While both cell types also express significant levels of PC2, it is likely that the actions of PC1 and PC5 are critical in the activation of pro-vasopressin and pro-oxytocin, respectively.

It is interesting to note that furin expression is detected throughout the hippocampal formation. When considering potential substrates within this region, the neurotrophic factors are candidates, since their precursors require cleavage at pairs of basic residues to generate biologically active growth factors, a subject which has been partially resolved with recent molecular and cellular analyses (see below).

In conclusion, the sum of the data obtained so far (Marcinkiewicz et al., 1993a; Day et al., 1993; Schäfer et al., 1993) provides a wealth of information on the complex tissue localization of each convertase enzyme. This starting point should stimulate further work leading to elucidation of the tissue-specific role of each enzyme and to the definition of its potential substrates. The integration of this information within the global context of brain development and function should allow us to better understand the diverse interactions that this family of processing enzymes can achieve.

Processing of pro-opiomelanocortin

Among the known precursor molecules, pro-opiomelanocortin (POMC) is the best studied of all. It represents a good model for differential processing since this gene is expressed in three different tissues i.e. the anterior and the intermediate lobes of the pituitary and in the hypothalamus. The colocalization of PC1 and PC2 with POMC in the pituitary and in the arcuate nucleus is shown in Fig. 3 (Seidah et al., 1996).

A recombinant vaccinia virus vector was used to coexpress the two candidate mouse prohormone convertases PC1 and PC2 together with mouse POMC in the constitutively secreting cells, BSC-40 and in the endocrine-derived cell lines PC12 and AtT-20 which exhibit regulated secretion. Monitoring POMC processing demonstrated the distinct PC1 and PC2 cleavage specificities (Benjannet et al., 1991).

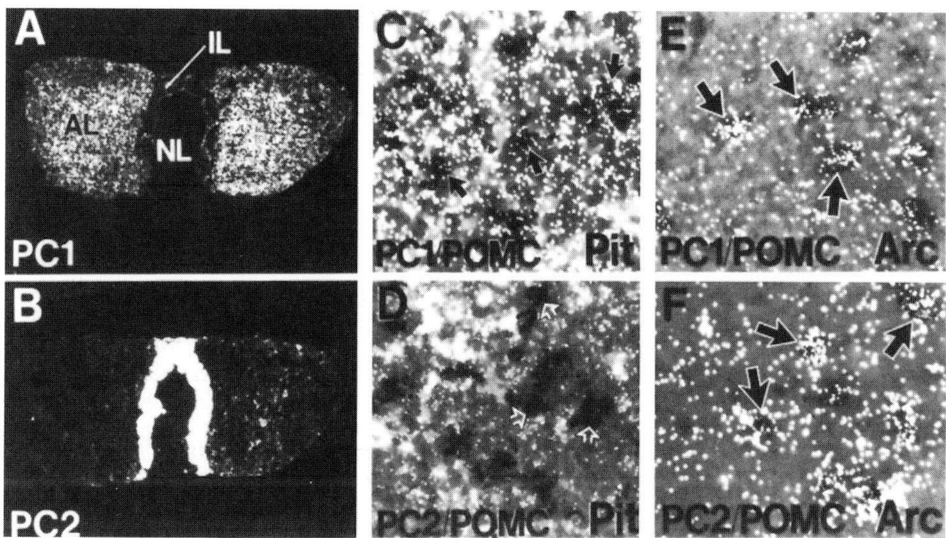

Fig. 3. *In situ* hybridization showing distribution of PC1 and PC2 mRNA, both known to cleave POMC in the pituitary gland and the arcuate nucleus. In (A) PC1 and (B) PC2 are shown to be distinctly expressed in the pituitary: while PC1 is highly expressed in the anterior lobe (AL), PC2 is almost exclusively present in the intermediate lobe (IL). At higher magnification of the anterior lobe in (C), dark arrows show the co-expression of PC1 (white grains) in the POMC cells (dark labeled corticotrophs). However, in (D), we observe that PC2 (white grains) is not co-expressed with POMC cells (white arrows). Finally in the arcuate nucleus (Arc), which also expresses POMC, we note the co-expression of PC1 and PC2 with POMC cells (E,F).
Magnifications A and B: x 16.5; C and D: x 280; E and F: x 500

The diagram in Fig. 4 shows that PC1 and PC2 (Benjannet et al., 1991; Seidah et al., 1990, 1991) are selectively cleaving POMC at distinct pairs of basic residues. Within this molecule, PC2 is capable of cleaving all the five pairs of basic residues analyzed, namely AlaGlnArgArg$_{76}\downarrow$, GluGlyLysArg$_{97}\downarrow$, GlyLysLysArg$_{114}\downarrow$, GluPheLysArg$_{138}\downarrow$ and LysAspLysArg$_{178}\downarrow$. PC1 preferentially cleaves POMC at the GluGlyLysArg$_{97}\downarrow$ and GluPheLysArg$_{138}\downarrow$ bonds, with minor cleavage at LysAspLysArg$_{178}\downarrow$ to produce β-endorphin (βEND) (Seidah et al., 1981). Noteworthy, PC1 principally generates adrenocorticotropin (ACTH) and β-lipotropic hormone (βLPH), while the efficient production of α-melanocyte-stimulating hormone (αMSH) and βEnd, requires the action of PC2.

M. Chrétien et al.

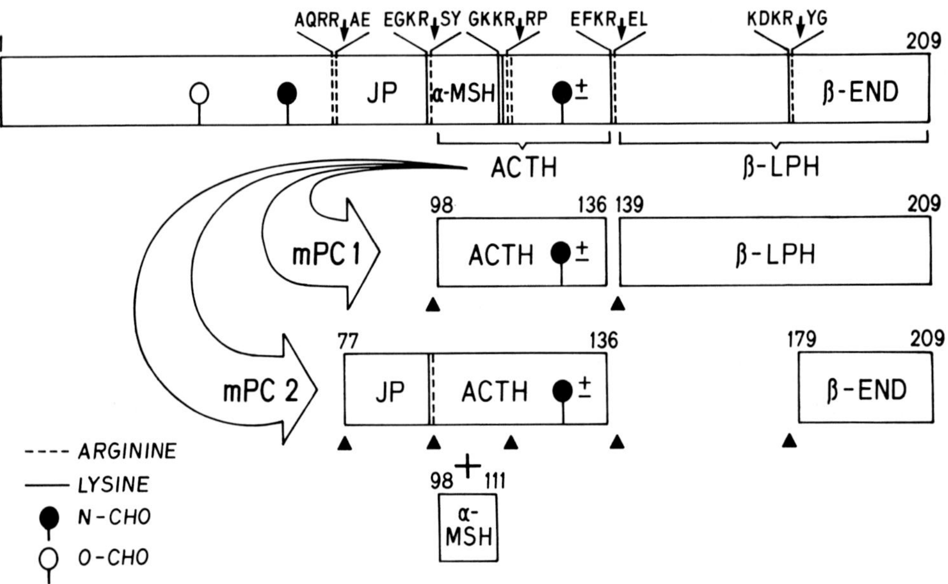

Fig. 4. Major end products of POMC processing by PC1 and PC2. Arrows represent the cleavage sites following pairs of basic residues (single-letter amino acid symbols). Black triangles tally the cleavage sites of each enzyme. Numbers represent the start and end positions of the processed peptides based on the POMC sequence. The pairs of basic residues as well as the N- and O-glycosylation (CHO) sites as well as are emphasized.

The data presented argue for a physiological role of PC1 in the processing of POMC in the anterior pituitary. In support of this hypothesis, PC1 generates ACTH and βLPH as the major final products of POMC (Fig. 4), both of which are known to represent the most abundant POMC-derived peptides in the anterior pituitary. Even though some βEnd was detected in the anterior pituitary, it is interesting that no significant production of αMSH was found. Therefore, it is possible that in the anterior pituitary, PC1 is the responsible convertase for the minor βEND production. In the pars intermedia, the major products of this precursor are αMSH and βEnd-derived peptides. In agreement we find that PC2 mRNA is much more abundant than PC1 mRNA in the numointermediate labe (Seidah et al., 1991; Marcinkiewicz et al., 1993b; Fig. 3). Therefore, our data are consistent with actions of both PC1 and PC2 in this tissue to efficiently process POMC into its final physiological products. Furthermore, the ontogeny of PC1 and PC2 constitutes a good example on the relationship between substrate and enzyme when they are colocalized in the same cells (Marcinkiewicz et al., 1993b).

Processing of proneurotrophins

The five known convertases which are expressed in either constitutively secreting and/or regulated cells (Seidah et al., 1993, 1994) and appear to be involved in the processing of proneurotrophins. We have recently carried out a detailed comparative analysis of the processing of different neutrophin precursors (proNGF, proBDNF and proNT3) by the convertases in both cell types. The results of this work show that in co-infected cells the candidate processing enzymes are furin, PACE4 and PC5/6-B. This conclusion was reached for both constitutive and regulated cells.

Furin is a likely candidate for processing proneurotrophins since this enzyme is ubiquitously expressed, including in cells that generate neurotrophins (Day et al., 1993; Schäfer et al., 1993) and it is produced early in embryonic development (Zheng et al., 1994). However, our data revealed that production of mature NGF and NT3 occurs to a small extent in the human colon adenocarcinoma cell line LoVo which are devoid of furin activity, suggesting that other mammalian convertases, in addition to furin, can process neurotrophin precursors. PACE4 which is produced by LoVo cells, seems a likely another candidate. Our data and the *in situ* hybridization histochemical localization of furin and PACE4 in glial cells of central nervous system and of furin in the hippocampus – both are sources of neurotrophins - support this hypothesis. The third candidate proneurotrophin convertase that we propose is PC5/6-B. This isoform is almost double of the size of the more widely expressed PC5 and has an extended Cys-rich domain and a putative transmembrane sequence close to the carboxy-terminus (Nakagawa et al ., 1993). Presumably it derives from a single gene (Mbikay et al., 1995) by alternative splicing. PC5/6-B is abundantly expressed in the epithelial cells of the small intestine and in the adrenal cortex. However, in order to substantiate the possible physiological involvement of PC5/6-B in proneurotrophin processing, it will be necessary to colocalize the protein in various tissues containing one or more members of the neurotrophin family. We should mention that the newly discovered PC7 (Seidah et al., 1996) could also be involved, since it is present in the brain and highly concentrated in the hippocampus.

Conclusion

In conclusion, the work presented here provides a framework for understanding the various steps involved in the biosynthesis of the neuropeptides and neurotrophins. Future studies of this complex phenomenon will afford many new insights into the mechanism of neurotrophin regulation and synthesis *in vivo*.

Over 25 years elapsed since the discovery that prohormones are activated at pairs of basic residues. The enzymes involved in the post-translational processing of protein and hormonal precursors were only recently identified and molecularly characterized. The cDNA cloning of the mRNA coding for the convertases PC1, PC2, furin, PACE4, PC4, PC5/PC6 and PC7 opens new avenues in our detailed understanding of the mechanism of action and specificity of these enzymes. Even though other PC-like enzymes may yet be discovered, the general principles which govern the zymogen activation, proteolytic specificity, gene expression and regulation and cellular localization of the convertases are being worked out in details in a number of laboratories.

Acknowledgments
The authors wish to thank Dr. W. Dong for his technical assistance. This work was supported by grants from the Medical Research Council of Canada (PG11474), the NeuroScience Network of Canada and J.A. De Sève Succession. The secretarial assistance of S. Emond is appreciated.

References

Benjannet, S., Rondeau, N., Day, R., Chrétien, M. and Seidah, N.G. (1991) PC1 and PC2 are pro-protein convertases capable of cleaving POMC at distinct pairs of basic residues. *Proc. Natl. Acad. Sci. USA* 88:3564-3568.

Chrétien, M. and Li, C.H. (1967) Isolation, purification and characterization of γ-lipotropic hormone from sheep pituitary glands. *Can. J. Biochem.* 45:1163-1174.

Chrétien, M., Benjannet, S., Gossard, F., Gianoulakis, C., Crine, P., Lis, M. and Seidah, N.G. (1979) From β-lipotropin to β-endorphin and proopiomelanocortin. *Can. J. Biochem.* 57:1111-1121.

Day, R., Schäfer, M.K.-H., Watson, S.J., Chrétien, M. and Seidah, N.G. (1992) Distribution and regulation of the prohormone convertases PC1 and PC2 in the rat pituitary. *Mol. Endocrinol.* 6:485-497.

Day, R., Schäfer, M.K.-H., Cullinan, W.E., Watson, S.J., Chrétien, M. and Seidah, N.G. (1993) Region specific expression of furin mRNA in the rat brain. *Neurosci. Lett.* 149:7-30.

Dong, W., Marcinkiewicz, M., Vieau, D., Chrétien, M., Seidah, N.G. and Day, R. (1995) Distinct mRNA expression of the highly homologous convertases PC5 and PACE4 in the rat brain and pituitary. *J. Neurosci.* 15:1778-1796.

Eipper, B., May, V., Cullen, E.I., Sato, S.M., Murthy, A.S.N. and Mains, R.E. (1987) Cotranslational and posttranslational processing in the production of bioactive peptides. *In*: H.Y. Meltzer (ed.): *Psychopharmacology: The Third Generation of Progress,* Raven Press, New York, pp. 385-400.

Feany, M.B. and Quinn, W.G. (1995) A neuropeptide gene defined by the *Drosophila* memory mutant *amnesiac. Science* 268:869-873.

Hökfelt, T., Johansson, O., Ljungdahl, A., Lundbert, J.M. and Schultzberg, M. (1980) Peptidergic neurons. *Nature* 284:515-521.

Lazure, C., Seidah, N.G., Pélaprat, D. and Chrétien, M. (1983) Proteases and post-translational processing of prohormones: A review. *Can. J. Biochem. Cell Biol.* 61:501-515.

Levi-Montalcini, R. (1987) The nerve growth factor 35 years later. *Science* 237, 1154-1162.

Maisonpierre, P.C., Belluscio, L., Squinto, S., Ip, N.Y., Furth, M.E., Lindsay, R.M. and Yancopoulos, G.D. (1990) Neurotrophin-3: a neurotrophic factor related to NGF and BDNF. *Science* 247:1446-1451.

Marcinkiewicz, M., Seidah, N.G. et Chrétien, M. (1993a) L'importance de nouvelles convertases des proprotéines et prohormones pour le système nerveux: données récentes. *Médecine/Sciences* 9:553-561.

Marcinkiewicz, M., Day, R., Seidah, N.G. and Chrétien, M. (1993b) Ontogeny of the prohormone convertases PC1 and PC2 in the mouse hypophysis and their colocalization with corticotropin and α-melanotropin. *Proc. Natl. Acad. Sci. USA* 90:4922-4926.

Mbikay, M., Seidah, N.G., Chrétien, M. and Simpson, E.M. (1995) Chromosomal assignment of the genes of proprotein convertases PC4, PC5 and PACE 4 in mouse and human. *Genomics* 26:123-129.

Nakagawa, T., Hosaka, M. and Nakayama, K. (1993) Identification of an isoform with an extremely large Cys-rich region of PC6, a kex-like processing endoprotease. *FEBS Lett.* 327:165-171.

Schäfer, M.K.-H., Day, R., Cullinan, W.E., Chrétien, M., Seidah, N.G. and Watson, S.J (1993) Gene expression of prohormone and proprotein convertases in the rat CNS: A comparative *in Situ* hybridization analysis. *J. Neurosci.* 13:1258-1279.

Seidah, N.G., Benjannet, S., Pareek, S., Savaria, D., Hamelin, J., Laliberté, J., Lazure, C., Chrétien, M. and Murphy, R.A. (1996) Cellular processing of the NGF precursor by the mammalian pro-protein convertases. *Biochem. J.*; in press.

Seidah, N.G. and Chrétien, M. (1992) Pro-protein and prohormone convertases of the subtilisin family - Recent developments and future perspectives. *Trends Endocrinol. Metabolism.* 3:131-138.

Seidah, N.G. and Chrétien, M. (1994) Pro-protein convertases of subtilisin/kexin family. *Methods Enzymol.* 244:175-188.

Seidah, N.G., Chrétien, M. and Day, R. (1994) The family of subtilisin/kexin like pro-protein and prohormone convertases: divergent or shared functions. *Biochimie* 76:197-209.

Seidah, N.G., Day, R. and Chrétien, M. (1993) The family of prohormone and pro-protein convertases. *Biochem. Soc. Trans.* 21:685-691.

Seidah, N.G., Gaspar, L., Mion, P., Marcinkiewicz, M., Mbilay, M. and Chrétien, M. (1990) cDNA sequence of two distinct pituitary proteins homologous to Kex2 and Furin gene products: Tissue-specific mRNAs encoding candidates for prohormone processing proteinases. *DNA* 9:415-424.

Seidah, N.G., Hamelin, J., Mamarbachi, M., Dong, W., Tadros, H., Mbikay, M., Chrétien, M. and Day, R. (1996) cDNA structure, tissue distribution and chromosomal localization of rat PC7: a novel mammalian proprotein convertase closest to yeast kexin-like proteinases. *Proc. Natl. Acad. Sci. USA; in press.*

Seidah, N.G., Marcinkiewicz, M., Benjannet, S., Gaspar, L., Beaubien, G., Mattei, M.G., Lazure, C., Mbikay, M. and Chrétien, M. (1991) Cloning and primary sequence of a mouse candidate prohormone convertase PC1 homologous to PC2, Furin, and Kex2: Distinct chromosomal localization and messenger RNA distribution in brain and pituitary compared to PC2. *Mol. Endocrinol.* 5:111-122.

Seidah, N.G., Rochemont, J., Hamelin, J., Benjannet, S. and Chrétien, M. (1981) The missing fragment of the pro-sequence of human proopiomelanocortin: sequence and evidence for C-terminal amidation. *Biochem. Biophys. Res. Commun.* 102:710-716.

Steiner, D.F., Cunningham, D., Spiegelman, L. and Aten, B. (1967) Insulin biosynthesis: evidence for a precursor. *Science* 157:697-700.

Zheng, M., Streck, R.D., Scott, R.E.M., Seidah, N.G. and Pintar, J.E. (1994) The developmental expression in rat of proteases furin, PC1, PC2, and carboxypeptidase E: Implications for early maturation of proteolytic processing capacity. *J. Neurosci.* 14:4656-4673.

The Peptidergic Neuron
B. Krisch and R. Mentlein (eds)
© 1996 Birkhäuser Verlag Basel/Switzerland

Biosynthesis of neuropeptides in cnidarians: evidence for unusual neuropeptide precursor processing enzymes

C.J.P. Grimmelikhuijzen and I. Leviev

Department of Cell Biology and Anatomy, University of Copenhagen, Universitetsparken 15, DK - 2100 Copenhagen Ø, Denmark

Summary. Evolutionarily "old" nervous systems, such as those of cnidarians, are strongly peptidergic. From a single sea anemone species, *Anthopleura elegantissima*, we and other collegues have recently isolated 17 different neuropeptides. Many of these neuropeptides are located in neuronal secretory vesicles and have excitatory or inhibitory actions on muscle preparations or isolated muscle cells, suggesting that they are neurotransmitters or neuromodulators. One of the sea anemone neuropeptides, <Glu-Gln-Pro-Gly-Leu-Trp-NH$_2$ (metamorphosin A, or MMA) induces metamorphosis in a hydroid planula larva to become a hydropolyp. This shows that cnidarian neuropeptides also can be neurohormones that control developmental processes, such as growth, differentiation and pattern formation. We have cloned the precursor proteins for most of the cnidarian neuropeptides. Here, we give the precursor protein for MMA as an example. Prepro-MMA contains 10 copies of authentic, immature MMA (Gln-Gln-Pro-Gly-Leu-Trp-Gly) and 27 other, closely related, putative neuropeptide sequences. The 10 copies of immature MMA are preceded by the sequence Xaa-Ala-Xaa-Pro, suggesting that dipeptidyl aminopeptidase IV plays a role in the final processing of Xaa-Ala-Xaa-Pro-MMA. This illustrates that primitive nervous systems use unusual processing enzymes for the maturation of their neuropeptide precursors.

Introduction

Neuropeptides are found throughout the animal kingdom and are especially abundant in evolutionarily "old" nervous systems, such as those of cnidarians, the lowest animal group having a nervous system. Using various radioimmunoassays, we have isolated 16 different neuropeptides from a single sea anemone species, *Anthopleura elegantissima*. These peptides are all structurally related and contain the C-terminal sequence Arg-Xaa-NH$_2$ or Lys-Xaa-NH$_2$, where Xaa is Ala, Asn, Ile, Phe, Pro, or Trp (for reviews see Grimmelikhuijzen et al., 1992, 1995). The peptides are located in neuronal secretory granules and have excitatory or inhibitory actions on muscle preparations or isolated muscle cells from sea anemones, suggesting that they are neurotransmitters or neuromodulators (McFarlane et al., 1991; Grimmelikhuijzen et al., 1992, 1995; Westfall and Grimmelik-huijzen, 1993; McFarlane et al., 1993; Westfall et al., 1995).

Cnidarians often have a life cycle including a polyp, a medusa and a planula larva stage. Recently, Leitz and coworkers (1994, 1995) have isolated a neuropeptide from *A. elegantissima* that induces metamorphosis in planula larvae of the marine hydroid *Hydractinia echinata*. This peptide, <Glu-Gln-Pro-Gly-Leu-Trp-NH$_2$ (metamorphosin A or MMA) has an interesting structure, because it does not belong to the large family of Arg-Xaa-NH$_2$ or Lys-Xaa-NH$_2$ neuropeptides present in sea anemones. The work of Leitz and coworkers is also interesting, because it shows that neuropeptides in cnidarians, in addition to being neurotransmitters or neuromodulators, also can be neurohormones that control developmental processes such as growth, differentiation and pattern formation.

We have cloned the precursor proteins for most of the cnidarian neuropeptides. The precursor protein for the neuropeptide Antho-RFamide (<Glu-Gly-Arg-Phe-NH$_2$) from *A. elegantissima*, for example, contains 13 copies of immature Antho-RFamide (Gln-Gly-Arg-Phe-Gly) and 20 copies of other, putative neuropeptides (Schmutzler et al., 1992). Each immature Antho-RFamide sequence was followed, at its C terminus, by one or two basic amino acid residues that are established cleavage sites for endoproteolytic precursor processing (Sossin et al., 1989). At the N terminus of the Antho-RFamide sequences, however, basic residues were lacking and, instead, acidic amino acid residues (Asp and Glu) occurred. This means that a novel prohormone processing enzyme cleaving at the C terminal sides of acidic residues must be present in cnidarian neurones. We have found these acidic cleavage sites in many other neuropeptide precursor proteins from sea anemones, sea pansies, *Hydra*, hydromedusae and scyphomedusae (Darmer et al., 1991; Grimmelikhuijzen et al., 1994, 1995; Reinscheid and Grimmelikhuijzen, 1994; Schmutzler et al., 1994). By cloning a large number of cnidarian neuropeptide precursor proteins, we also found evidence for cleavage at the C-terminal sides of Ser/Thr, Asn, Tyr, and Xaa-Ala or Xaa-Pro sequences (reviewed in Grimmelikhuijzen et al., 1994, 1995). Our data suggest, therefore, that primitive nervous systems use a large variety of unusual prohormone

processing enzymes, in addition to the number that we know so far from higher animals. In the following, we give an example of a cnidarian neuropeptide precursor protein, in order to illustrate that (1) neuropeptide biosynthesis is very efficient in cnidarians and, (2) unusual processing enzymes play an important role in neuropeptide precursor processing.

```
CAATGAACTGAGTGGAACACAAGTAATACATATTCTTCACTTCGGTTGATA ATG GCC CTC AAG TGT CAT CTA GTT CTA CTG    81
                                                   Met Ala Leu Lys Cys His Leu Val Leu Leu    10

GCC ATT ACT TTA CTA TTA GCA CAG TGT TCA GGG TCA GTA GAC AAG AAG GAT AGT ACG ACG AAT CAC TTA   150
Ala Ile Thr Leu Leu Leu Ala Gln Cys Ser Gly Ser Val Asp Lys Lys Asp Ser Thr Thr Asn His Leu    33

GAT GAG AAG AAA ACA GAT TCC ACA GAA GCA CAT ATT GTA CAA GAA ACA GAC GCG TTA AAA GAA AAT TCT   219
Asp Glu Lys Lys Thr Asp Ser Thr Glu Ala His Ile Val Gln Glu Thr Asp Ala Leu Lys Glu Asn Ser    56

TAT CTT GGC GCC GAG GAG GAA TCT AAA GAA GAA GAC AAG AAG AGA TCC GCC GCT CCT CAG CAG CCT GGC   288
Tyr Leu Gly Ala Glu Glu Glu Ser Lys Glu Glu Asp Lys Lys Arg Ser Ala Ala Pro **Gln Gln Pro Gly**    79

CTC TGG GGG AAA CGC CAG AAA ATA GGA CTA TGG GGA AGA TCC GCT GAC GCA GGA CAG CCA GGC CTC TGG   357
**Leu Trp Gly** Lys Arg Gln Lys Ile Gly Leu Trp Gly Arg Ser Ala Asp Ala Gly Gln Pro Gly Leu Trp   102

GGC AAA CGA CAA AGT CCC GGA TTA TGG GGA AGA TCC GCT GAC GCA GGA CAG CCA GGC CTC TGG GGC AAA   426
Gly Lys Arg Gln Ser Pro Gly Leu Trp Gly Arg Ser Ala Asp Ala Gly Gln Pro Gly Leu Trp Gly Lys   125

CGT CAA AAT CCC GGA TTA TGG GGA AGA TCC GCT GAC GCA GGA CAG CCA GGC CTC TGG GGC AAA CGT CAA   495
Arg Gln Asn Pro Gly Leu Trp Gly Arg Ser Ala Asp Ala Gly Gln Pro Gly Leu Trp Gly Lys Arg Gln   148

AAT CCC GGA TTA TGG GGA AGA TCG GCT GAC GCA GGA CAG CCA GGC CTC TGG GGC AAA CGT CAA AAT CCC   564
Asn Pro Gly Leu Trp Gly Arg Ser Ala Asp Ala Gly Gln Pro Gly Leu Trp Gly Lys Arg Gln Asn Pro   171

GGA TTA TGG GGA AGG TCC GCT GAC GCA AGA CAA CCC GGA CTC TGG GGC AAA CGT CAA AAT CCC GGA TTA   633
Gly Leu Trp Gly Arg Ser Ala Asp Ala Arg Gln Pro Gly Leu Trp Gly Lys Arg Glu Ile Tyr Ala Leu   194

TGG GGA GGA AAA CGT CAA AAT CCC GGA CTT TGG GGA AGA TCC GCT GAT CCA GGA CAG CCC GGC CTC TGG   702
Trp Gly Gly Lys Arg Gln Asn Pro Gly Leu Trp Gly Arg Ser Ala Asp Pro Gly Gln Pro Gly Leu Trp   217

GGC AAA CGT GAA CTC GTC GGA TTA TGG GGG GGA AAA CGT CAA AAC CCC GGA TTG TGG GGA AGA TCG GCT   771
Gly Lys Arg Glu Leu Val Gly Leu Trp Gly Gly Lys Arg Gln Asn Pro Gly Leu Trp Gly Arg Ser Ala   240

GAA GCA GGA CAG CCA GGA CTT TGG GGA AAA CGC CAA AAA ATA GGA TTG TGG GGA CGT TCG GCT GAC CCA   840
Glu Ala Gly Gln Pro Gly Leu Trp Gly Lys Arg Gln Lys Ile Gly Leu Trp Gly Arg Ser Ala Asp Pro   263

CTT CAG CCT GGC CTC TGG GGC AAA CGT CAA AAT CCC GGA TTA TGG GGA AGA TCT GCT GAC CCG CAG CAG   909
Leu Gln Pro Gly Leu Trp Gly Lys Arg Gln Asn Pro Gly Leu Trp Gly Arg Ser Ala Asp Pro **Gln Gln**   286

CCT GGC CTC TGG GGC AAA CGT CAA AAT CCC GGA TTA TGG GGA AGA TCT GCT GAC CCG CAG CAG CCT GGC   978
**Pro Gly Leu Trp Gly** Lys Arg Gln Asn Pro Gly Leu Trp Gly Arg Ser Ala Asp Pro **Gln Gln Pro Gly**   309

CTC TGG GGC AAA CGT CAA AAT CCC GGA TTA TGG GGA AGA TCT GCT GAC CCG CAG CAG CCT GGC CTC TGG  1047
**Leu Trp Gly** Lys Arg Gln Asn Pro Gly Leu Trp Gly Arg Ser Ala Asp Pro **Gln Gln Pro Gly Leu Trp**   332

GGC AAA CGT CAA AAT CCC GGA TTA TGG GGA AGA TCT GCT GAC CCG CAG CAA CCT GGC CTC TGG GGC AAA  1116
**Gly** Lys Arg Gln Asn Pro Gly Leu Trp Gly Arg Ser Ala Asp Pro **Gln Gln Pro Gly Leu Trp Gly** Lys   355

AGC CCC GGT TTA TGG GGA CGA TCC GCT GAC CCA CAA CAG CCT GGA CTT TGG GGG AAA CGC CAA AAT CCC  1185
Ser Pro Gly Leu Trp Gly Arg Ser Ala Asp Pro **Gln Gln Pro Gly Leu Trp Gly** Lys Arg Gln Asn Pro   378

GGA TTT TGG GGA AGA TCT GCT GAC CCG CAG CAG CCT GGC CTC TGG GGC AAA CGT CAA AAT CCC GGA TTA  1254
Gly Phe Trp Gly Arg Ser Ala Asp Pro **Gln Gln Pro Gly Leu Trp Gly** Lys Arg Gln Asn Pro Gly Leu   401

TGG GGA AGA TCT GCT GAC CCG CAG CAA CCT GGC CTC TGG GGC AAA CGT CAA AAT CCC GGA TTA TGG GGA  1323
Trp Gly Arg Ser Ala Asp Pro **Gln Gln Pro Gly Leu Trp Gly** Lys Arg Gln Asn Pro Gly Leu Trp Gly   424

AGA TCT GCT GAC CCG CAG CAA CCT GGC CTC TGG GGC AAA CGT CAA AAC CCC GGT TTA TGG GGA CGA TCC  1392
Arg Ser Ala Asp Pro **Gln Gln Pro Gly Leu Trp Gly** Lys Arg Gln Asn Pro Gly Leu Trp Gly Arg Ser   447

GCT GAC CCA CAA CAG CCT GGA CTT TGG GGG AAA CGC CAA AAT CCA GGA CTA TGG GGA AGA AGT GCT GGC  1461
Ala Asp Pro **Gln Gln Pro Gly Leu Trp Gly** Lys Arg Gln Asn Pro Gly Leu Trp Gly Arg Ser Ala Gly   470

TCC GGT CAA CTC GGA CTT TGG GGT AAA AGG CAA TCA CGC ATT GGA TTA TGG GGA AGA TCT GCC GAG CCT  1530
Ser Gly Gln Leu Gly Leu Trp Gly Lys Arg Gln Ser Arg Ile Gly Leu Trp Gly Arg Ser Ala Glu Pro   493

CCA CAA TTT GAA GAT TTA GAA GAT TTA AAG AAA AAA TCA GCA ATT CCC CAA CCA AAA GGA CAA TGA TAA  1599
Pro Gln Phe Glu Asp Leu Glu Asp Leu Lys Lys Lys Ser Ala Ile Pro Gln Pro Lys Gly Gln Stop Stop   514
```

Fig. 1. cDNA and deduced amino acid sequence of the MMA precursor from *A. elegantissima*. Only the coding sequence of the cDNA is shown (for the complete sequence see Leviev and Grimmelikhuijzen, 1995). Authentic MMA sequences are underlined and printed bold-faced type. Highly likely, but putative neuropeptide sequences are underlined only. Amino acid sequences that have an uncertain status are underlined by a stippled line. Modified from Leviev and Grimmelikhuijzen, 1995.

Biosynthesis of MMA

We have cloned the precursor protein for the metamorphosis inducing neuropeptide MMA from *A. elegantissima* (Fig.1). This precursor contains 10 copies of immature MMA (Gln-Gln-Pro-Gly-Leu-Trp-Gly), of which 9 copies are regularly distributed in the C-terminal half of the protein. All immature MMA copies are followed and preceded by basic cleavage sites (arrows in the upper 3 lines of Table I). After cleavage has taken place at these sites, however, the immature MMA sequences are still N-terminally elongated by the sequence Ser-Ala-Xaa-Pro. As authentic MMA has been isolated from *A. elegantissima*, this clearly proves that there must be processing at the C-terminal sides of Pro residues. The most likely enzyme catalyzing this cleavage is dipeptidyl aminopeptidase IV (DPP) that cleaves at the C-terminal sides of N terminally located Xaa-Ala and Xaa-Pro sequences (Kreil, 1990; Mentlein, 1988). DPP would remove the N-terminal elongations of MMA in two steps, first cleaving after Ser-Ala and then after the Xaa-Pro sequence.

DPP is known to be involved in the final processing of yeast α-mating factor, honey bee melittin, moth cecropin and the frog skin peptides xenopsin and caerulein (reviewed in Kreil, 1990). All these peptides are biologically active, but they are not neuropeptides. The involvement of DPP in the final processing of immature MMA would be one of the first examples that DPP plays a role in the processing of a neuropeptide precursor. Recently, we have proposed that DPP plays a similar role in the processing of two other cnidarian neuropeptide precursors (Grimmelikhuijzen et al., 1994).

In addition to the 10 copies of immature MMA, there is a large number of other, putative neuropeptide sequences that are closely related to authentic MMA (Table I). For reasons of simplicity, we have named the most frequent, putative neuropeptide sequence Antho-LWamide I (14 copies), the authentic metamorphosis inducing peptide (MMA) isolated by Leitz et al. (1994) Antho-LWamide II (10 copies), a third peptide occurring in high copy number Antho-LWamide III (6 copies) and other, closely related peptides Antho-LWamides IV-IX (Table I). The 14 immature Antho-LWamide I sequences are flanked by basic amino acid residues, so it is likely that they will be processed and released from the precursor protein. The Antho-LWamides IV, VII and VIII have exactly the same

Table I. N- and C-terminal extensions of MMA and related neuropeptide sequences.

N- and C-terminal extensions and neuropeptide sequence	Copy number	Name
Arg↓Ser-Ala-Asp-Pro-**Gln-Gln-Pro-Gly-Leu-Trp-Gly**-Lys-Arg↓ [a]	8	MMA (Antho-LWamide II)
Arg↓Ser-Ala-Asp-Pro-**Gln-Gln-Pro-Gly-Leu-Trp-Gly**-Lys↓ [b]	1	MMA (Antho-LWamide II)
Arg↓Ser-Ala-Ala-Pro-**Gln-Gln-Pro-Gly-Leu-Trp-Gly**-Lys-Arg↓ [c]	1	MMA (Antho-LWamide II)
Lys-Arg↓Gln-Asn-Pro-Gly-Leu-Trp-Gly-Arg↓ [d]	14	Antho-LWamide I
Lys-Arg↓Gln-Ser-Pro-Gly-Leu-Trp-Gly-Arg↓ [e]	1	Antho-LWamide VII
Lys-Arg↓Gln-Lys-Ile-Gly-Leu-Trp-Gly-Arg↓ [f]	2	Antho-LWamide IV
Lys-Arg↓Gln-Ser-Arg-Ile-Gly-Leu-Trp-Gly-Arg↓ [g]	1	Antho-LWamide VIII
Arg↓Ser-Ala-Gly-Ser-Gly-Gln-Leu-Gly-Leu-Trp-Gly-Lys-Arg↓ [h]	1	Antho-LWamide IX
Arg↓Ser-Ala-Asp-Ala-Gly-Gln-Pro-Gly-Leu-Trp-Gly-Lys-Arg↓ [i]	4	Antho-LWamide III
Arg↓Ser-Ala-Glu-Ala-Gly-Gln-Pro-Gly-Leu-Trp-Gly-Lys-Arg↓ [j]	1	Antho-LWamide III
Arg↓Ser-Ala-Asp-Pro-Gly-Gln-Pro-Gly-Leu-Trp-Gly-Lys-Arg↓ [k]	1	Antho-LWamide III
Arg↓Ser-Ala-Asp-Pro-Leu-Gln-Pro-Gly-Leu-Trp-Gly-Lys-Arg↓ [l]	1	Antho-LWamide V
Arg↓Ser-Ala-Asp-Ala-Arg-Gln-Pro-Gly-Leu-Trp-Gly-Lys-Arg↓ [m]	1	Antho-LWamide VI
Lys↓Ser-Pro-Gly-Leu-Trp-Gly-Arg↓ [n]	1	
Lys-Arg↓Glu-Leu-Val-Gly-Leu-Trp-Gly-Gly-Lys-Arg↓ [o]	1	
Lys-Arg↓Glu-Ile-Tyr-Ala-Leu-Trp-Gly-Gly-Lys-Arg↓ [p]	1	
Arg↓Ser-Ala-Glu-Pro-Pro-Gln-Phe-Glu-Asp-Leu-Glu-Asp-Leu-Lys-Lys-Lys↓ [q]	1	

The sites of initial cleavage at dibasic or monobasic residues are indicated by arrows. MMA copies are underlined and printed boldface type. Highly likely, but putative peptide sequences are underlined only. Uncertain mature sequences or residues are underlined by a stippled line. The neuropeptide sequences given in this table can be found back in Fig.1 at the following amino acid positions: a: 285, 306, 327, 367, 388, 409, 430, 451; b: 348; c: 76; d: 127, 148, 169, 200, 231 273, 294, 315, 336, 376, 397, 418, 439, 460; e: 106; f: 85, 252; g: 481; h: 470; i: 97, 118, 139, 160; j: 243; k: 212; l: 264; m: 181; n: 358; o: 221; p: 190; q: 492. Modified from Leview and Grimmelikhuijzen, 1995.

processing sites as Antho-LWamide I, and it is likely that these peptides will also be produced from the Antho-LWamide precursor. All mature Antho-LWamides I, II, IV, VII and VIII will have an N-terminal <Glu residue and the C-terminal sequence Gly-Leu-Trp-NH_2 (Table I).

There are other, putative neuropeptide sequences, the Antho-LWamides III, V, VI and IX that are very similar to MMA (Table I). These immature sequences are flanked by basic residues and are N-terminally elongated by Xaa-Ala or Xaa-Pro sequences, so that they will probably be released from the precursor and processed. Their final structures, however, are still uncertain (see Leviev and Grimmelikhuijzen, 1995, for a detailed discussion). Finally, there are 4 other, putative peptide sequences flanked by basic residues (underlined by stippled lines in Fig. 1 and Table I), but it is not certain whether these will yield intact peptides (Leviev and Grimmelikhuijzen, 1995).

In summary, the Antho-LWamide precursor contains 37 immature neuropeptide sequences that are likely to yield 9 different, mature neuropeptides: the Antho-LWamides I-IX. Table II shows the putative structure of these mature neuropeptides and the extent to which these peptides are structurally related. All neuropeptides have the C-terminal sequence Gly-Leu-Trp-NH_2 in common, and also in the N-terminal regions are many sequence similarities. Leitz and coworkers (1994) have found that the metamorphosis inducing potency of MMA resides in the C-terminal part of the peptide, as N-terminal variants are still able to induce metamorphosis in planula larvae of *Hydractinia*. Thus, it is possible that many, or perhaps all of the peptides of Table II have a metamorphosis inducing capacity.

Table II. Established and putative, mature neuropeptides that could be produced from the MMA precursor protein.

Copies	Structure	Name
14	<Glu-Asn-Pro-Gly-Leu-Trp-NH$_2$	Antho-LWamide I
10	<Glu-Gln-Pro-Gly-Leu-Trp-NH$_2$	MMA (Antho-LWamide II)
6	Gly-Gln-Pro-Gly-Leu-Trp-NH$_2$	Antho-LWamide III
1	Leu-Gln-Pro-Gly-Leu-Trp-NH$_2$	Antho-LWamide V
1	Arg-Gln-Pro-Gly-Leu-Trp-NH$_2$	Antho-LWamide VI
1	<Glu-Ser-Pro-Gly-Leu-Trp-NH$_2$	Antho-LWamide VII
2	<Glu-Lys-Ile-Gly-Leu-Trp-NH$_2$	Antho-LWamide IV
1	<Glu-Ser-Arg-Ile-Gly-Leu-Trp-NH$_2$	Antho-LWamide VIII
1	Gly-Ser-Gly-Gln-Leu-Gly-Leu-Trp NH$_2$	Antho-LWamide IX

———

37

Amino acid residues that are identical with those of MMA are highlighted by boxes. Residues that might possibly not be contained in the mature peptides are underlined with a stippled line. Modified from Leviev and Grimmelikhuijzen, 1995.

Acknowledgements
We thank Astrid Juel Jensen for typing the manuscript and the Danish Natural Science Research Council for financial support.

References

Darmer, D., Schmutzler, C., Diekhoff, D. and Grimmelikhuijzen, C.J.P. (1991) Primary structure of the precursor for the sea anemone neuropeptide Antho-RFamide (<Glu-Gly-Alg-Phe-NH$_2$). *Proc. Natl. Acad. Sci. USA* 88: 2555-2559.

Grimmelikhuijzen, C.J.P., Carstensen, K., Darmer, D., Moosler, A., Nothacker, H.-P., Reinscheid, R.K., Schmutzler, C., Vollert, H., McFarlane, I.D. and Rinehart, K.L. (1992) Coelenterate neuropeptides: structure, action and biosynthesis. *Amer. Zool.* 32: 1-12.

Grimmelikhuijzen, C.J.P., Darmer, D., Schmutzler, C., Reinscheid, R.K. and Carstensen, K. (1994) Biosynthesis of neuropeptides in the Cnidaria: new discoveries of old principles. *In*: K.G. Davey, R.E. Peter and S.S. Tobe (eds.): *Perspectives in Comparative Endocrinology*, National Research Council Canada, Ottawa: 97-108.

Grimmelikhuijzen, C.J.P., Leviev, I., and Carstensen, K. (1995) Peptides in the nervous systems of cnidarians: structure, function and biosynthesis. *Int. Rev. Cytol.; in press.*

Kreil, G. (1990) Processing of precursors by dipeptidylaminopeptidases: A case of molecular ticketing. *Trends Biochem. Sci.* 15: 23-26.

Leitz, T., Morand, K. and Mann, M. (1994) Metamorphosin A: A novel peptide controlling development of the lower metazoan *Hydractinia echinata* (Coelenterata, Hydrozoa). *Dev. Biol.* 163: 440-446.

Leitz, T. and Lay, M. (1995). Metamorphosin A is a neuropeptide. *Roux. Arch. Dev. Biol.* 204: 276-279.

Leviev, I. and Grimmelikhuijzen, C.J.P. (1995) Molecular cloning of a preprohormone from sea anemones containing numerous copies of a metamorphosis inducing neuropeptide: a likely role for dipeptidyl aminopeptidase in neuropeptide precursor processing. *Proc. Natl. Acad. Sci. USA; in press.*

McFarlane, I.D., Anderson, P.A.V. and Grimmelikhuijzen, C.J.P. (1991) Effects of three anthozoan neuropeptides, Antho-RWamide I, Antho-RWamide II and Antho-RFamide, on slow muscles from sea anemones. *J. Exp. Biol.* 156: 419-431.

McFarlane, I.D., Hudman, D., Nothacker, H.-P. and Grimmelikhuijzen, C.J.P. (1993) The expansion behaviour of sea anemones may be coordinated by two inhibitory neuropeptides, Antho-KAamide and Antho-RIamide. *Proc. Roy. Soc. B (London)* 253: 183-188.

Mentlein, R. (1988) Proline residues in the maturation and degradation of peptide hormones and neuropeptides. *FEBS Lett.* 234: 251-256.

Reinscheid, R.K. and Grimmelikhuijzen, C.J.P. (1994) Primary structure of the precursor for the anthozoan neuropeptide Antho-RFamide from *Renilla köllikeri*: evidence for unusual processing enzymes. *J. Neurochem.* 62: 1214-1222.

Schmutzler, C., Darmer, D., Diekhoff, D. and Grimmelikhuijzen, C.J.P. (1992) Identification of a novel type of processing sites in the precursor for the sea anemone neuropeptide Antho-RFamide (<Glu-Gly-Arg-Phe-NH$_2$) from *Anthopleura elegantissima. J. Biol. Chem.* 267: 22534-22541.

Schmutzler, C., Diekhoff, D. and Grimmelikhuijzen, C.J.P. (1994) The primary structure of the Pol-RFamide neuropeptide precursor protein from the hydromedusa *Polyorchis penicillatus* indicates a novel processing proteinase activity. *Biochem. J.* 299: 431-436.

Sossin, W.S., Fisher, J.M. and Scheller, R.H. (1989) Cellular and molecular biology of neuropeptide processing and packaging. *Neuron* 2:1407-1417.

Westfall, J.A. and Grimmelikhuijzen, C.J.P. (1993) Antho-RFamide immunoreactivity in neuronal synaptic and nonsynaptic vesicles of sea anemones. *Biol. Bull.* 185: 109-114.

Westfall, J.A., Sayyar, K.L., Elliott, C.F. and Grimmelikhuijzen, C.J.P. (1995) Ultrastructural localization of Antho-RWamide I and II at neuromuscular synapses in the gastrodermis and oral sphincter muscle of the sea anemone *Calliactis parasitica. Biol. Bull.; in press.*

The Peptidergic Neuron
B. Krisch and R. Mentlein (eds)
© 1996 Birkhäuser Verlag Basel/Switzerland

Evidence for a receptor-mediated mechanism for sorting pro-opiomelanocortin to the regulated secretory pathway

Y.P. Loh, C.R. Snell[1] and D.R. Cool

Section on Cellular Neurobiology, Laboratory of Developmental Neurobiology, National Institute of Child Health and Human Development, National Institutes of Health, Bethesda, MD 20892, USA
[1] *The Sandoz Institute for Medical Research, 5, Gower Place, London WCIE 6BN, UK*

Summary: A sorting signal which directs the prohormone pro-opiomelanocortin (POMC) to the regulated secretory pathway was identified. This sorting signal motif consists of a 13 amino acid ($POMC_{8-20}$) amphipathic loop stabilized by a disulfide bridge (Cys_8-Cys_{20}) located at the N-terminus of pro-opiomelanocortin. It binds at an acidic pH specifically to an ~64 kD sorting receptor protein present in Golgi and secretory granule membranes of neuroendocrine cells. These findings support the hypothesis that pro-opiomelanocortin is targeted to the regulated secretory pathway by a receptor-mediated mechanism.

Introduction

Neuropeptides and peptide hormones are synthesized in neuroendocrine cells as larger precursors which are subsequently proteolytically processed at paired/mono-basic residues to yield the biologically active peptide products (Loh, 1993). The proneuropeptides and prohormones, like other secretory, lysosomal and plasma membrane proteins, traverse a common pathway in the cell from the site of synthesis at the rough endoplasmic reticulum, through its cisternae and into the Golgi apparatus. At the *trans*-Golgi network (TGN), the intracellular routing of these proteins diverge. Lysosomal proteins are sorted into lysosomes while proteins destined for the plasma membrane (e.g. receptors) or for secretion in a non-regulated manner are routed to the cell surface via the constitutive "default" pathway, common to all cells. Secretory proteins that are destined for release in a regulated Ca^{2+}-dependent manner, such as pro-neuropeptides, are sorted and packaged into immature secretory granules of the regulated secretory pathway (RSP) and processed (Burgess and Kelly, 1987; Orci et al., 1987).

Our laboratory has focussed on elucidating the mechanisms by which proneuropeptides and prohormones are sorted at the *trans*-Golgi network into the granules of the regulated secretory pathwork of neuroendocrine cells. The ACTH-endorphin prohormone, pro-opiomelanocortin (POMC, see Fig. 1) which is synthesized in the pituitary and brain, is used as the model system for these studies.

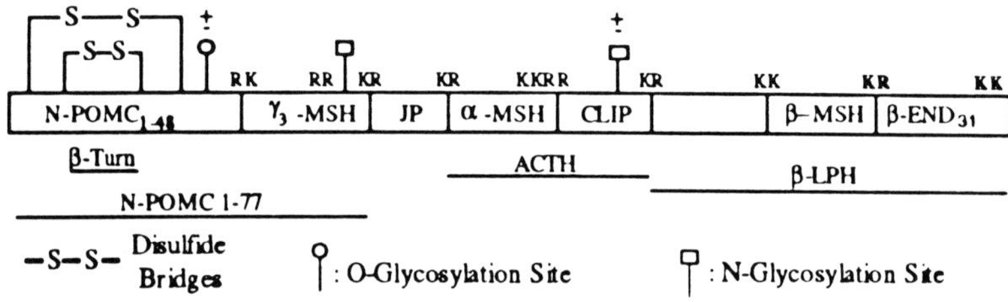

Fig. 1. Structure of bovine pro-opiomelanocortin. Disulfide bridges and the amphipathic loop are shown.

Identification of a sorting signal for the regulated secretory pathway

Proteins which are sorted into or retained in specific subcellular compartments often contain a unique molecular signal directing them to their final cellular destination. Most notable of these are the KDEL signals Lys-Asp-Glu-Leu found in the primary amino acid sequence of proteins retained in the rough endoplasmatic reticulum, and the mannose-6-phosphate signal for targeting proteins to the lysosomes (Munro and Pelham, 1987). However, comparison of the amino acid sequences of numerous proteins destined for the regulated secretory pathway has failed to reveal a consensus sequence that may act as a sorting signal for this route. This raises the possibility that such sorting signals may instead be encoded by a specific conformation-dependent motif within the secondary structure of the protein. Several studies have been done to identify a sorting signal for the regulated secretory pathway. The N-terminal 78 amino acids of pro-somatostatin (Sevarino et al., 1989; Stoller and Shields, 1989) and the loop domain between Cys_{16} and Cys_{37} of human chromogranin B (Chanat et al., 1993) were found to contain sorting information for the regulated secretory pathway. However, neither of these studies have defined the regulated secretory pathway sorting signal motif.

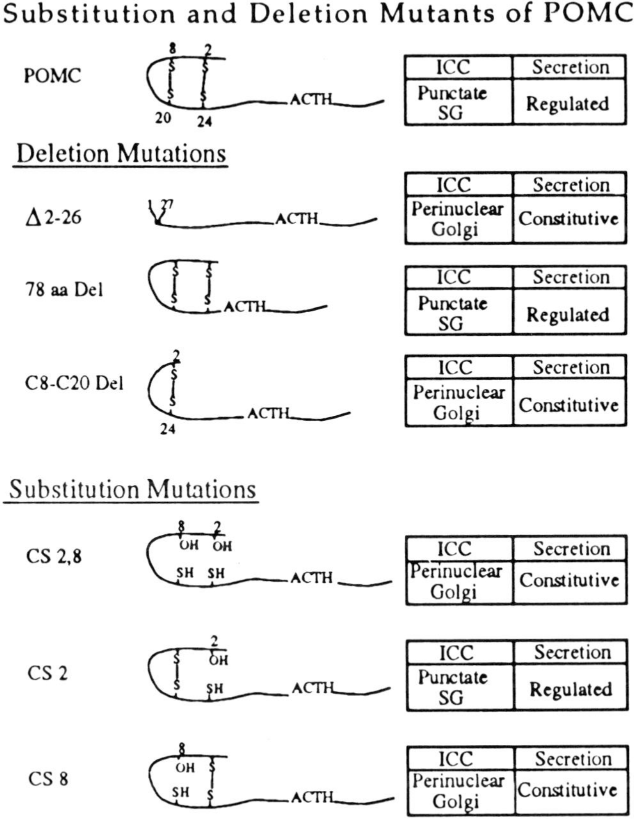

Fig. 2. Diagrammatic summary of POMC and the various substitution and deletion mutants.

Pro-opiomelanocortin contains a highly conserved domain within the N-terminal 26 amino acids which is represented by a unique hairpin loop structure stabilized by two disulfide bridges, Cys_2-Cys_{24} and Cys_8-Cys_{20} (Fig. 2). We have carried out deletion and substitution mutation studies and showed that this region of Pro-optiomelanocortin contains a sorting signal motif for the regulated secretory pathway. Three deletion mutations (see Fig.2) were made as follows: i) amino acids 2-26 were deleted (Δ2-26); ii) 78 amino acids from Lys_{24} to Arg_{101} were deleted (78 aa Del); iii) amino acids 8-20 were deleted to remove the loop (Cys_8-Cys_{20} Del). In a second set of three mutations, one or both disulfide bridges were disrupted by substituting a serine residue for a cysteine residue at positions 2 or 8 or both 2 and 8 (CS2; CS8; CS2,8, respectively) to determine whether the disulfide bridges were necessary for sorting, and if so, which bridge was more important (see Fig. 2). The constructs were transfected into the mouse neuroblastoma cell line, Neuro-2a, which does not endogenously synthesize pro-opiomelanocortin, but does have both a regulated and a constitutive secretory pathway. Immunocytochemistry and release studies were carried out on the transfected cells to assay for regulated secretion of the mutated Pro-opiomelanocortin molecules (Cool and Loh, 1994; Cool et al., 1995). When expressed in these cells, wild type pro-opiomelanocortin and the CS2 and 78 aa Del mutants showed secretory granules punctate immunopositive for adrenocorticotropic hormone (ACTH) and stimulated secretion with high K^+. However, when Δ2-26, Cys_8-Cys_{20} Del, CS8 or CS2,8 mutants were expressed in the Neuro-2a cells, immunoreactive ACTH was not found in punctate granules but rather in the Golgi and perinuclear region. Only constitutive secretion of unprocessed pro-opiomelanocortin was found for Neuro-2a cells expressing these mutants. This data showed that a motif containing residues 8-20 and one disulfide bridge (Cys_8-Cys_{20}) was sufficient and necessary for sorting pro-opiomelanocortin to the regulated secretory pathway. Further evidence that the N-terminal 26 amino acids were sufficient for targeting pro-opiomelanocortin to the regulated secretory pathway was provided by experiments showing that signal peptide + N-$POMC_{1-26}$, when fused to chloramphenicol acetyltransferase, directed this bacterial protein into the regulated secretory pathway in AtT20 cells (Tam et al., 1993).

The hairpin loop region of N-POMC (Cys_8-Cys_{20}) was modeled and identified as an amphipathic loop with four highly conserved amino acid residues ($Asp_{10}Leu_{11}Glu_{14}Leu_{18}$) exposed (Fig. 3) and stabilized by the disulfide bridge, Cys_8-Cys_{20}. This model is supported by recent NMR studies of the N-$POMC_{1-26}$ peptide (A. Bird, personal communication).

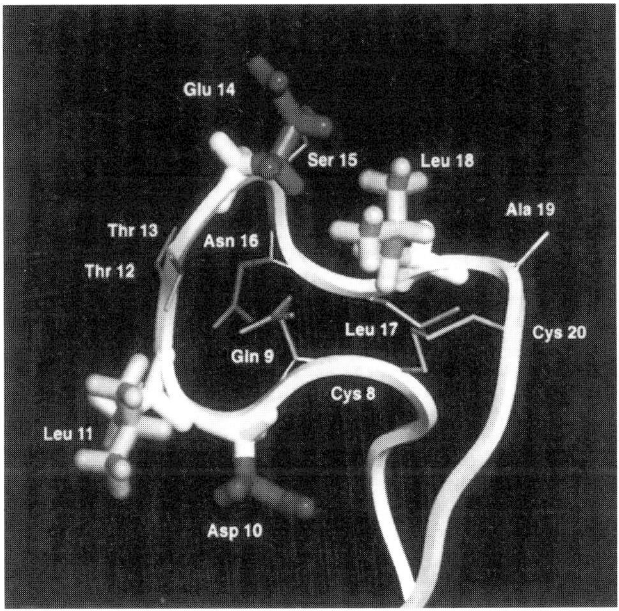

Fig. 3. Three-dimensional computer diagram of N-POMC$_{1-26}$ showing the Cys$_8$/Cys$_{20}$ disulfide bridge and the amphipathic loop portion of the lowest energy structure for POMC$_{1-26}$.

Characterization of a binding protein for the sorting signal of the regulated secretory pathway

We hypothesized that the mechanism of sorting pro-opiomelanocortin to the regulated secretory pathway involves binding of the sorting signal to a membrane receptor at the *trans*-Golgi network, which then buds off to form an immature secretory granule. Iodinated N-POMC$_{1-26}$, with the 2 disulfide bridges intact, was used as a ligand, and lysed Golgi and secretory granule membranes derived from bovine intermediate lobe cells were used as the source of membrane proteins to assay for putative sorting receptors (Cool et al., manuscript in preparation).

[^{125}I]N-POMC$_{1-26}$ was found to bind to Golgi and secretory granule membranes in a pH-dependent manner, with the optimal between pH 5.5 and 6.5. Moreover, the binding was trypsin-sensitive, indicative of binding to a protein. Fig. 4 shows that the bound [^{125}I]N-POMC$_{1-26}$ was displaced by unlabeled N-POMC$_{1-26}$ with an IC$_{50}$ = 65 µM at pH 5.5 in secretory granule membranes. Ca^{2+} had no effect on binding. The specificity of the binding of $[^{125}I]$N-POMC$_{1-26}$ was further confirmed by the lack of displacement by other parts of the pro-opiomelanocortin molecule such as ACTH and β-endorphin. Finally, cross-linking experiments revealed binding of N-POMC$_{1-26}$ to a ~ 64 kD protein. Thus, all the data indicate the existence of a protein in the Golgi and secretory granule membranes which specifically binds the pro-opiomelanocortin sorting signal. We refer to this binding protein as the regulated secretory pathway sorting receptor.

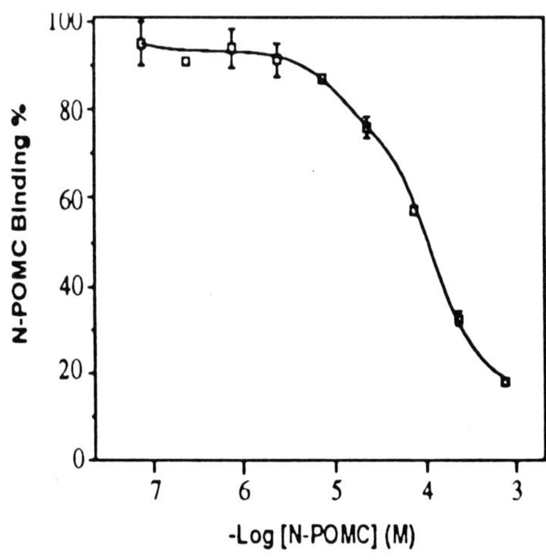

Fig. 4. Concentration-dependent inhibition of $[^{125}I]$N-POMC binding by N-POMC$_{1-26}$. Binding of 7 nM $[^{125}I]$N-POMC$_{1-26}$ to bovine secretory granule membranes was determined in the presence and absence of various concentrations of N-POMC$_{1-26}$. Specific binding in the anbsence of N-POMC$_{1-26}$ was 25.2 ± 1.3 nmoles/mg of protein and was taken as 100% binding. Values are means ± SE from 2 experiments.

A receptor-mediated mechanism for sorting pro-opiomelanocortin to the regulated secretory pathway

Fig. 5 shows our working model for the sorting of pro-opiomelanocortin to the regulated secretory pathway. This model involves 3 steps:

1) When pro-opiomelanocortin molecules arrive at the *trans*-Golgi-network, which has an acidic environment of pH 6.2 (Seksek et al., 1995), these molecules bind optimally to the sorting receptor via the sorting signal motif.

2) Since a 1:1 binding stoichiometry of pro-opiomelanocortin to the receptor would result in an inefficient mechanism for sorting and packaging large quantities of pro-opiomelanocortin, we propose that there is intermolecular aggregation of pro-opiomelanocortin. While only one pro-opiomelanocortin molecule needs to bind to the receptor, the aggregation of pro-opiomelanocortin, whether prior to binding to the receptor or subsequently, remains to be investigated. Indeed, another protein which is sorted to the regulated secretory pathway, chromogranin B, has been shown to undergo aggregation at the *trans*-Golgi network (Chanat and Huttner, 1991).

3) The *trans*-Golgi network then buds off to form an immature secretory granule containing aggregates bound to the membrane receptors. Such a sorting signal/receptor mediated mechanism for targeting pro-opiomelanocortin to the regulated secretory pathway is not only consistent with our data but also with studies on intact, or semi-intact cells which show that sorting to the regulated secretory pathway requires an acidic environment (Burgess and Kelly, 1987) and that Ca^{2+} is not required for this sorting to occur (Carnell and Moore, 1994).

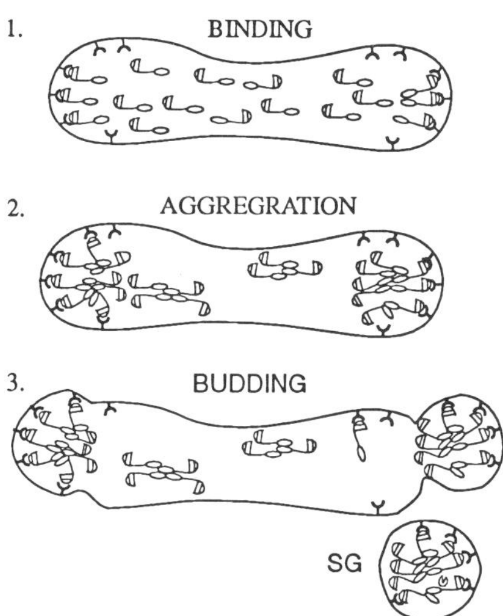

Fig. 5. Working model for the RSP sorting mechanism.
1. POMC in the TGN binds to the membrane bound receptor.
2. Aggregation of POMC.
3. Packaging of POMC and budding of immature secretory granules (SG) with processsing enzymes.

References

Burgess, T.L. and Kelly, R.B. (1987) Constitutive and regulated secretion of proteins. *Ann. Rev. Cell Biol.* 3: 243-293.

Carnell, L. and Moore, H.-P.H. (1994) Transport via the regulated secretory pathway in semi-intact PC12 cells: role of intra-cisternal calcium and pH in the transport and sorting of secretorgranin II. *J. Cell Biol.* 127: 693-705.

Chanat, E. and Huttner, W.B. (1991) Milieu-induced, selective aggregation of regulated secretory proteins in the *trans*-Golgi network *J. Cell Biol.* 115: 1505-19.

Chanat, E., Weiss, U., Huttner, W.B. and Tooze, S.A. (1993) Reduction of the disulfide bond of chromogranin B (secretogranin I) in the *trans*-Golgi network causes its missorting to the constitutive secretory pathway. *EMBO J.* 12: 2159-2168.

Cool, D.R., Fenger, M., Snell, C.R. and Loh, Y.P. (1995) Identification of the sorting signal motif within Pro-opiomelanocortin for the regulated secretory pathway. *J. Biol. Chem.* 270: 8723-8729.

Cool, D.R. and Loh, Y.P. (1994) Identification of a sorting signal for the regulated secretory pathway at the N-terminus of pro-opiomelanocortin. *Biochimie* 76: 265-270.

Loh, Y.P., Beinfeld, M. and Birch, N. P. (1993) Proteolytic processing of prohormones and pro-neuropeptides. *In*: Y.P. Loh (ed.): *Mechanisms of intracellular trafficking and processing of pro-proteins*, CRC Press, Boca Raton, Florida, pp. 179-224.

Munro, S. and Pelham, H. (1987) A C-terminal signal prevents secretion of luminal ER proteins. *Cell* 48: 899-907.

Orci, L., Ravazzola, M., Storch, M.-J., Anderson, R.G.W., Vassalli, J.-D. and Perrelet, A. (1987) Proteolytic maturation of insulin is a post-Golgi event which occurs in acidifying clathrin-coated secretory vesicles. *Cell* 49: 865-868.

Seksek, O., Biwersi, J. and Verkman, A.S. (1995) Direct measurement of *trans*-Golgi pH in living cells and regulation by second messengers. *J. Biol. Chem.* 270: 4967-4970.

Sevarino, K.A., Stock, P., Ventimiglia, R., Mandel, G. and Goodman, R. (1989) Amino-terminal sequences of prosomatostatin direct intracellular targeting but not processing specificity. *Cell* 57: 11-19.

Stoller, T.J. and Shields, D. (1989) The propeptide of preprosomatostatin mediates intracellular transport and secretion of α-globin from mammalian cells. *J. Cell Biol.* 108: 1647-1655.

Tam, W.H.H., Andreasson, K.A. and Loh, Y.P. (1993) The amino-terminal sequence of pro-opiomelanocortin directs intracellular targeting to the regulated secretory pathway. *Eur. J. Cell Biol.* 62: 294-306.

Molecular mechanisms of neurotransmitter and neuropeptide release

J. Pevsner

Department of Neuroscience, The Johns Hopkins School of Medicine and Department of Neurology, The Kennedy Krieger Institute, 707 N. Broadway, Baltimore, MD 21205 USA

Summary. In synaptic transmission, calcium entry into the presynaptic nerve terminal causes docked synaptic vesicles to fuse with the plasma membrane and release their neurotransmitter contents across the synapse. A biochemical pathway for synaptic vesicle docking and fusion is currently being elucidated. The proteins implicated in this pathway are conserved between eukaryotes from yeast to mammals, and isoforms of these proteins mediate vesicle trafficking in a variety of constitutive and regulated transport steps. It is likely that the molecular mechanisms of synaptic vesicle exocytosis also apply to the release of neuropeptides from large dense-core vesicles. We have characterized n-sec1, a cytosolic protein of the nerve terminal that binds syntaxin. We have also identified three mammalian homologues of n-sec1 that may regulate vesicle trafficking between the Golgi apparatus and lysosomes.

Introduction

In synaptic transmission, an action potential is propagated down an axon to the presynaptic nerve terminal where voltage-gated calcium channels cause a rapid, transient influx of calcium ions into the cytosol. Synaptic vesicles which have been loaded with neurotransmitter and have docked at the presynaptic plasma membrane then fuse with the cell surface to release their neurotransmitter contents across the synapse (Scheller, 1995; Südhof, 1995; Volknandt, 1995). This process of vesicle-mediated exocytosis underlies the ability of neurons to communicate with postsynaptic cells and hence to perform one of their most basic functions, information transfer.

Synaptic vesicle exocytosis represents a highly specialized case of regulated secretion, and it may serve as a valuable model system for the understanding of intracellular vesicle trafficking in general. Eukaryotic cells are organized into distinct intracellular compartments that depend on specific and efficient sorting mechanisms to deliver both membrane and soluble cargo to the appropriate destination. This sorting is generally achieved by vesicular transport of the cargo (reviewed in Pryer et al., 1992).

Three experimental approaches have recently yielded information on the molecular mechanisms of vesicle trafficking: (1) genetic analysis of organisms that are defective in a sorting pathway, such as the vacuolar protein sorting (vps) defective mutants or the secretory (sec) mutants in the yeast *Saccharomyces cerevisiae*; (2) *in vitro* reconstitution of membrane trafficking in mammalian cells; and (3) biochemical analyses of synaptic vesicles in mammalian nerve terminal (Bennett and Scheller, 1994).

List of Abbreviations

b0303.9	a *Caenorhabditis elegans* hypothetical 68 kDa protein related sec1 protein
h-vps45	human homologue of yeast Vps45p
mSec1	see n-sec-1
Munc-18	see n-sec-1
n-sec1	neural-specific homologue of yeast Sec1p; also called Munc-18 (mammalian unc-18), rbSec1A (rat brain Sec1A), rbSec1B (an alternatively spliced isoform) and mSec1 (mammalian sec1). Related mammalian isoforms include Munc-18b and Munc-18c (also called Munc-18-2).
NSF	N-ethylmaleimide sensitive factor
r-vps33a / r-vps33b	rat homologue of yeast Vps33p
rop	ras2-opposite gene, a *Drosophila* sec1 homologue
sec	secretion mutant (identified in genetic screens in the yeast *Saccharomyces cerivisiae*
Sec1p	yeast proteins essential for definite parts of intracellular trafficking
SLY	suppressor of lethality in yeast
Sly1p	suppressor of lethality of ypt1; a yeast sec1 - related protein
SNAP	synaptosome-associated protein
SNAP-25	SNAP of 25 kDa
α SNAP	soluble NSF attachment protein
SNARE	SNAP receptor
SSO1	suppressor of Sec1
SV2	synaptic vesicle antigen 2
t-SNARE	SNARE located on the target membrane (such as syntaxin)
UNC-18	uncoordinated mutant 18, a *Caenorhabditis elegans* sec1 homologue
VAMP	vesicle associated membrane protein; also called synaptobrevin
VPS	vacuolar protein-sorting mutant
Vps6p	VPS mutant 6; a yeast syntaxin-related protein; also called PEP12
Vps33p	vacuolar protein sorting mutant 33; a yeast sec1-related protein; also as Slp 1p
Vps45p	see above, mutant 45
v-SNARE	SNARE located on the vesicle membrane

These diverse approaches have led to a general model for vesicle trafficking throughout the secretory and endocytic pathways. According to the SNARE hypothesis, the specificity of vesicle targeting is guaranteed by the binding of vesicle proteins (v-SNAREs, such as the vesicle-associated membrane protein VAMP) with proteins located on the target membrane (t-SNAREs, such as syntaxin and the synaptosomal-associated protein of 25 kDa, SNAP-25)(reviewed in Rothman and Warren, 1994; Scheller, 1995). A family of VAMP and syntaxin isoforms exists, with varying binding affinities for each other, and the particular combination of v-SNAREs and t-SNAREs that interact may determine where and how a vesicle fuses with its target membrane. In a model of the SNARE hypothesis, synaptic vesicle docking and fusion at the target membrane occur through the formation and dissociation of protein complexes (Fig. 1; Söllner and Rothman, 1994; Scheller, 1995; Südhof, 1995). Initially, syntaxin is bound to the soluble protein n-sec1, essential for constitutive secretion in mammals. After n-sec1 dissociates (Fig. 1, step i), a complex forms having a sedimentation coefficient of 7S and consisting of the t-SNAREs syntaxin and SNAP-25 and the v-SNAREs VAMP and synaptotagmin. With the subsequent addition of the soluble NSF attachment protein (abbreviated αSNAP) and the N-ethylmaleimide sensitive factor (NSF)(Fig. 1, steps ii and iii), a 20S complex forms. When NSF hydrolyzes ATP (Fig. 1, step iv) this complex dissociates and the vesicle fuses with the plasma membrane by an unknown mechanism (step v).

In addition to synaptic vesicles that mediate rapid neurotransmission, nerve terminals and endocrine cells contain large dense-core vesicles. These two vesicle classes are distinguished by several prominent differences in structure and function (Cameron et al., 1993; Martin, 1994).

(1) Transmitter content: synaptic vesicles contain classical neurotransmitters such as acetylcholine, GABA, and glutamate, while large dense-core vesicles contain neuropeptides that are synthesized in the cell body, and catecholamines.

(2) Size: synaptic vesicles are homogeneous and typically 50-80 nm in diameter, while large dense-core vesicles are heterogeneous in size but typically 100-150 nm in diameter.

(3) Morphology: by electron microscopy synaptic vesicles are clear, while large dense-core vesicles are opaque.

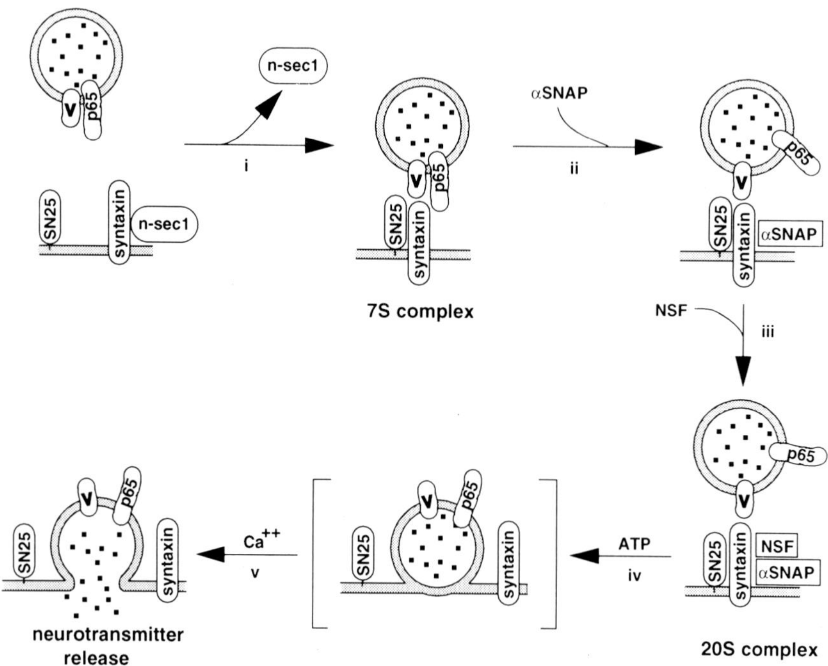

Fig. 1. A pathway for synaptic vesicle docking and fusion based on the formation and dissociation of several protein complexes. Initially, the presynaptic plasma membrane protein syntaxin is bound tightly to n-sec1, a soluble neural-enriched homologue of the yeast protein Sec1p. After dissociation of n-sec1 from syntaxin, the vesicle docks to the plasma membrane (step 1). The synaptic vesicle proteins VAMP (vesicle-associated membrane protein; abbreviated V) and synaptotagmin (abbreviated p65) bind to the plasma membrane proteins syntaxin and SNAP-25 (synaptosomal-associated protein of 25 kDa) to form a complex with an *in vitro* sedimentation coefficient of 7S. Subsequently, the two soluble proteins αSNAP (soluble NSF attachment protein) and NSF (N-ethyl maleimide sensitive factor) are added while synaptotagmin dissociates, resulting in the formation of a protein complex with a sedimentation coefficient of 20S (steps ii, iii). After ATP hydrolysis by NSF (step iv), the 20S particle dissociates and vesicle fusion occurs (step v).

(4) Vesicle protein content: synaptic vesicles contain several integral or peripheral membrane proteins that are not found on large dense-core vesicles including synapsins and synaptic vesicle antigen 2. Conversely, large dense-core vesicles contain proteins not found on synaptic vesicles including calpactin and the chromogranins (e.g. O'Connor et al., 1994).

(5) Release site: synaptic vesicles release neurotransmitters from specialized regions of the presynaptic plasma membrane called active zones. In contrast, neuropeptides are released from sites dispersed throughout the nerve terminal.

(6) Time course of release: after synaptic vesicles have docked at the plasma membrane, the fusion event and neurotransmitter release occur in less than 1 millisecond. Large dense-core vesicle -mediated release is slower, perhaps because calcium channels and large dense-core vesicles are not closely colocalized (Chow et al., 1994).

(7) Calcium dependence of release: while each vesicle type requires elevated cytosolic calcium levels at multiple stages, including both low-affinity (20-100 mM) and high-affinity calcium sensors, synaptic vesicles and large dense-core vesicles differ in their dependence on calcium for exocytosis (Thomas et al., 1993).

Despite these fundamental differences between synaptic vesicles and large dense-core vesicles, it is likely that similar molecular mechanisms underlie their function (Golding, 1994; Martin, 1994). The two vesicle types share several proteins in common including synaptotagmin, VAMP or a VAMP homologue, SNAP25, rab3A, and n-sec1 (Hodel et al., 1994; but see Garcia et al., 1995). Synaptotagmin and syntaxin are required for large dense-core vesicle exocytosis in PC12 cells (Bennett et al., 1993), and antibodies against syntaxin inhibit calcium-dependent secretion from chromaffin cells (Gutierrez et al., 1995). In addition, synaptic-like microvesicles in peptide-secreting endocrine cells contain synaptic vesicle proteins (Thomas-Reetz and de Camilli, 1994). Finally, botulinum and tetanus toxins, which selectively cleave VAMP, syntaxin and SNAP-25 to inhibit neurotransmission, also inhibit exocytosis of catecholamines from chromaffin cells.

We have focused our studies on n-sec1, a mammalian homologue of the yeast Sec1p-related proteins that are essential for constitutive secretion in yeast. Genetic screens in yeast have identified four homologous proteins that are required for vesicular transport in distinct intracellular pathways: Sec1p (Aalto et al., 1991), Vps33p (Banta et al., 1990; Wada et al., 1990), Vps45p (Cowles et al., 1994; Piper et al., 1994), and Sly1p (Ossig et al., 1991; Dascher et al., 1991). These four hydrophilic proteins are approximately 80 kDa and share 20 to 24% amino acid identity (Aalto et al., 1992). Each is essential for vesicular trafficking at distinct stages of the secretory pathway, as disruption of the gene causes an

                    J. Pevsner

abnormal accumulation of transport vesicles between endoplasmatic reticulum and Golgi (*sly1*), between Golgi and vacuole (lysosome)(*vps45* and *vps33*), or between Golgi and plasma membrane (*sec1*)(Fig. 2, top panel).

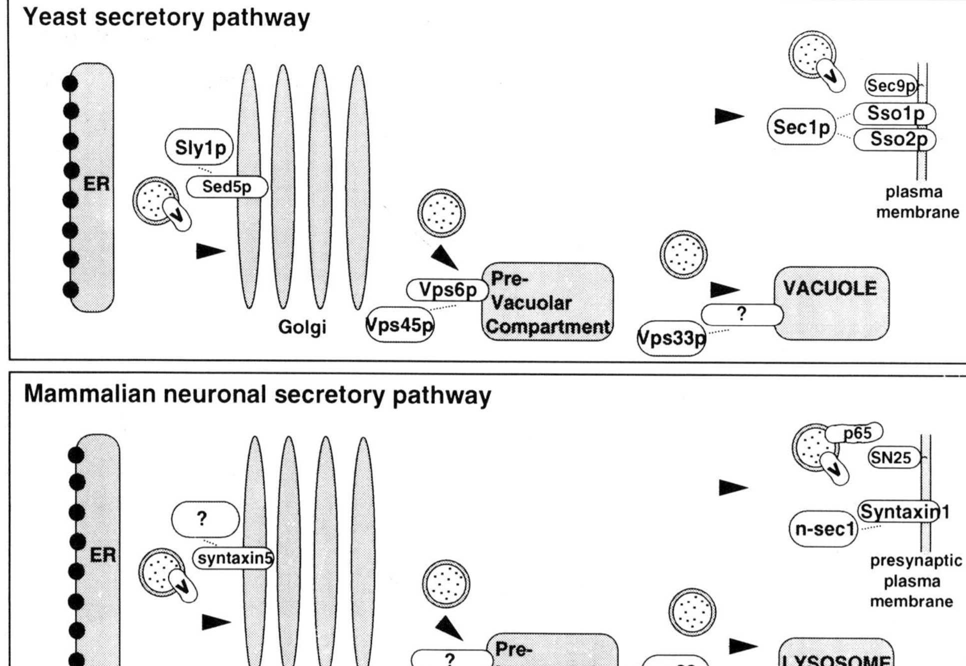

Fig. 2. Schematic of the Sec1-related proteins present in the constitutive secretory pathway in yeast (top panel) and in mammalian neurons (bottom panel). Vesicles are indicated by circles filled with dots (cargo) at multiple stages of each secretory pathway. The four yeast Sec1-related proteins are each thought to bind to a syntaxin-like protein: Sly1p (suppressor of lethality of ypt1) binds to the syntaxin-like Sed5p at the *cis*-Golgi apparatus, Vps45p may bind the syntaxin homologue Vps6p/PEP12 at a prevacuolar compartment, and Sec1p interacts with the plasma membrane syntaxins SSO1p and SSO2p (suppressor of Sec one). Abbreviations: ER, endoplasmic reticulum; V, VAMP-like protein (VAMP is the vesicle-associated membrane protein/synaptobrevin); r-vps33a and r-vps33b, rat brain homologues of yeast Vps33p; h-vps45, human brain homologue of yeast Vps45p; vps, vacuolar protein sorting mutant.

A common feature of the sec1-related proteins is their interaction with syntaxins (Fig. 2, top panel). The yeast SEC1 gene mutation is suppressed by suppressors of sec1 (SSO1 and SSO2), which encode two syntaxin homologues localized to the plasma membrane (Aalto et al., 1993). Yeast SLY1 may interact with a syntaxin homologue localized to the *cis*-Golgi apparatus, SED5 (Hardwick and Pelham, 1992); and VPS45 may interact with another syntaxin homologue, VPS6/PEP12. This association of sec1-related proteins with syntaxins extends to mammalian cells. The neural-specific Sec1p homologue n-sec1, also called Munc18 (Hata et al., 1993) and RbSec1 (Garcia et al., 1994), binds syntaxins 1a, 2 and 3 with nanomolar affinity (Hata et al., 1993; Pevsner et al., 1994a & b; Garcia et al., 1994, 1995). The ability of n-sec1 to bind syntaxin with nanomolar affinity, and its ability to inhibit VAMP and SNAP-25 binding to syntaxin, suggest that n-sec1 may regulate the activity of syntaxin (Pevsner et al., 1994b). Other Sec1p homologues include rop in *Drosophila* (Harrison et al., 1994; Schulze et al., 1994) and unc-18 in *Caenorhabditis elegans* (Hosono et al., 1992). Disruption of these genes impairs synaptic transmission, consistent with the hypothesis that the sec1-related proteins modulate synaptic vesicle docking.

Biochemical studies on n-sec1 suggest that it may regulate the ability of synaptic vesicles to dock. This report describes brain cDNA clones encoding three novel mammalian proteins related to n-sec1. These proteins are homologues of Vps33p from rat (called r-vps33a and r-vps33b) and a homologue of Vps45p from human (called h-vps45). They may participate in vesicle trafficking between the Golgi apparatus and lysosomes (Fig. 2, bottom panel).

Materials and methods

cDNA cloning of r-vps33a. Two oligonucleotides were synthesized (5'-AAAAGCACTCGTTGATCAAGGTG-3' and 5'-TCTGCGGTTTCAGCAGGCCGGCC-3') corresponding to clone hbc1194, a human expressed sequence tag (GenBank accession number T11336) predicted to have a protein product homologous to yeast vps33p. The oligonucleotides were used to isolate a 141 base pair cDNA fragment by the polymerase chain reaction (PCR) using 1 ng of adult male human pancreatic cDNA (Clontech, Palo Alto, CA) as a template. The PCR was performed with 30 cycles of denaturation (94°C, 1 min), annealing (50°C, 1 min) and extension (72°C, 2 min) with a final extension at 72°C for 10 min in a DNA thermal cycler (Perkin Elmer model 480). The 141 base pair product was purified (Geneclean, Bio101), radiolabeled with [a^{32}P]dCTP by random primer labeling, and used to probe 360,000 plaques from a rat brain cDNA library in lZAPII (Stratagene). Hybridization was performed at 55°C for 12 h using a probe concentration of 500,000 dpm/ml hybridization solution, and filters were washed at 55°C in 0.3 M NaCl/30 mM sodium citrate/0.1% sodium dodecyl sulfate. Following secondary screening, four clones (BO1 to BO4) were positive and plasmid DNA was excised with helper phage (Stratagene). R-vps33a clone FLB3 was subsequently isolated. All clones were sequenced on both strands in the coding region by the dideoxynucleotide method (USBiochemicals).

Cloning of r-vps33b. Two oligonucleotides (5'-CAGTAACAGGCTCTCCTCTG-3' and 5'-CTTGCTCACTTTAACTCTCC-3') were synthesized corresponding to human clone HCETC59R (Human Genome Sciences, Inc., Rockville, MD; GenBank accession number M34474) and used to isolate a 309 base pair PCR product using the conditions described above. This product was radiolabeled and used to screen 360,000 plaques from a rat brain cDNA library (5'-Stretch Plus library prepared by oligo(dT) and random priming; Clontech Laboratories, Palo Alto). Hybridization and wash conditions were as described above. Two positive clones (RS1 and RS2) were identified, and the EcoRI fragments were purified by cesium chloride DNA preparation, subcloned into bluescript KSII+ and sequenced. R-vps33b is closely related to a short human cDNA sequence in GenBank (accession number D20750).

Cloning of h-vps45. The oligonucleotides 5'-GTAGGGTACTTGTCAATTCG-3' and 5'-TTCCACGTCACTCTTGCTGG-3' corresponding to human clone GCEMC05R (Human Genome Sciences, Inc., Rockville, MD) were used to generate a 349 base pair fragment by PCR as described above using the clone as a template. This DNA was radiolabeled and used to screen a human brain cDNA library in λZAPII (Stratagene) under conditions as described above. A single 2.1 kilobase cDNA clone was isolated in Bluescript SK- vector and sequenced. H-vps45 is homologous to a human expressed sequence tag (accession number R12336).

Sequence analyses. DNA sequences were analyzed using the Wisconsin Package, version 8 (Genetics Computer Group, Madison, WI). Percent amino acid identities were obtained using the Bestfit program (gap penalty 3.00, gap extension penalty 0.10).

Isoelectric points were predicted with the peptide sort program of the GCG suite. Database searches were of GenBank release 88.0 and/or Swiss protein database release 31.0. The three novel sequences were deposited in GenBank with the accession numbers U35244 (r-vps33a), U35245 (r-vps33b), and U35246 (h-vps45).

Northern analysis. R-vps33a clone BO2 was digested with the restriction enzyme PstI to isolate an 800 base pair fragment. For r-vps33b, fragmentation of clone RS1 by the restriction enzyme EcoRI resulted in a 2.0 kilobase fragment. For h-vps45, an EcoRI fragment (nucleotides 1-851) was generated. Each of these three cDNA fragments was random primer labeled and used to probe a Northern blot consisting of 2 μg each of seven rat tissues (Clontech Laboratories Inc., Palo Alto, CA). Prehybridization and hybridization were performed in 0.75 M NaCl/50 mM sodium phosphate/6 mM EDTA/10x Denhardt's solution/50% formamide/0.1 mg per ml denatured, sheared salmon sperm DNA/2% sodium dodecyl sulfate, and hybridization was at 42°C for 12 h. Blots were washed twice in 0.3 M NaCl/30 mM sodium citrate/0.05% sodium dodecyl sulfate at 42°C then twice in 15 mM NaCl/1.5mM sodium citrate/0.1% sodium dodecyl sulfate at 50°C. The radioactive signal was quantitated by phosphorimaging (Molecular Dynamics).

In vitro translation and binding assay. Soluble n-sec1, r-vps33a, r-vps33b and h-vps45 were obtained by *in vitro* translation. N-sec1, r-vps33a, r-vps33b and h-vps45 cDNAs in bluescript plasmid were transcribed with T3 or T7 polymerase and *in vitro* translated using rabbit reticulocyte lysate (Amersham) in the presence of [^{35}S]methionine (New England Nuclear/DuPont). Syntaxins 1a, 2, 3 and 4 glutathione-S-transferase fusion proteins immobilized on beads were made as previously described (Pevsner et al., 1994a) except bacteria were lysed by sonication (25 pulses of 1 second in a Branson model 250 Sonifier, setting 4) instead of French pressing. Binding experiments were conducted by incubating 1-2 μl radiolabeled soluble protein with 15 μl beads containing equal amounts of fusion protein (approximately 0.5 μg) with phosphate buffered saline (PBS) in a total volume of 30 μl for 1 h at 22°C. Following incubation beads were washed twice with 0.5 ml PBS/0.1% Tween-20 and once with 0.5 ml PBS. Bound proteins were solubilized from the beads in 15 μl sample buffer, and electrophoresed on 10% polyacrylamide sodium dodecyl sulfate gels. The gels were fixed for 30 min in 50% methanol/10% acetic acid, incubated in an autoradiography enhancer (Amplify, Amersham) for 15 min, dried and exposed to film for 12-36 h at -70°C.

Results

r-vps33a, r-vps33b and h-vps45. We have characterized cDNA clones encoding three additional mammalian homologues belonging to the sec1-related family (Pevsner et al., submitted). Three human partial cDNA clones (expressed sequence tags) were identified in DNA databases corresponding to VPS33 and VPS45. For each, oligonucleotides were used to amplify cDNA fragments of approximately 300 base pairs. These fragments were radiolabeled and used to screen mammalian brain cDNA libraries. In this way clones r-vps33a and r-vps33b were isolated from rat brain. The predicted r-vps33a protein has 597 amino acids, and has a molecular mass of 67,513 Da with an isoelectric point of 7.0. The protein product r-vps33b consists of 617 amino acids with a molecular mass of 70,693 Da and an isoelectric point of 6.8. R-vps33a and r-vps33b share 30% and 26% amino acid identity with yeast vps33p, respectively. The two rat proteins share 35% amino acid identity with each other (Table I). The regions of similarity are dispersed across the lengths of the proteins (data not shown).

We also isolated a cDNA clone encoding h-vps45 from a human brain cDNA library. The predicted h-vps45 protein has 570 amino acid residues, a molecular mass of 64,951 daltons and an isoelectric point of 8.1. H-vps45 has 38% amino acid identity with yeast vps45p. This is the highest degree of homology between any higher eukaryotic member of the sec1 family and a yeast member (Table I). The regions of similarity between h-vps45 and vps45p are dispersed along the lengths of the proteins.

A search of the current DNA and protein databases and subsequent protein homology comparisons revealed that r-vps33a, r-vps33b and h-vps45 are members of the sec1 family of hydrophilic proteins (Table I). This group includes two additional members of the family not previously reported. A nematode sly1 has 34% amino acid identity to yeast sly1p and also significant identity (25%) to r-vps33a. A nematode hypothetical 68 kDa protein, b0303.9, shares 29% amino acid identity with r-vps33a, 25% amino acid identity with r-vps33b, and less identity with all other members of the family. It is likely that mammalian homologues of nematode sly1 and b0303.9 exist and are distinct from those mammalian family members that are currently known.

Table I. Percent amino acid identity of members of the sec1 family. Identities were calculated with the Bestfit program. The GenBank accession numbers for these proteins are: Sec1, X62541; Vps33, M34638; Vps45, U07972; Sly1, X54323; unc-18, S34207; b0303.9, P34260; *Caenorhabditis elegans* sly1, Z35640; rop, X67218; Munc-18b, U19520; Munc-18c, U19521; n-sec1, L26264. Not included in the table are RbSec1B, U21116 (97% identity to n-sec1; Garcia et al., 1995) and bovine unc-18, L26088 (99.8% identity to n-sec1; Hata et al., 1993).

Species	protein	Sec1	vps33	vps45	Sly1	unc18	b0303	sly1	rop	munc-18b	munc-18c	n-sec1	r-vps 33a	r-vps 33b	h-vps 45
S. cerevisiae	Sec1p	100	20	21	23	26	19	19	27	28	24	26	18	18	22
	Vps33p		100	22	19	21	20	23	20	20	19	20	30	26	21
	Vps45p			100	24	22	21	23	22	22	22	22	21	20	38
	Sly1p				100	23	24	34	22	24	24	22	21	18	24
C. elegans	unc-18					100	24	22	58	53	45	59	21	20	25
	b0303.9						100	18	22	22	20	21	29	25	21
	sly1							100	19	23	20	19	25	17	23
Drosoph. mel.	rop								100	56	45	65	19	21	25
Mus musculis	Munc-18b									100	47	63	18	20	24
	Munc-18c										100	53	18	18	24
Rattus norv.	n-sec1											100	21	18	22
	r-vps33a												100	35	20
	r-vps33b													100	22
Homo sapiens	h-vps45														100

Regional distribution of gene expression. By RNA blot analysis, r-vps33a mRNA was expressed in all tissues assayed (brain, spleen, lung, liver, skeletal muscle, kidney, testis) with transcript sizes of 2.3 kb, 3.5 kb, 4.2 kb and 7.0 kb (Fig. 3A). The relative abundance of each transcript was assessed by phosphor imaging. In brain and skeletal muscle, the 4.4 kb transcript was predominant, while in spleen, lung, liver, kidney and testis the 2.5 kb transcript was most abundant. The presence of multiple transcript sizes is likely due in part to differential polyadenylation sites, based on sequence analysis of the cDNAs. R-vps33b mRNA was observed to have a single transcript size of 2.8 kb in all tissues (Fig. 3B).

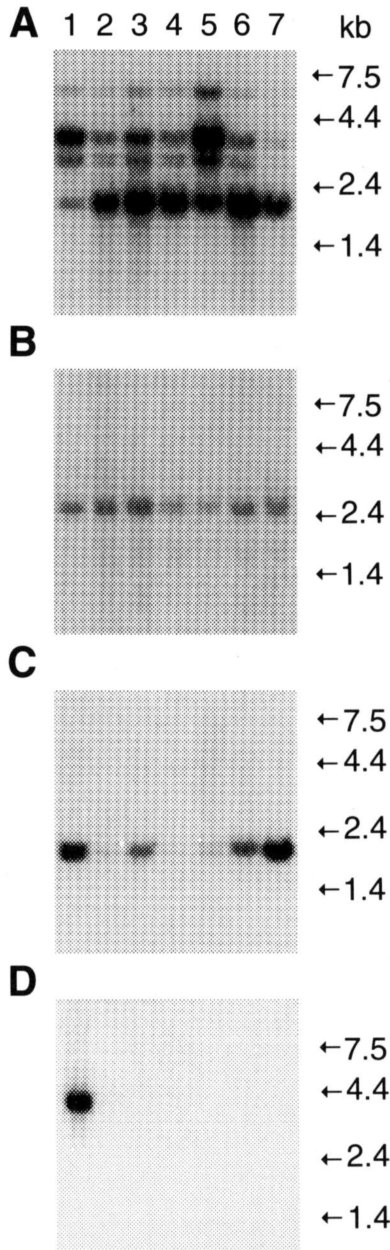

Fig. 3. Regional RNA blot analysis of r-vps33a, r-vps33b, h-vps45 and n-sec1. Each lane contained 2μg of poly(A)$^+$ RNA from a male rat. Molecular size markers are indicated in kilobases. Blots were probed with random primer labeled r-vps33a (panel A), r-vps33b (panel B), h-vps45 (panel C), and n-sec1 (panel D). Lanes: 1, brain; 2, spleen; 3, lung; 4, liver; 5, skeletal muscle; 6, kidney; 7, testis.

H-vps45 mRNA was present in all tissues with a 2.3 kb transcript present in testis > brain > kidney > lung > skeletal muscle > spleen > liver (Fig. 3C). In contrast to the broad tissue distribution of these three genes, n-sec1 expression was highly enriched in brain (Fig. 3D; see also Pevsner et al., 1994a).

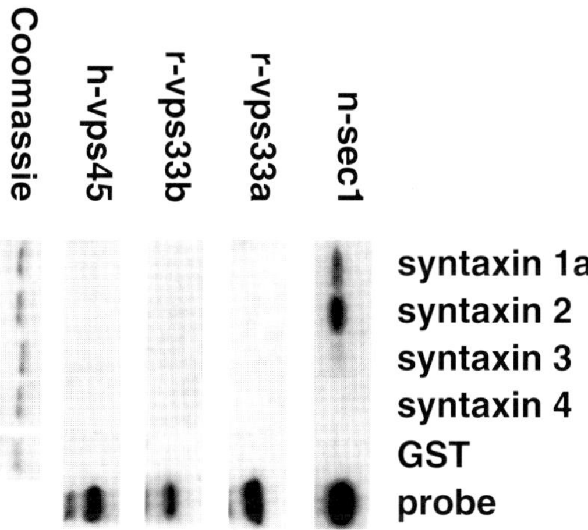

Fig. 4. Binding of *in vitro* translated sec1 homologues to rat syntaxin isoforms. Translated, radioactively labeled n-sec1, r-vps33a, r-vps33b, and h-vps45 were incubated with syntaxin 1, 2, 3, and 4 glutathione-S-transferase constructs or glutathione-S-transferase alone immobilized on glutathione agarose beads. Bound proteins were eluted, electrophoresed, and visualized by autoradiography. Probe lane indicates the signal from 1 µl of each radiolabelled protein (approximate molecular weight 70 kDa). Coomassie indicates staining of the syntaxin GST constructs in the amounts used in each experiment. Data are from a typical experiment performed independently three times with similar results.

It is likely that each sec1-related protein interacts with at least one syntaxin-like molecule. We determined that r-vps33a, r-vps33b and h-vps45 do not bind the known syntaxins 1a, 2, 3 or 4, based upon *in vitro* assays. The syntaxins were prepared as glutathione-*S*-transferase fusion proteins and immobilized on glutathione agarose beads. As a positive control for binding, [^{35}S]methionine-labeled, *in vitro* translated n-sec1 bound to syntaxins 1, 2, and 3 but not 4 as previously described (Fig. 4; compare Pevsner et al., 1994b). *In vitro* translated r-vps33a, r-vps33b and h-vps45 did not bind to the syntaxins (Fig. 4). Recombinant syntaxin 5 is poorly soluble, remaining in bacterial inclusion bodies, and was not tested. This result is consistent with the idea that syntaxins 1, 2, 3, and 4 are localized to the plasma membrane, and bind n-sec1 or proteins closely related to n-sec1. In contrast, other intracellularly localized syntaxins would likely bind r-vps33a, r-vps33b and h-vps45 or other sec1-related proteins localized to intracellular organelles (Table II). Staining of the syntaxin and glutathione-*S*-tranferase proteins with Coomassie Blue confirmed that equal amounts of protein were incubated with radiolabeled probes (Fig. 4).

Discussion

Vesicle trafficking is mediated in part by several families of proteins: vesicular proteins (v-SNAREs) such as VAMP and synaptotagmin, target membrane proteins (t-SNAREs) such as syntaxin and SNAP-25, and soluble proteins such as rab3A, αSNAP and NSF. The molecular mechanisms of vesicle trafficking have been particularly well characterized for synaptic vesicles, due to their great abundance in the nerve terminal (Pevsner and Scheller, 1994). A remarkable feature of a model for synaptic vesicles docking and fusion (Fig. 1) is that most of the proteins implicated in this regulated secretory pathway in a mammalian system have homologues in the constitutive secretory pathway of yeast. Furthermore, these proteins have homologues that are functional in distinct intracellular pathways (Fig. 2). Thus it is likely that a single mechanism of vesicle movement from a donor compartment to a target or acceptor compartment underlies all vesicular trafficking. This mechanism requires important variations to account for the specialized needs of particular trafficking steps. For example, regulated secretory pathways require a „brake" to allow selective trafficking upon the arrival of an appropriate signal. For synaptic vesicles, synaptotagmin has been proposed to assume this role, based in part upon the impairments in neurotransmission observed in mouse, *Drosophila* and *Caenorhabditis elegans* synaptotagmin mutants (Geppert et al., 1994; Popov and Poo, 1993; DeBello et al., 1993).

It is likely that neuropeptide-containing large dense-core vesicles utilize similar mechanisms for transmitter release as small clear synaptic vesicles. Both vesicle types contain synaptotagmin, VAMP and rab3. Large dense-core vesicles are heterogeneous in size and in composition, and may utilize multiple release mechanisms (Seward et al.,

1995). For example, in *Aplysia* neuromuscular synapses, neuropeptide release from identified motor neurons is elicited by a broad range of stimulation patterns (Whim et al., 1993). However, it is likely that multiple isoforms of proteins involved in exocytosis can account for the diversity of release mechanisms (reviewed in Bark and Wilson, 1994).

The sec1-related family of proteins consists of cytosolic proteins that are required for vesicle function at specific intracellular pathways in yeast, and each sec1-related protein is likely to interact with a syntaxin or with multiple syntaxins (Fig. 2, top panel; Table II) (Pfeffer, 1994; Aalto et al., 1992). N-sec1 binds syntaxin 1a with nanomolar affinity, as well as syntaxins 2 and 3 (but not 4) (Pevsner et al., 1994a). Other mammalian sec1-related proteins are Munc-18b, which has not been shown to bind syntaxins, and Munc-18c, which binds syntaxins 1a and 4 (Tellam et al., 1995; Hata and Südhof, 1995; Katagiri et al., 1995). The function and intracellular locations of Munc-18b and c are not known.

This paper describes the identification of three additional sec1-related proteins: r-vps33a, r-vps33b, and h-vps45. These proteins share limited homology with n-sec1, and based upon amino acid identities they are likely to be mammalian homologues of yeast Vps33p and Vps45p. The vacuolar protein-sorting proteins are required for the proper sorting of hydrolases and other enzymes to the lysosome-like vacuole in yeast (Raymond et al., 1992). Thus, r-vps33a, r-vps33b, and h-vps45 may function in vesicular trafficking from the Golgi to a prelysosomal compartment (i.e. a late endosomal compartment) and to lysosomes. If their functions are analogous to n-sec1, they may bind lysosomal or prelysosomal syntaxin-like proteins. A further understanding of the intracellular localizations, biochemical interactions, and physiological functions of these three proteins may help elucidate the role of all sec1-like proteins in vesicle trafficking. Conversely, the principles of vesicle trafficking that are being characterized in the nerve terminal may be applied to the pathway of Golgi to lysosomal vesicle trafficking, further demonstrating the generality of vesicle trafficking mechanisms.

J. Pevsner

Table II. Summary of interactions between mammalian syntaxins and sec1-related proteins. The Table highlights that syntaxins have not been identified that bind r-vps33a, r-vps33b, and h-vps45. +, binding interaction detected; -, binding not detected; ?, not done.

Sec1-like protein	Mammalian syntaxin									Notes
	1A	**1B**	**2**	**3A**	**3B**	**3C**	**3D**	**4**	**5**	**Notes**
n-sec1	+	?	+	+	+	-	-	-	-	1
rbSec1B	+	?	?	?	?	?	?	?	?	2
Munc-18b	?	?	?	?	?	?	?	?	?	3
Munc-18c	+	?	?	?	?	?	?	+	?	4
r-vps33a	-	?	-	-	?	?	?	-	?	5
r-vps33b	-	?	-	-	?	?	?	-	?	5
h-vps45	-	?	-	-	?	?	?	-	?	5

1. Pevsner et al., 1994a,b; also called Munc-18 (Hata et al., 1993), rbSec1A (Garcia et al., 1994) and mSec1 (Hodel et al., 1994). Studies on syntaxin 3 isoforms are by Ibaraki et al. (1995).
2. Garcia et al. (1995).
3. Tellam et al. (1995).
4. Tellam et al. (1995); also called Munc-18-2 (Hata and Südhof, 1995) and muSec1 (Katagiri et al., 1995)
5. Pevsner et al., submitted.

Acknowledgments
The experiments described in this study were performed in Richard Scheller's laboratory. I thank A. Fitzgerald for helpful discussions and N. Varg for manuscript preparation.

References

Aalto, M.K., Ruohonen, L., Hosono, K. and Keränen, S. (1991) Cloning and sequencing of the yeast *Saccharomyces cerevisiae SEC1* gene localized on chromosome IV. *Yeast* 7: 643-650.

Aalto, M., Keränen, S. and Ronne, H. (1992) A family of proteins involved in intracellular transport. *Cell* 68: 181-182.

Aalto, M.K., Ronne, H. and Keränen, S. (1993) Yeast syntaxins Sso1p and Sso2p belong to a family of related membrane proteins that function in vesicular transport. *EMBO J.* 12: 4095-4104.

Banta, L.M., Vida, T.A., Herman, P.K. and Emr, S.D. (1990) Characterization of yeast vps33p, a protein required for vacuolar protein sorting and vacuole biogenesis. *Mol. Cell. Biol.* 10: 4638-4649.

Bark, I.C. and Wilson, M.C. (1994) Regulated vesicular fusion in neurons: snapping together the details. *Proc. Natl. Acad. Sci. USA* 91:4621-4624.

Bennett, M.K., Garcia-Arrarás, J.E., Elferink, L.A., Peterson, K., Fleming, A.M., Hazuka, C.D. and Scheller, R.H. (1993) The syntaxin family of vesicular transport receptors. *Cell* 74: 1-20.

Bennett, M.K. and Scheller, R.H. (1994) A molecular description of synaptic vesicle membrane trafficking. *Annu. Rev. Biochem.* 63: 63-100.

Cameron, P., Mundigl, O. and De Camilli, P. (1993) Traffic of synaptic vesicle proteins in polarized and nonpolarized cells. *J. Cell Sci.* Supplement 17:93-100

Chow, R.H., Klingauf, J. and Neher, E. (1994) Time course of Ca2+ concentration triggering exocytosis in neuroendocrine cells. *Proc. Natl. Acad. Sci. USA* 91:12765-12769.

Cowles, C.R., Emr, S.D. and Horazdovsky, B.F. (1994) Mutations in the *VPS45* gene, a *SEC1* homologue, result in vacuolar protein sorting defects and accumulation of membrane vesicles. *J. Cell Sci.* 107: 3449-3459.

Dascher, C., Ossig, R., Gallwitz, D. and Schmitt, H.D. (1991) Identification and structure of four yeast genes (*SLY*) that are able to suppress the functional loss of *YPT1*, a member of the *RAS* superfamily. *Mol. Cell. Biol.* 11: 872-885.

DeBello, W.M., Betz, H. and Augustine, G.J. (1993) Synaptotagmin and neurotransmitter release. *Cell* 74: 947-950.

Garcia, E.P., Gatti, E., Butler, M., Burton, J. and de Camilli, P. (1994) A rat brain Sec1 homologue related to Rop and UNC18 interacts with syntaxin. *Proc. Natl. Acad. Sci. USA* 91: 2003-2007.

Garcia, E.P., McPherson, P.S., Chilcote, T.J., Takei, K. and De Camilli, P. (1995) rbSec1A and B colocalize with syntaxin 1 and SNAP-25 throughout the axon, but are not in a stable complex with syntaxin. *J. Cell Biol.* 129: 105-120.

Geppert, M.., Goda, Y., Hammer, R.E., Li, C., Rosahl, T.W., Stevens, C.F. and Südhof, T.C. (1994) Synaptotagmin I: a major Ca2+ sensor for transmitter release at a central synapse. *Cell* 79:717-727.

Golding, D.W. (1994) A pattern confirmed and refined – synaptic, nonsynaptic and parasynaptic exocytosis. *Bioessays* 16:503-508.

Guttierrez, L.M., Quintanar, J.L., Viniegra, S., Salinas, E., Moya, F. and Reig, J.A. (1995) Anti-syntaxin antibodies inhibit calcium-dependent catecholamine secretion from permeabilized chromaffin cells. *Biochem. Biophys. Res. Commun.* 206: 1-7.

Hardwick, K.G. and Pelham, H.R.B. (1992) SED5 encodes a 39-kD integral membrane protein required for vesicular transport between the ER and the Golgi complex. *J. Cell Biol.* 119: 513-521.

Harrison, S.D., Broadie, K., van de Goor, J. and Rubin, G.M. (1994) Mutations in the Drosophila *Rop* gene suggest a function in general secretion and synaptic transmission. *Neuron* 13: 555-566.

Hata, Y., Slaughter, C.A. and Südhof, T.C. (1993) Synaptic vesicle fusion complex contains *unc-18* homologue bound to syntaxin. *Nature* 366: 347-351.

Hata, Y. and Südhof, T.C. (1995) A novel ubiquitous form of Munc-18 interacts with multiple syntaxins. *J. Biol. Chem.* 270:13022-13028.

Hodel, A., Schäfer, T., Gerosa, D. and Burger, M.M. (1994) In chromaffin cells, the mammalian Sec1p homologue is a syntaxin 1A-binding protein associated with chromaffin granules. *J. Biol. Chem.* 269: 8623-8626.

Hosono, R., Hekimi, S., Kamiya, Y., Sassa, T., Murakami, S., Nishiwaki, K., Miwa, J., Taketo, A. and Koaira, K.-I. (1992) The *unc-18* gene encodes a novel protein affecting the kinetics of acetylcholine metabolism in the nematode *Caenorhabditis elegans*. *J. Neurochem.* 58: 1517-1525.

Ibaraki, K., Horikawa, H.P., Morita, T., Mori, H., Sakimura, K., Mishina, M., Saisu, H. and Abe, T. (1995) Identification of four different forms of syntaxin 3. *Biochem. Biophys. Res. Commun.* 211: 997-1005.

Katagiri, H., Terasaki, J., Murata, T., Ishihara, H., Ogihara, T., Inukai, K., Fukushima, Y., Anai, M., Kikuchi, M., Miyazaki, J.-i., Yazaki, Y. and Oka, Y. (1995) A novel isoform of syntaxin-binding protein homologous to yeast Sec1 expressed ubiquitously in mammalian cells. *J. Biol. Chem.* 270: 4963-4966.

Martin, T.F.J. (1994) The molecular machinery for fast and slow neurosecretion. *Curr. Op. Neurobiol.* 4:626-632.

O'Connor, D.T., Wu, H., Gill, B.M., Rozansky, D.J., Tang, K., Mahata, S.K., Mahata, M.M., Eskeland, N.L., Videen, J.S., Zhang, X., Takiyyudin and Palmer, R.J. (1994) Hormone storage vesicle proteins: Transcriptional basis of the widespread neuroendocrine expression of chromogranin A, and evidence of its diverse biological actions, intracellular and extracellular. *Ann. N. Y. Acad. Sci.* 733:36-45.

Ossig, R., Dascher, C., Trepte, H.-H., Schmitt, H.D. and Gallwitz, D. (1991) The yeast *SLY* gene products, suppressors of defects in the essential GTP-binding Ypt1 protein, may act in endoplasmic reticulum-to-Golgi transport. *Mol. Cell. Biol.* 11: 2980-2993.

Pevsner, J. and Scheller, R.H. (1994) Mechanisms of vesicle docking and fusion: insights from the nervous system. *Curr. Op. Cell Biol.* 6: 555-560.

Pevsner, J. Hsu, S.-C., and Scheller, R. H. (1994a) n-Sec1: a neural-specific syntaxin-binding protein. *Proc. Natl. Acad. Sci. USA* 91: 1445-1449.

Pevsner, J., Hsu, S.-C., Braun, J.E.A., Calakos, N., Ting, A.E., Bennett, M.K. and Scheller, R.H. (1994b) Specificity and regulation of a synaptic vesicle docking complex. *Neuron* 13: 353-361.

Pevsner, J., Hsu, S.-C., Hyde, P.S. and Scheller, R.H. (submitted) Mammalian homologues of yeast vacuolar protein sorting (vps) genes implicated in Golgi-to-lysosome trafficking.

Pfeffer, S.R. (1994) Clues to brain function from baker's yeast. *Proc. Natl. Acad. Sci. USA* 91:1987-1988.

Piper, R.C., Whitters, E.A. and Stevens, T.H. (1994) Yeast Vps45p is a Sec1p-like protein required for the consumption of vacuole-targeted, post-Golgi transport vesicles. *Eur. J. Cell Biol.* 65: 305-318.

Popov, S.V. and Poo, M.-M. (1993) Synaptotagmin: a calcium-sensitive inhibitor of exocytosis? *Cell* 73: 1247-1249.

Pryer, N.K., Wuestehube, L.J. and Schekman, R. (1992) Vesicle-mediated sorting. *Annu. Rev. Biochem.* 61: 471-516.

Raymond, C.K., Howald-Stevenson, I., Vater, C.A. and Stevens, T.H. (1992) Morphological classification of the yeast vacuolar protein sorting mutants: evidence for a prevacuolar compartment in class E *vps* mutants. *Mol. Biol. Cell* 3: 1389-1402.

Rothman, J.E. and Warren, G. (1994) Implications of the SNARE hypothesis for intracellular membrane topolology and dynamics. *Curr. Biol.* 4: 220-233.

Scheller, R. (1995) Membrane trafficking in the presynaptic nerve terminal. *Neuron* 14: 1-20.

Schulze, K.L., Littleton, J.T., Salzberg, A., Halachmi, N., Stern, M., Lev, Z. and Bellen, H.J. (1994) rop, a Drosophila homolog of yeast sec1 and vertebrate n-sec1/munc-18 proteins, is a negative regulator of neurotransmitter release in vivo. *Neuron* 13: 1099-1108.

Seward, E.P., Chernevskaya, N.I. and Nowycky, M.C. (1995) Exocytosis in peptidergic nerve terminals exhibits two calcium-sensitive phases during pulsatile calcium entry. *J. Neurosci.* 15:3390-3399.

Söllner, T. and Rothman, J.E. (1994) Neurotransmissioning fusion machinery at the synapse. *Trends Neurosci.* 17: 344-353.

Südhof, T.C. (1995) The synaptic vesicle cycle: a cascade of protein-protein interactions. *Nature* 375:645-653.

Tellam, J.T., McIntosh, S. and James, D.E. (1995) Molecular identification of two novel Munc-18 isoforms expressed in non-neuronal tissues. *J. Biol. Chem.* 270: 5857-5863.

Thomas, P., Wong, J.G., Lee, A.K. and Almers, W. (1993) A low affinity Ca^{2+} receptor controls the final steps in peptide secretion from pituitary melanotrophs. *Neuron* 11:93-104.

Thomas-Reetz, A.C. and De Camilli, P. (1994) A role for synaptic vesicles in non-neuronal cells: clues from pancreatic β cells and from chromaffin cells. *FASEB J.* 8:209-216.

Volknandt, W. (1995) The synaptic vesicle and its targets. *Neuroscience* 64:277-300.

Wada, Y., Kitamoto, K., Kanbe, T., Tanaka, K. and Anraku, Y. (1990) The *SLP1* gene of *Saccharomyces cerevisiae* is essential for vacuolar morphogenesis and function. *Mol. Cell. Biol.* 10: 2214-2223.

Whim, M.D., Church, P.J. and Lloyd, P.E. (1994) Functional roles of peptide cotransmitters at neuromuscular synapses in *Aplysia*. *Mol. Neurobiol.* 7:335-347.

The Peptidergic Neuron
B. Krisch and R. Mentlein (eds)
© 1996 Birkhäuser Verlag Basel/Switzerland

Scanning electron microscopy of an active neurohaemal area, the cockroach (*Periplaneta americana*) corpora cardiaca: looking at neurosecretion from an unprecedented viewpoint

P.D. Verhaert and A.J. De Loof

Zoological Institute of the University, Naamsestraat 59, B-3000 LEUVEN, Belgium

Summary. Unlike neurosecretory release organs of vertebrates, equivalent sites in the body of invertebrates with an open circulatory system, are well suited for scanning electron microscopy (SEM). The present paper describes such a SEM study of the principal neurohaemal organ of insects, i.e., the corpus cardiacum. Besides a classical fixation protocol using glutaraldehyde, two additional - uncommon - methods of tissue preparation were examined, one of which employed Bouin-Hollande's 10% sublimate as fixative. The various preparations all yielded different scanning electron micrographs. The images observed are correlated with previously described ultrastructural data on various types of (invertebrate) neurosecretory release profiles and with the "fixing" strength (speed) of the chemical solutions used.

Introduction

Microscopic investigation of the process of neurosecretion is commonly done with both electron and light microscopy. Typical electron microscopy viewing includes transmission electron microscopy, such as the tannic acid method (e.g. Khan and Saleuddin, 1992), and freeze fracture or quick-freeze, deep-etch electron microscopy (e.g. Senda et al., 1994). Everyday light microscopic imaging techniques in neuroendocrinology are Gomori staining and related classical histochemical methods (see review by Rowell, 1976), as well as more specific and selective immunocytochemistry (e.g. Verhaert et al., 1984; Verhaert and De Loof, 1993).

One feature which all of the above type studies have in common is that they examine merely a section/fraction of the neurosecretory nerve ending, hereby concentrating on the interior of the cell. The exterior of the neuroendocrine fiber has rarely been focused upon in microscopy. One of the most probable reasons may be the tight bond between the nerve ending and the wall of a blood vessel/capillary. This is undoubtedly so for most vertebrates. Some selected invertebrates, however, such as arthropods and molluscs, possess an open circulation, which implies that their neurohaemal organs are ending blindly in the body cavity filled with haemolymph (called haemocoel). Hence they should be much more easily accessible for scanning electron microscopy viewing. With this in mind we undertook a scanning electron microscopy study of corpora cardiaca, the insect main neurohemal organ (equivalent to the vertebrate pituitary).

Material and methods

The American cockroach, *Periplaneta americana*, was the insect selected for this study. The cockroach corpus cardiacum was one of the first (if not the very first) invertebrate neurohaemal organ ever to be studied (see Scharrer and Scharrer, 1944). Consequently transmission electron microscopy data with respect to the neurosecretory activity of the retrocerebral complex of the present species have been around for a long time (Scharrer, 1968; Scharrer and Kater, 1969; Scharrer and Wurzelmann, 1974). Moreover, extensive light microscopic data on this neurohemal organ are likewise available, e.g. from various of our own neuropeptide immunohistochemical stainings (see Verhaert et al., 1984; Verhaert and De Loof, 1993).

Initially corpora cardiaca were prepared for scanning electron microscopy the routine way: fixation for approximately 20 h (4°C) in 2% glutardialdehyde (in 0.05 M sodium cacodylate, pH 7.3; GA), followed by dehydration, critical-point drying and gold coating. Alternatively, we also substituted the glutardialdehyde fixative by Bouin-Hollande's 10% sublimate fixative solution (BHS) or insect saline (Ringer). Imaging was done with a Philips SEM 515 microscope.

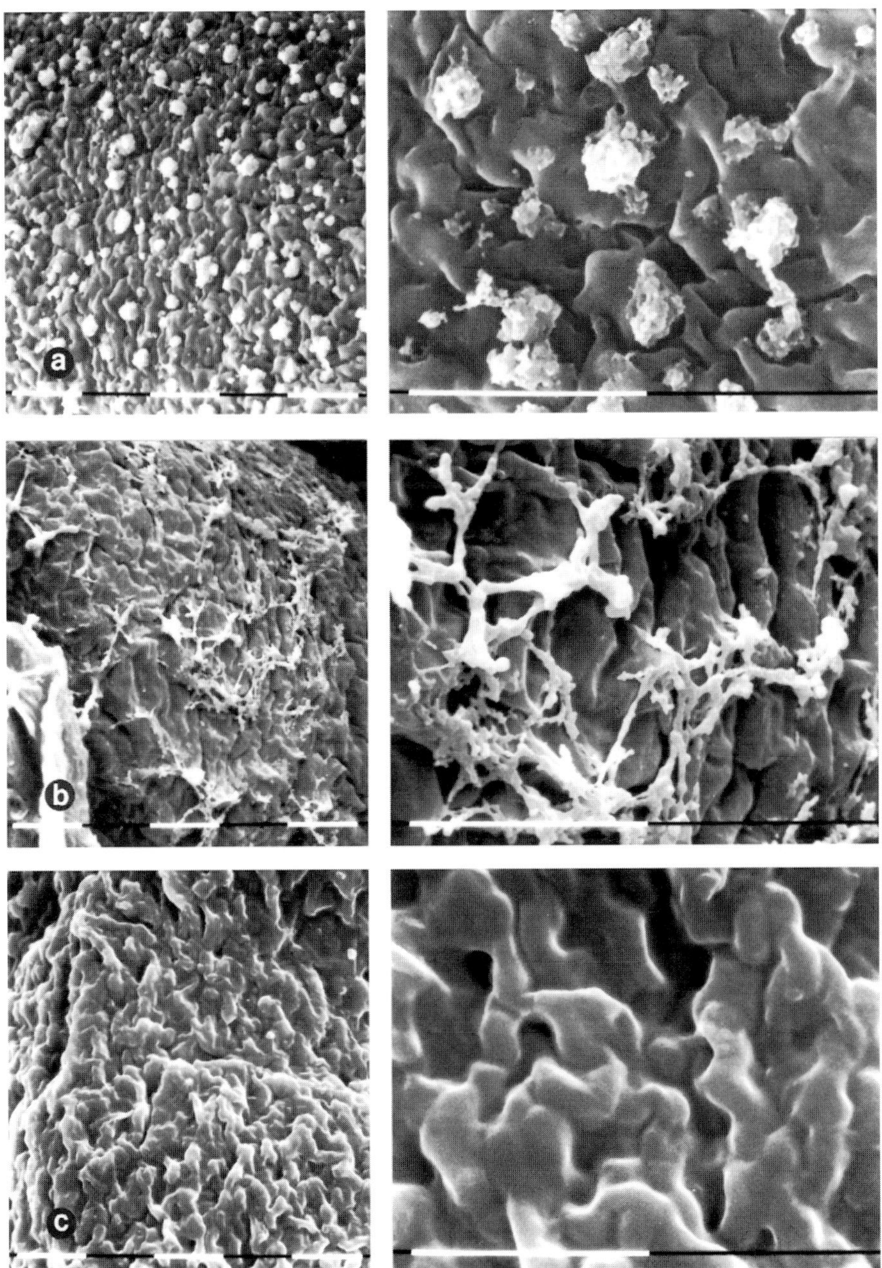

Fig. 1. Scanning electron micrograph recordings of cockroach corpora cardiaca treated according to "BHS", "GA" and "Ringer" protocol (panels a,b and c respectively). Right pictures are magnifications of left ones. Scale bar = 10 μm.

Results

Neurohaemal organs from the "GA" and "BHS" protocols yield a conspicuous secretion-like substance on their folded surface (Figs. 1a,b). The secretion-type material apparent on corpora cardiaca from the "GA" protocol (Fig. 1b) appears as extended strands, which at some point seem to form a network-like secretion around the exterior of the organ. The Bouin-Hollande-sublimate fixed neurosecretion-like material (Fig. 1a) appears as small chunks/drops, of which the different sizes and shapes are reminiscent of the various types of (invertebrate) neurosecretion described in transmission electron microscopy. The secretion is clearly confined to those surface areas of the corpora cardiaca which are effectively involved in neuroendocrine release. Furthermore, in none of the cases any secretion-like material is observed on the corpora allata or on the cerebral surface. Moreover, the density of the droplets at the surface of the corpora cardiaca agrees with the density of neurosecretory nerve endings which can be stained with neuropeptide antibodies in light microscopy (see e.g. Verhaert et al., 1984; Verhaert and De Loof, 1993). Corpora cardiaca prepared according to the "Ringer" protocol (Fig. 1c) exhibit a perfectly clean surface.

Discussion

In order to interpret the diversity of the scanning electron microscopy images generated from the differently prepared corpora cardiaca, one can take the available transmission electron microscopic data as a starting point. Diverse types of neuroendocrine release have been recognized in invertebrates (see Khan and Saleuddin, 1992). These include so-called simple omega-figures ("type I"), multiple ("type II") and complex exocytosis-profiles ("type III") (resp. Figs. 2 I, II, III). Careful examination of previously published transmission electron microscopic pictures of cockroach corpora cardiaca (see Scharrer and co-workers, quoted above) indicate that similar phenomena may occur in the present species as well. On the other hand, when examining the fixation process during dissection (binocular magnifying optics), one observes the following. The tissue submerged in Bouin-Hollande-sublimate stains pale white-green nearly instantaneously. The color of corpora cardiaca immersed in glutardialdehyde fixative solution changes from the typical opalescent "Tyndall blue" (see Orchard, 1983) into flat white (not transparent) in about a minute. Corpora cardiaca incubated in isotonic Ringer solution for 12 h appear to maintain their characteristic iridescent blue-white appearance for at least 1 h, whereafter they slowly loose their brightness, eventually becoming nearly as pale as the surrounding tissues.

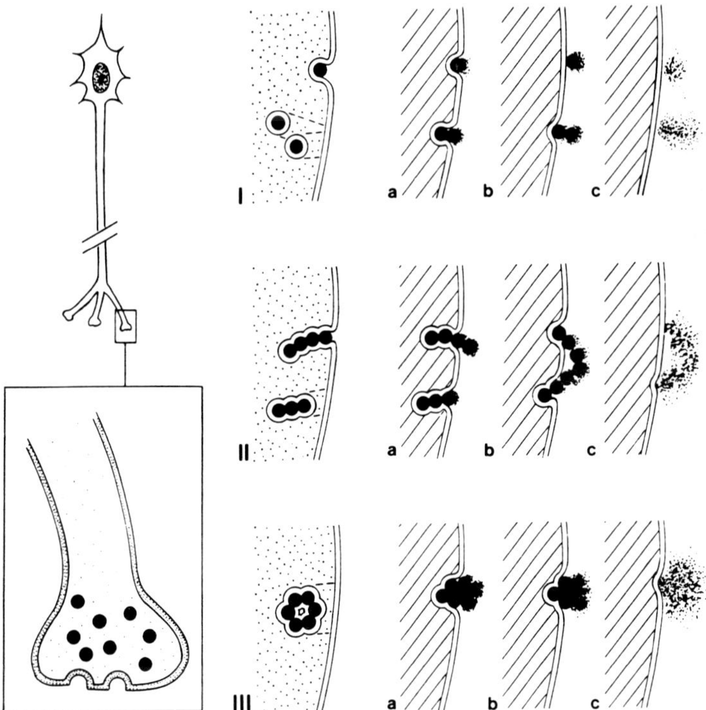

Fig. 2. Cartoon depicting a two-dimensional view on the process of neuroendocrine release from a neurosecretory nerve terminal. Panels I-III represent type I-III exocytosis profiles. Far left columns are modified transmission electron microscopic diagrams from Khan and Saleuddin (1992), other columns illustrate the respective result on the surface of the cell from "BHS" (a), "GA" (b) and "Ringer" (c) protocol (see text for details).

Both considerations help in trying to explain our scanning electron microscopic images. This is illustrated in Fig. 2. Our observations suggest that the tissues dissected under Bouin-Hollande-sublimate apparently are fixed most rapidly. Exteriorization of secretory

Cell-surface peptidases involved in neuropeptide metabolism: an overview

A.J. Kenny

Le Presbytère, St. Etienne d'Albagnan, 34390-France   (formerly at Department of Biochemistry and Molecular Biology, University of Leeds, Leeds, LS2 9JT, United Kingdom)

Summary. There is general agreement that secreted neuropeptides are not degraded within the cell, but are hydrolysed at the cell-surface by ectoenzymes. An earlier view implied the existence of a battery of peptide-specific enzymes; it is now clear that most neuropeptides are degraded by a relatively small number of peptidases which are not special to the nervous system, but have a broad distribution on the surfaces of many different cell types. The kidney provides an ideal source for the isolation and study of most of the cell-surface peptidases. There are now more than a dozen well-characterized enzymes in this group (most have been cloned). Three of these: endopeptidase–24.11 (E–24.11), endopeptidase–24.18 (E–24.18) and aminopeptidase-W (AP–W) were discovered in the course of my research in Leeds. Over the last 20 years some unpredictable developments ensured that endopeptidase–24.11 grew in scientific and medical popularity. First came the proof that it was responsible for the "enkephalinase" activity in brain; second, the realization of its crucial role in inactivating atrial natriuretic peptides and third, the discovery (as a result of molecular cloning) of its identity to CD-10 (CALLA). endopeptidase–24.11 seems to be the principal enzyme in the brain and elsewhere for initiating the degradation of many, but not all neuropeptides. Endopeptidase–24.18 has, by contrast, a limited distribution, essentially restricted to brush borders, with none so far revealed in the nervous system. Kinetically, it is much less efficient in hydrolysing neuropeptides than is endopeptidase–24.11, but capable of attacking some more complex molecules. Aminopeptidase W has a strong preference for dipeptides in which the second residue is aromatic. It is a neuronal marker in peripheral nerves, contrasting with endopeptidase–24.11, which is neuronal in the central but located on Schwann cells in the peripheral nervous system.

Introduction

This overview is concerned with the hydrolysis of neuropeptides after their release from their cell of origin. In the majority of instances the hydrolytic attack leads to inactivation, so that, for a peptidergic neuron, the initial cleavage can serve to terminate the signal. Early investigations of these processes assumed that each peptide required a specific peptidase and the terminology for the putative enzymes ("enkephalinase" etc.) outstripped their biochemical characterization. We now know that most of these enzymes are neither peptide–, nor tissue–specific. Indeed fundamental enzymological studies are more effective with tissues containing the peptidase in abundance and using simple assay systems, than are attempts to isolate the same peptidase from the brain, where it may be three orders of magnitude less abundant.

Table I. Cell-surface peptidases. Transmembrane proteins, anchored by a hydrophobic peptide domain are designated (where known) C or N (types I or II) indicating attachment near the N- or C-terminus.

Enzyme	Class	Anchor
Endopeptidase–24.11 (E–24.11)	metallo (Zn)	peptide (N)
Endopeptidase–24.18 (E–24.18)	metallo (Zn)	peptide (C)
Endothelin converting enzyme (ECE)	metallo (Zn)	peptide (N)
Angiotensin converting enzyme (ACE)	metallo (Zn)	peptide (C)
Dipeptidyl peptidase IV (DPP-IV)	serine	peptide (N)
Aminopeptidase A (AP-A)	metallo (Ca)	peptide (N)
Aminopeptidase N (AP-N)	metallo (Zn)	peptide (N)
Aminopeptidase P (AP-P)	metallo (Zn)	glycolipid
Aminopeptidase W (AP-W)	metallo (Zn)	peptide
Carboxypeptidase M (CP-M)	metallo (Zn)	glycolipid
Carboxypeptidase P (CP-P)	metallo (Zn)	peptide
Membrane dipeptidase (MDP)	metallo (Zn)	glycolipid
TRH-degrading ectoenzyme	metallo (Zn)	peptide (N)
γ-Glutamyl transpeptidase (γ-GTP)		peptide (N)

The surfaces of most mammalian cells are endowed with a battery of hydrolases. They are *ectoenzymes*, a term implying that they are integral plasma membrane proteins with a highly asymmetric topology, such that the bulk of the protein is at the external surface and linked to the membrane through a stalk peptide. Most of them are transmembrane proteins, possessing a peptide anchor and a very small cytoplasmic domain, but a few have glycolipid anchors comprising a phosphatidylinositol moiety attached at the C-terminus and thus lack any cytoplasmic domain. Peptidases form the major group of ectoenzymes (Kenny et al., 1987). Table I is a list of some well-characterized peptidases, most of which are located at the cell surface of the renal proximal tubule cell (the exceptions are endothelin-converting enzyme and TRH-degrading ectoenzyme). Of the enzymes shown, three are obligatory endopeptidases in contrast to the rest, which are exopeptidases. The latter group require an unsubstituted N– or C–terminus and are thus hindered in attacking those neuropeptides with blocked termini, a restriction not encountered by the endopeptidases.

There are several reviews of this group of peptidases (e.g. Checler 1993; Kenny and Hooper, 1991; Kenny et al., 1987; Turner et al., 1989) and only an outline of their actions (other than those described below in more detail) need be given here. Endothelin-converting enzyme is highly specific in hydrolysing the Trp^{21}–Val^{22} bond in 'big' endothelin to generate the highly active 21–residue endothelin and thus a rare example of a peptide–specific enzyme (Yorimitsu et al., 1995). Angiotensin-converting enzyme removes the C-terminal dipeptide from angiotensin I to form the active angiotensin II, but in generic terms it is a peptidyl dipeptidase capable of attacking a wide range of peptides. Dipeptidyl peptidase IV attacks peptides with X–Pro or X–Ala at their N-termini, releasing the dipeptide. Of the aminopeptidases, aminopeptidase N has a very broad specificity, while aminopeptidase A prefers Glu and Asp amino-termini, aminopeptidase P attacks an X-Pro bond and aminopeptidase W prefers short peptides with Trp (or other aromatics) as the second residue. Carboxypeptidase P removes X from peptides terminating with Pro-X and carboxypeptidase M is specific for peptides terminating with Lys or Arg. Membrane dipeptidase is a true dipeptidase, with the broadest of specificities. Thyrotropin (TRH)–degrading ectoenzyme is located on brain synaptic membranes and in the anterior pituitary. It is relatively specific for this peptide, which it cleaves to release the N-terminal pyroglutamic acid residue (Bauer 1994, 1995; Schauder et al., 1994).

Historical background

More than a quarter of a century has passed since we noticed the existence of a new membrane peptidase in kidney microvilli (Wong-Leung and Kenny, 1968). A few years later it was characterized in my laboratory as a zinc metallo-endopeptidase with a primary specificity similar to that of a group of microbial enzymes, typified by thermolysin, i.e. it cleaved bonds involving the amino function of hydrophobic amino acid residues (Kerr and Kenny, 1974b). Like thermolysin, this endopeptidase is specifically inhibited ($K_i = 2$ nM) by phosphoramidon (Kenny, 1977). Later, it became clear that it was widely distributed and topological studies showed it to be an ectoenzyme (Kenny et al., 1983) with a hydrophobic anchor near the N–terminus (Fulcher et al., 1986), in line with an earlier speculation (Kenny and Booth, 1978), that some microvillar proteins might achieve this topology in the plasma membrane (now known as a Type II membrane protein) by the retention of the signal peptide.

About 20 years ago a burst of interest in the metabolism of the enkephalins introduced this enzyme to neuropharmacologists. One of the enkephalin bonds hydrolysed by crude brain extracts is Gly^3-Phe^4, releasing the C-terminal dipeptide (Craves et al., 1978; Malfroy et al., 1978), but there was some confusion as to whether this 'enkephalinase' activity was a peptidyl dipeptidase or an endopeptidase. The identity of this activity with the renal brush border endopeptidase became clear only as a result of some studies exploiting inhibitors and antibodies (Fulcher et al., 1982; Matsas et al., 1983; Relton et al., 1983; Almenoff and Orlowski, 1984) and, to avoid confusion for an enzyme with many potential substrates, the name endopeptidase–24.11 was adopted (Matsas et al., 1983). Thus, the enzyme became a focus of major interest for pharmaceutical chemists searching for a new class of analgesics.

This pattern of events was repeated a few years later in regard to the metabolism of atrial natriuretic peptides. Renal brush border membranes had been shown to hydrolyse atriopeptin III at several bonds, including Cys-Phe (Olins et al., 1986). The enzyme initiating the attack on human atrial natriuretic peptide was then shown to be endopeptidase–24.11 (Stephenson and Kenny, 1987a) and the mode of attack was defined in detail (Kenny and Stephenson, 1988; Vanneste et al., 1988). Once again endopeptidase–24.11 became a therapeutic target, this time in developing inhibitors of endopeptidase–24.11 (Table II) that might be effective in cardiovascular diseases (for review, see Wilkins et al., 1993).

Table II. Two distinct membrane metallo-endopeptidase families.

Name:	Endopeptidase–24.11	Endopeptidase–24.18
Related enzymes:		
Mammalian	Endothelin-converting enzyme	Meprin (mouse) PABA-peptide hydrolase (human)
Microbial	Thermolysin	
Crustacean		Astacin
Inhibitors:	Phosphoramidon Thiorphan Retrothiorphan Acetorphan Candoxatrilat etc.	Actinonin

The early availability of a specific inhibitor (phosphoramidon) of endopeptidase–24.11 was crucial to the discovery of the another endopeptidase in rat kidney brush borders. The hydrolysis of insulin B chain by rat microvilli (as distinct from microvilli prepared from rabbit, pig, human) was inhibited by phosphoramidon to a maximum of 50% and the residual activity was shown to be a distinct endopeptidase (Kenny et al., 1981). The activity, given the provisional name of 'endopeptidase-2', was later characterized in some detail, (Kenny and Ingram, 1987; Stephenson and Kenny, 1988; Barnes et al., 1989) and, when classified as EC 3.4.24.18, the more informative name, endopeptidase–24.18 was adopted (Table II). Endopeptidase–24.18 is one of a distinct family of membrane peptidases (Stephenson and Kenny, 1989) which includes mouse kidney 'meprin' (Beynon et al., 1981) and the human intestinal 'PABA–peptide hydrolase' (Sterchi et al., 1983).

Following an immunization of a mouse with pig renal brush border membranes a clone, GK5C1, was obtained that secreted an IgG which bound to the microvillar membrane of kidney and intestine. The antigen was a 130 kDa subunit glycoprotein with all the structural and toplogical properties of an ectoenzyme, but it failed to hydrolyse any of the substrates for the known hydrolases. Success was finally achieved with Leu-Trp as substrate and the antigen was shown to be a previously unreported aminopeptidase (Gee and Kenny, 1985). This is probably the first, and still one of the few instances to date, of a previously unknown enzyme being revealed by hybridoma technology. It is a zinc metallopeptidase unusually insensitive to inhibition by chelating agents, though readily inhibited by amastatin (Gee and Kenny, 1987) and by some sulphydryl inhibitors such as Rentiapril (Tieku and Hooper, 1992). Glu-Trp is the best substrate so far known and provides the basis of a fluorometric assay (Jackson et al., 1988). Aminopeptidase W was identified in the choroid plexus (Bourne et al., 1989), but no immunochemical staining was observed in the central nervous system. However, it was demonstrated to be an axonal antigen in peripheral nerves (Barnes et al., 1991). In Wallerian degeneration following section of a nerve, aminopeptidase W exhibited an altered location, now being seen in the invading macrophages (Kenny and Bourne, 1991). The functional significance of aminopeptidase W is at present obscure. Enzymologically it has a strong preference for short peptides, Leu-Trp being hydrolysed twice as rapidly as Leu-Trp-Leu and 8-times faster than Leu-Trp-Met-Arg and it is difficult to guess what biologically important peptides of this size might be targets in those locations where the peptidase is found.

Which peptidases hydrolyse which peptides *in vivo*?

Simple *in vitro* experiments are important to show the potential of a peptidase to attack a given peptide, but a positive result does not necessarily indicate that this is a physiological event. The fate of a peptide will depend crucially on the peptidases it encounters in the vicinity of its release. The subcellular localization of peptidases is fundamental, hence the emphasis in this review on *cell-surface* enzymes. When a peptide encounters a battery of such peptidases it becomes important to know which enzyme initiates the inactivating cleavage. Ultimately many may play a role in the total degradation of the substrate, but these later steps are of little physiological interest. Inhibitors provide a valuable tool in such investigations, making it possible to perform a biochemical dissection of a complex system, such as a membrane preparation from renal brush border. Thus, for renal microvilli, endopeptidase–24.11 was shown to be the initiator of the attack on angiotensins I, II and III, bradykinin, substance P, atrial natriuretic peptide and oxytocin (Stephenson and Kenny, 1987a,b). Endopeptidase–24.11 was also the dominant peptidase in membranes from choroid plexus for the hydrolysis of brain and atrial natriuretic peptides (Bourne and Kenny, 1990) as it was in pig striatal synaptic membranes for the hydrolysis of substance P, [D-Ala^2] Leuenkephalin and cholecystokinin-(CCK)-8 (Matsas et al., 1983, 1984b). Specific inhibitors have also played a role in experiments with tissue slices from brain in which the recovery of released neuropeptide could be shown to be significantly enhanced in the presence of an appropriate inhibitor, e.g. CCK-8 and substance P by inhibitors of endopeptidase–24.11 on superfused slices of substantia nigra (Mauborgne et al., 1987; Littlewood et al., 1988). Biochemical approaches together with immunohistochemical evidence of colocalization of peptidase and putative peptide substrate can go some way to defining a function, especially where the latter data are at an ultrastructural level.

Structural and enzymic comparisons of endopeptidase–24.11 and endopeptidase–24.18

Endopeptidase–24.11 is a 94 kDa ectoenzyme, a type II integral membrane protein with a short intracellular N-terminal domain, a hydrophobic membrane–associated domain, corresponding to the retained signal sequence, and a large extracytoplasmic domain which includes the active site (Fig. 1). It has been cloned and sequenced in rat (Malfroy et al., 1987), rabbit (Devault et al., 1987), mouse (Chen et al., 1992) and human (Letarte et al., 1988).

Endopeptidase–24.18 is a disulphide-linked tetramer composed of two nonidentical subunits (α and β, Fig. 1) and like endopeptidase–24.11 it contains Zn at the active site and is strongly inhibited by chelating agents (Kenny and Ingram, 1987). Both subunits of the rat endopeptidase–24.18 have been cloned and sequenced (Corbeil et al., 1992; Johnson and Hersh, 1992). The α-subunit cDNA revealed a single putative hydrophobic membrane anchor located near the C-terminus of the protein, but a processing step leads to the loss of this domain and, when the α-subunit was expressed in COS-1 cells, it emerged as a secreted protein, lacking activity unless trypsin-treated (Corbeil et al., 1993).

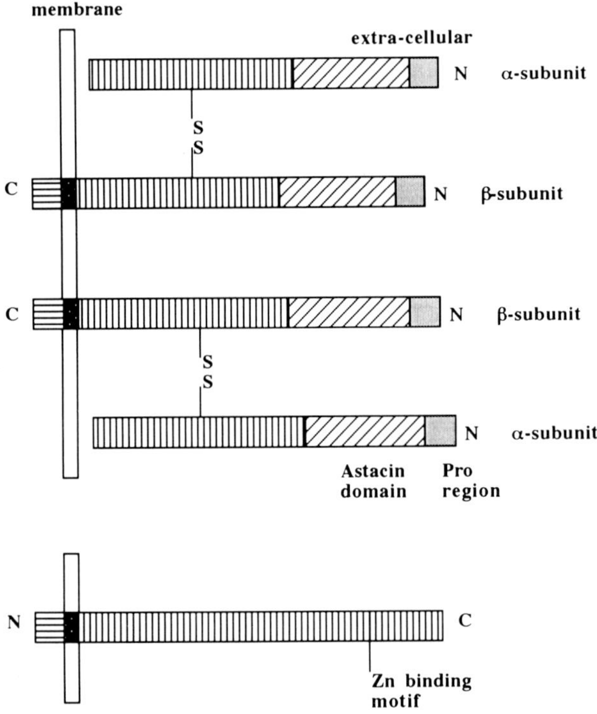

Fig. 1. Primary structures of two cell-surface endopeptidases. (See text for references).
(a) Endopeptidase–24.18, a tetramer of which only the β-subunits are transmembrane, being anchored near the C-terminus. The astacin domain contains the Zn-binding motif; full activity requires the proteolytic removal of the N-terminal domains.
(b) Endopeptidase–24.11, a transmembrane protein, anchored near the N-terminus. In most species it exists as a noncovalently-linked dimer.

The mature tetrameric protein is anchored only by the β–subunits, the α-subunits being attached by SS bridge(s) (Milhiet et al., 1994). Endopeptidase–24.18 is a member of the astacin family, showing homology to the 200 residue zinc metalloproteinase isolated from the crayfish *Astacus fluviatilis* (Titani et al., 1987).

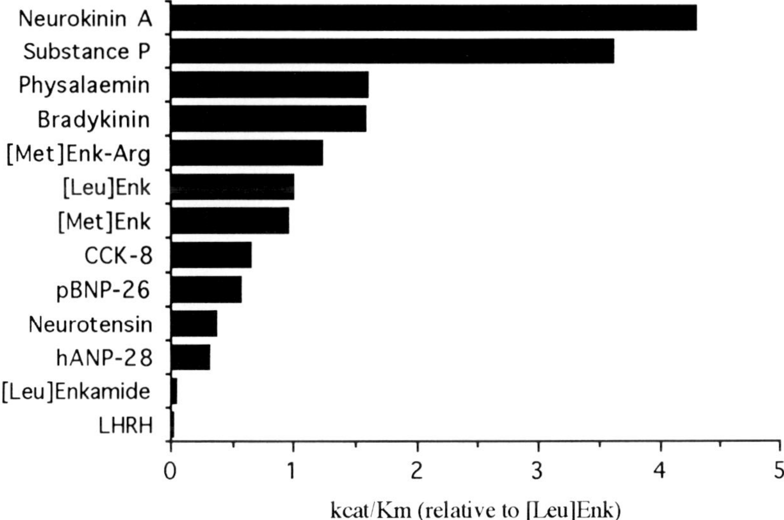

Fig. 2. Efficiency of hydrolysis of some neuropeptides by endopeptidase–24.11. Data expressed as k_{cat}/K_m (specificity constant) relative to [Leu]enkephalin which is taken as unity. In absolute terms the specificity constant for this peptide is 44 $min^{-1}\mu M^{-1}$.

The primary specificity of endopeptidase–24.11 is to hydrolyse peptide bonds involving the amino function of hydrophobic amino acid residues, similar to thermolysin (Kerr and Kenny, 1974a). However subsite interactions and conformational factors greatly influence the activity, so that the efficiency of hydrolysis of a group of neuropeptides may span two orders of magnitude (Fig. 2). The specificity constant (k_{cat}/K_m ratio) for substance P is 100 times greater than that for luliberin (LHRH), though both contain the hydrolysable Gly–Leu bond (Matsas et al., 1984a). Simple peptides are preferred: the B chain of insulin is hydrolysed, but the intact insulin molecule is wholly resistant to hydrolysis (Kenny, 1986; Stephenson and Kenny, 1987b). Larger substrates are generally not hydrolysed, but in a limited survey of some cytokines, hydrolysis of interleukin–1β (IL–1β) and IL-6 was observed (by sodium dodecylsulfate polyacrylamide gel electrophoresis) but only with a very high enzyme : substrate ratio (5:1) and long incubation. No effect on IL-1α, tumour necrosis factor-α, transforming growth factor-α (TGF-α), epidermal growth factor, nerve growth factor and insulin-like growth factor could be shown in these conditions (Choudry and Kenny, 1991).

The bond specificity of endopeptidase–24.18 is more complex. There is a preference for aromatic residues, but these may occupy either the P1 or P1' positions. Various neuropeptides were hydrolysed by the purified rat enzyme, but generally the efficiency of hydrolysis by endopeptidase–24.18 was much inferior to that by endopeptidase–24.11. For substance P and bradykinin, both good substrates for endopeptidase-24.11, the K_m values were 3-fold higher and the k_{cat}/K_m ratios about two orders of magnitude lower for endopeptidase–24.18. Only with luliberin (LHRH) as substrate was endopeptidase–24.18 slightly better than endopeptidase–24.11 (Stephenson and Kenny, 1988) and neuropeptide Y appears to be another exception to the general rule that endopeptidase–24.11 is the dominant neuropeptide-hydrolysing enzyme (Price et al., 1991). However, endopeptidase–24.18 may have a wider repertoire in hydrolysing more complex molecules, e.g., it was very effective in hydrolysing transforming growth factor-α, a peptide with three cystines (Choudry and Kenny, 1991). Endopeptidase–24.18 has a rather limited tissue distribution (essentially brush borders of kidney and intestine) and it is difficult to speculate on the physiological significance of these *in vitro* observations. In contrast, we have a better understanding of the function of endopeptidase–24.11, at least in certain locations and systems.

The role of endopeptidase–24.11 in the metabolism of natriuretic peptides

The main forms of natriuretic peptides (NP) in the human (h) are: α-hANP (atrial), hBNP-32 (brain), hCNP-22 (C-type) and urodilatin. They are derived from different precursor molecules (except for urodilatin, which is derived from the α-hANP precursor), but all contain a disulphide-linked loop of 17 residues, the integrity of which is essential for their biological activity in the regulation of sodium and water homeostasis and blood pressure. ANP is abundant in cardiac atria, but is also found in several other tissues including the brain, adrenal gland and testis. BNP is present in brain and heart, the highest concentration being in the atrium, but a greater quantity is present in the ventricles. CNP is essentially confined to the nervous system. Studies on the hydrolysis of these peptides (Kenny et al., 1993), have revealed some notable differences in rates, thus hBNP-32 was hydrolysed about 15 times more slowly than hANP. Moreover, hBNP-32 was hydrolysed differently from any of the other peptides, in particular the initial cleavage was not within the ring, but in the head peptide at Met^4-Val^5. The pattern of attack on bonds within the disulphide-linked ring of human BNP differs substantially from that of porcine BNP-26 (Bourne and Kenny, 1990). Kinetic studies also place hBNP-32 in a different category, the K_i for which was 5-fold higher than that for α–hANP. If hBNP-32 was unusually resistant to hydrolysis by endopeptidase–24.11, CNP was, by contrast, hydrolysed more rapidly than any of the other peptides in this group, almost twice as rapidly as α-hANP. CNP is mainly found in the brain and might therefore be a target for endopeptidase–24.11 in the central nervous system.

The role of endopeptidase–24.11 and other peptidases in the central nervous system

The surfaces in contact with the cerebrospinal fluid are relatively rich in cell-surface peptidases. Membrane preparations of choroid plexus contain endopeptidase–24.ll, angiotensin-converting enzyme, dipeptidyl peptidase IV, aminopeptidase–N, aminopeptidase–A, aminopeptidase–W and carboxypeptidase–M, but appear to lack aminopeptidase–P, carboxypeptidase–P and membrane dipeptidase. These preparations are not free of endothelial membranes and part of the activity of angiotensin-converting enzyme and aminopeptidase–N is known, from immunohistochemistry, to include a vascular component. The pial membrane enveloping the entire central nervous system is rich in endopeptidase–24.11 and aminopeptidase–N, while that covering the brain also has angiotensin-converting enzyme (Bourne et al., 1989). It seems likely that this distribution of peptidase activity, especially that of endopeptidase–24.11, serves to clear the cerebrospinal fluid of many neuropeptides, consistent with reports that the concentrations in cerebrospinal fluid of atrial natriuretic factor and substance P, both of which are targets for endopeptidase–24.11, are about ten-fold less than in plasma.

Endopeptidase–24.11 has been purified from pig brain (Relton et al., 1983) where it is localized in the neuropil, being concentrated in the corpus striatum, and is striosomally ordered (Matsas et al., 1986; Waksman et al., 1986; Pollard et al., 1987; Barnes et al., 1988a). Aminopeptidase–N on the other hand has a very diffuse distribution throughout the brain neuropil and, like angiotensin-converting enzyme, is also present on the endothelium of vessels (Barnes et al., 1988a). Neither is striosomally ordered. Indeed several lines of evidence, including primary cell cultures and immunogold labelling of ultrathin cryosections, suggest that aminopeptidase–N has a major location on glial membranes and predominates on the basal (rather than the apical) surface of endothelial cells. A clear ultrastructural demonstration that aminopeptidase–N is also present on neuronal membranes has so far been elusive (Barnes et al., 1994).

The neuronal localization of endopeptidase-24.11 is supported by a number of independent observations. It was demonstrated by immunoperoxidase staining in a population of central cortex neurons cultured from pig striatal tissue (Matsas and Kenny, 1989) and immunoperoxidase electron microscopy of pig brain has shown it to be associated with some axonal membranes and synapses in the globus pallidus (Barnes et al., 1988b). Electron microscopic studies on rat brain have revealed immunoperoxidase reactivity in the nucleus solitarius (Lasher et al., 1990) and another using radio-iodinated monoclonal antibody to endopeptidase-24.11 showed a location predominantly, but not exclusively, on neuronal membranes (Marcel et al., 1990). Electron microscopy and Percoll-gradient fractionation have shown that endopeptidase-24.11 in the pig substantia nigra is present on pre- and postsynaptic membranes (Barnes et al., 1992) and is thus well placed to terminate some neuropeptide signals. At the light microscopic level immunostaining of adjacent sections of pig striatum for endopeptidase-24.11 and neuropeptide revealed colocalization of enzyme with both SP and Leu-enkephalin in the caudate-putamen and in bundles of efferent fibres passing through the globus pallidus (Matsas et al., 1986). More significantly, ultrastructural studies of the substantia nigra have provided support for a role in the hydrolysis of substance P (Barnes et al., 1993). These double-labelling studies, using immunogold and immunoperoxidase methods at the ultrastructural level, have confirmed the presence of substance P in a number of synapses that also stained positively for endopeptidase–24.11. This was the first evidence of colocalization of a target neuropeptide with its putative degrading enzyme. Substance P is often found in association with other peptides and hence it may not be the only target peptide at these synapses and, because substance P was the only neuropeptide to be studied, the results do not imply that this is an exclusive relationship: colocalization of endopeptidase–24.11 with other neuropeptides, including enkephalins, cannot be excluded.

References

Almenoff, J. and Orlowski, M. (1984) Biochemical and immunological properties of a membrane-bound brain metalloendopeptidase: comparison with thermolysin-like kidney neutral metalloendopeptidase. *J. Neurochem.* 42: 151-157.

Barnes, K., Bourne, A., Cook, P. A., Turner, A. J. and Kenny, A. J. (1991) Membrane peptidases in the peripheral nervous system of the pig: their localization by immunohistochemistry at light and electron microscopic levels. *Neuroscience* 44: 245-261.

Barnes, K., Ingram, J. and Kenny, A. J. (1989) Proteins of the kidney microvillar membrane. Structure and immunochemical properties of rat endopeptidase-2 and its immunohistochemical localization in tissues of rat and mouse. *Biochem. J.* 264: 335-346.

Barnes, K., Kenny, A. J. and Turner, A. J. (1994) Localization of aminopeptidase N and dipeptidyl peptidase IV in pig striatum and in neuronal and glial cell cultures. *Eur. J. Neurosci.* 6: 531-537.

Barnes, K., Matsas, R., Hooper, N. M., Turner, A. J. and Kenny, A. J. (1988a) Endopeptidase-24.11 is striosomally ordered in pig brain and, in contrast to aminopeptidase N and peptidyl dipeptidase A ('angiotensin converting enzyme'), is a marker for a set of striatal efferent fibres. *Neuroscience* 27: 799-817.

Barnes, K., Turner, A. J. and Kenny, A. J. (1988b) Electronmicroscopic immunocytochemistry of pig brain shows that endopeptidase-24.11 is localized in neuronal membranes. *Neurosci. Lett.* 94: 64-69.

Barnes, K., Turner, A. J. and Kenny, A. J. (1992) Membrane localization of endopeptidase-24.11 and peptidyl dipeptidase A (angiotensin converting enzyme) in the pig brain: a study using subcellular fractionation and electron microscopic immunocytochemistry. *J. Neurochem.* 58: 2088-2096.

Barnes, K., Turner, A. J. and Kenny, A. J. (1993) An immuno-electron microscopic study of pig substantia nigra shows colocalization of endopeptidase-24.11 with substance P. *Neuroscience* 53: 1073-1082.

Bauer, K. (1994) Purification and characterization of the thyrotropin-releasing-hormone-degrading ectoenzyme. *Eur. J. Biochem.* 224: 387-396.

Bauer, K. (1995) Inactivation of thyrotropin-releasing hormone (TRH) by the hormonally regulated TRH-degrading enzyme. *TEM* 6: 101-105.

Beynon, R. J., Shannon, J. D. and Bond, J. S. (1981) Purification and characterization of a metallo-endopeptidase from mouse kidney. *Biochem. J.* 199: 591-598.

Bourne, A., Barnes, K., Taylor, B. A., Turner, A. J. and Kenny, A. J. (1989) Membrane peptidases in the porcine choroid plexus and on other cell surfaces in contact with the cerebral spinal fluid. *Biochem. J.* 259: 69-80.

Bourne, A. and Kenny, A. J. (1990) The hydrolysis of brain and atrial natriuretic peptides by porcine choroid plexus is attributable to endopeptidase-24.11. *Biochem. J.* 271: 381-385.

Checler, F. (1993) Neuropeptide-degrading peptidases. *In:* S. H. Parvez, M. Naoi, T. Nagatsu and S. Parvez (eds.): *Methods in Neurotransmitter and Neuropeptide Research,* Elsevier Science Publishers B.V., Amsterdam, pp. 375-418.

Chen, C.-Y., Salles, G., Seldin, M. F., Kister, A. E., Reinherz, E. L. and Shipp, M. A. (1992) Murine common acute lymohoblastic leukemia antigen (CD10, neutral endopeptidase 24.11). Molecular characterization, chromosomal localization and modeling of the active site. *J. Immunol.* 148: 2817-2825.

Choudry, Y. and Kenny, A. J. (1991) Hydrolysis of transforming growth factor-alpha by cell surface peptidases *in vitro. Biochem. J.* 280: 57-60.

Corbeil, D., Gaudoux, F., Wainwright, S., Ingram, J., Kenny, A. J., Boileau, G. and Crine, P. (1992) Molecular cloning of the alpha-subunit of rat endopeptidase-24.18 (endopeptidase-2) and colocalization with endopeptidase-24.11 in rat kidney by *in situ* hybridization. *FEBS Lett.* 309: 203-208.

Corbeil, D., Milhiet, P.-E., Simon, V., Ingram, J., Kenny, A. J., Boileau, G. and Crine, P. (1993) Rat endopeptidase-24.18 α–subunit is secreted into the culture medium as a zymogen when expressed by COS-1 cells. *FEBS Lett.* 335: 361-366.

Craves, F. B., Law, P. Y., Hunt, C. A. and Loh, H. H. (1978) The metabolic disposition of radiolabeled enkephalins *in vitro* and *in situ. J. Pharmacol. Exper. Ther.* 206: 492-506.

Devault, A., Lazure, C., Nault, C., Le Moual, H., Seidah, N. G., Chretien, M., Kahn, P., Powell, J., Mallet, J., Beaumont, A. (1987) Amino acid sequence of kidney neutral endopeptidase-24.11 (enkephalinase) deduced from a complementary DNA. *EMBO J.* 6: 1317-1322.

Fulcher, I. S., Matsas, R., Turner, A. J. and Kenny, A. J. (1982) Kidney neutral endopeptidase and the hydrolysis of enkephalin by synaptic membranes show similar sensitivity to inhibitors. *Biochem. J.* 203: 519-522.

Fulcher, I. S., Pappin, D. J. C. and Kenny, A. J. (1986) The N-terminal amino acid sequence of pig kidney endopeptidase-24.11 shows homology with pro-sucrase-isomaltase. *Biochem. J.* 240: 305-308.

Gee, N. S. and Kenny, A. J. (1985) Proteins of the kidney microvillar membrane. The 130 kDa protein in pig kidney, recognised by monoclonal antibody GK5C1, is an ectoenzyme with aminopeptidase activity. *Biochem. J.* 230: 753-764.

Gee, N. S. and Kenny, A. J. (1987) Proteins of the kidney microvillar membrane. Enzymic and molecular properties of aminopeptidase W. *Biochem. J.* 246: 97-102.

Jackson, M. C., Choudry, Y., Bourne, A., Woodley, J. F. and Kenny, A. J. (1988) A fluorimetric assay for aminopeptidase W. *Biochem. J.* 253: 299-302.

Johnson, G. D. and Hersh, L. B. (1992) Cloning of rat meprin reveals the enzyme is a heterodimer. *J. Biol. Chem.* 267: 13505-13512.

Kenny, A. J. (1977) Proteinases associated with cell membranes. *In:* A. J. Barrett (ed.): *Proteinases in Mammalian Cells and Tissues,* Elsevier/North Holland Biomedical Press, Amsterdam, pp. 393-444.

Kenny, A. J. (1986) Regulatory peptide metabolism at cell surfaces: the key role of endopeptidase-24.11. *Biomed. Biochim. Acta* 45: 1503-1513.

Kenny, A. J. and Booth, A. G. (1978) Microvilli, their ultrastructure, enzymology and molecular organisation. *Essays Biochem.* 14: 1-44.

Kenny, A. J. and Bourne, A. (1991) Cellular reorganization of membrane peptidases in Wallerian degeneration of pig peripheral nerve. *J. Neurocytol.* 20: 875-885.

Kenny, A. J., Bourne, A. and Ingram, J. (1993) Hydrolysis of natriuretic peptides and C-receptor ligands by endopeptidase-24.11. *Biochem. J.* 191: 83-88.

Kenny, A. J., Fulcher, I. S., McGill, K. A. and Kershaw, D. (1983). Proteins of the kidney microvillar membrane. Reconstitution of endopeptidase in liposomes shows that it is a short-stalked protein. *Biochem. J.* 211: 755-762.

Kenny, A. J., Fulcher, I. S., Ridgwell, K. and Ingram, J. (1981) Microvillar membrane neutral endopeptidases. *Acta Biol. Med. Germ.* 40: 1465-1471.

Kenny, A. J. and Hooper, N. M. (1991) Peptidases involved in the metabolism of bioactive peptides. *In:* J. H. Henriksen (ed.): *Degradation of Bioactive Substances. Physiology and Pathology,* CRC Press Inc., Boca Ratton, pp. 47-79.

Kenny, A. J. and Ingram, J. (1987) Proteins of the kidney microvillar membrane. Purification and properties of the phosphoramidon-insensitive endopeptidase ('endopeptidase-2') from rat kidney. *Biochem. J.* 245: 515-524.

Kenny, A. J. and Stephenson, S. L. (1988) Role of endopeptidase-24.11 in the inactivation of atrial natriuretic peptide. *FEBS Letts* 232: 1-8.

Kenny, A. J., Stephenson, S. L. and Turner, A. J. (1987) Cell surface peptidases. *In:* A. J. Kenny and A. J. Turner (eds.): *Mammalian Ectoenzymes,* Elsevier, Amsterdam, pp. 169-210.

Kerr, M. A. and Kenny, A. J. (1974a) The purification and specificity of a neutral endopeptidase from rabbit kidney brush border. *Biochem. J.* 137: 477-488.

Kerr, M. A. and Kenny, A. J. (1974b) The molecular weight and properties of a neutral metallo-endopeptidase from rabbit kidney brush border. *Biochem. J.* 137: 489-495.

Lasher, R. S., Lutz, E., M., Mulholland, F., Sanderson, R., Stewart, J. M. and Bublitz, C. (1990) Immunocytochemical localization of endopeptidase-24.11 in the nucleus tractus solitarius of the rat brain. *Neurosci. Lett.* 117: 43-49.

Letarte, M., Vera, S., Tran, R., Addis, J. B. L., Onizuka, R. J., Quackenbush, E. J., Jongeneel, C. V. and McInnes, R. R. (1988) Common acute lymphocytic leukemia antigen is identical to neutral endopeptidase. *J. Exp. Med* 168: 1247-1253.

Littlewood, G. M., Iversen, L. L. and Turner, A. J. (1988) Neuropeptides and their peptidases: Functional considerations. *Neurochem. Int.* 12: 383-389.

Malfroy, B., Schofield, P. R., Kuang, W. J., Seeburg, P. H., Mason, A. J. and Henzel, W. J. (1987) Molecular cloning and amino acid sequence of rat enkephalinase. *Biochem. Biophys. Res. Commun.* 144: 59-66.

Malfroy, B., Swerts, J. P., Guyon, A., Roques, B. P. and Schwartz, J.-C. (1978) High affinity enkephalin-degrading peptidase in brain is increased after morphine. *Nature* 276: 523-526.

Marcel, D., Pollard, H., Verroust, P., Schwartz, J. C. and Beaudet, A. (1990) Electron microscopic localization of immunoreactive enkephalinase (EC 3.4.24.11) in the neostriatum of the rat. *J. Neurosci.* 10: 2804-2817.

Matsas, R., Fulcher, I. S., Kenny, A. J. and Turner, A. J. (1983) Substance P and [Leu]enkephalin are hydrolysed by an enzyme in pig caudate synaptic membranes that is identical with the endopeptidase of kidney microvilli. *Proc. Natl. Acad. Sci. USA* 80: 3111-3115.

Matsas, R. and Kenny, A. J. (1989). Immunocytochemical localization of endopeptidase-24.11 in cultured neurons. *Neuroscience* 31: 237-246.

Matsas, R., Kenny, A. J. and Turner, A. J. (1984a) The metabolism of neuropeptides. The hydrolysis of peptides, including enkephalins, tachykinins and their analogues, by endopeptidase-24.11. *Biochem. J.* 223: 433-440.

Matsas, R., Kenny, A. J. and Turner, A. J. (1986) An immunohistochemical study of endopeptidase-24.11 ("enkephalinase") in the pig nervous system. *Neuroscience* 18: 991-1012.

Matsas, R., Turner, A. J. and Kenny, A. J. (1984b) Endopeptidase-24.11 and aminopeptidase activity in brain synaptic membranes are jointly responsible for the hydrolysis of cholescystokinin octapeptide (CCK-8). *FEBS Lett.* 175: 124-128.

Mauborgne, A., Bourgoin, S., Benoliel, J. J., Hirsch, M., Berthier, J. L., Mamon, M. and Cesselin, F. (1987) Enkephalinase is involved in the degradation of endogenous substance P released from slices of rat substantia nigra. *J. Pharmacol. Pharmacol. Exp. Ther.* 243: 674-680.

Milhiet, P.-E., Corbeil, D., Simon, V., Kenny, A. J., Crine, P., Boileau, G. and . (1994) Expression of rat endopeptidase-24.18 in COS-1 cells: membrane topology and activity. *Biochem. J.* 300: 37-43.

Olins, G. M., Spear, K. L., Siegel, N. R., Zurcher-Neely, H. A. and Smith, C. E. (1986) *Fed. Proc. Fed. Am. Soc. Exp. Biol.* 45: 427.

Pollard, H., Llorens-Cortes, C., Couraud, J. Y., Ronco, P., Verrroust, P. and Schwartz, J. C. (1987) Enkephalinase (EC 3.4.24.11) is highly localized to a striatonigral pathway in rat brain. *Neurosci. Lett.* 77: 267-271.

Price, J. S., Kenny, A. J., Huskisson, N. S. and Brown, M. J. (1991) Neuropeptide Y (NPY) metabolism by endopeptidase-2 hinders characterization of NPY receptors in rat kidney. *Brit. J. Pharmacol.* 104: 321-326.

Relton, J. M., Gee, N. S., Matsas, R., Turner, A. J. and Kenny, A. J. (1983) Purification of endopeptidase-24.11 ('enkephalinase') from pig brain by immunoadsorbent chromatography. *Biochem. J.* 215: 519-523.

Schauder, S., Schomburg, L., Köhrle, J. and Bauer, K. (1994) Cloning of a cDNA encoding an ectoenzyme that degrades thyrotropin-releasing hormone. *Proc. Natl. Acad. Sci. USA* 91: 9534-9538.

Stephenson, S. L. and Kenny, A. J. (1987a) The hydrolysis of alpha-human atrial natriuretic peptide by pig kidney microvillar membranes is initiated by endopeptidase-24.11. *Biochem. J.* 243: 183-187.

Stephenson, S. L. and Kenny, A. J. (1987b) Metabolism of neuropeptides. Hydrolysis of angiotensins, bradykinin, substance P and oxytocin by pig microvillar membranes. *Biochem. J.* 241: 237-247.

Stephenson, S. L. and Kenny, A. J. (1988) The metabolism of neuropeptides. Hydrolysis of peptides by the phosphoramidon-insensitive rat kidney enzyme, 'endopeptidase-2', and by rat kidney microvillar membranes. *Biochem. J.* 255: 45-51.

Stephenson, S. L. and Kenny, A. J. (1989) Sorry, not one of the family. *Biochem. J.* 259: 622-623.

Sterchi, E. E., Green, J. R. and Lentze, M. J. (1983) Non-pancreatic hydrolysis of N-benzoyl-L-tyrosyl-p-aminobenzoic acid (PABA peptide) in the rat small intestine. *J. Ped. Gastroenterol. Nutrition* 2: 539-547.

Tieku, S. and Hooper, N. M. (1992) Inhibition of aminopeptidases N, A and W. A re-evaluation of the actions of bestatin and inhibitors of angiotensin converting enzyme. *Biochem. Pharmacol.* 44: 1725-1730.

Titani, K., Torff, H.-J., Hormel, S., Kumar, S., Walsh, K. A., Rodl, J., Neurath, H. and Zwilling, R. (1987) Amino acid sequence of a unique protease from the crayfish *Astacus fluviatilis*. *Biochemistry* 26: 222-226.

Turner, A. J., Hooper, N. M. and Kenny, A. J. (1989) Neuropeptide-degrading enzymes. *In:* G. Finkand , A. J. Harmar (eds.): *Neuropeptides, a Methodology,* John Wiley & Sons, London, pp. 189-223.

Vanneste, Y., Michel, A., Dimaline, R., Najdovski, T. and Deschodt-Lanckman, M. (1988) Hydrolysis of a-human atrial natriuretic peptide *in vitro* by human kidney membranes and purified endopeptidase-24.11. *Biochem. J.* 254: 531-537.

Waksman, G., Hamel, E., Delay-Goyet, P. and Roques, B. P. (1986) Neuronal localization of the neutral endopeptidase 'enkephalinase' in rat brain revealed by lesions and autoradiography. *EMBO J.* 5: 3163-3166.

Wilkins, M. R., Unwin, R. J. and Kenny, A. J. (1993) Endopeptidase-24.11 and its inhibitors: potential therapeutic agents for edematous disorders and hypertension. *Kidney Int.* 43: 273-285.

Wong-Leung, Y. L. and Kenny , A. J. (1968) The intracellular location of some particulate peptidases in the kidney of the rat and the rabbit. *Biochem. J.* 110: 5P-6P.

Yorimitsu, K., Moroi, K., Inagaki, N., Saito, T., Masuda, Y., Masaki, T., Seino, S. and Kimura, S. (1995) Cloning and sequencing of human endothelin converting enzyme in renal adenocarcinoma (ACHN) cells producing endothelin-2. *Biochem. Biophys. Res. Commun.* 208: 721-727.

The TRH-degrading ectoenzyme: a putative signal-terminator within the central nervous system and adenohypophyseal regulator of hormone secretion

L. Schomburg and K. Bauer

Max-Planck-Institut für experimentelle Endokrinologie, 30603 Hannover, Germany

Summary. The extracellular inactivation of TRH signals is catalyzed by specific TRH-degrading enzymes. The analysis of the primary structure of the TRH-degrading ectoenzyme reveals that it belongs to the family of Zn-dependent metallopeptidases. The tissue-specific regulation of the adenohypophyseal enzyme by peripheral hormones suggests that it may serve integrative functions within endocrine regulatory systems. In the brain, the high activity and the exclusive localization on neurons imply that this peptidase might also act as a terminator of neurotrophic TRH signals.

Introduction

Thyrotropin-releasing hormone (TRH, pyroGlu-His-Pro-NH$_2$), the first hypothalamic neuropeptide hormone structurally elucidated, has been isolated as a hypophysiotropic releasing factor that stimulates the release of adenohypophyseal hormones (Guillemin, 1978; Schally, 1978). Subsequently, TRH and high affinity TRH-receptors were shown to be widely distributed within the central nervous system, where the tripeptideamide most likely acts as a neurotransmitter and/or neuromodulator. Binding to its receptor stimulates phosphatidylinositol turnover, mobilization of intracellular Ca^{2+} and activation of protein kinase C.

The function of TRH as a potent signal substance necessarily implies the existence of a highly efficient inactivation system. There is considerable evidence that secreted TRH is enzymatically inactivated by a particulate enzyme that hydrolyzes the pyroGlu-His bond. Subcellular fractionation studies and other investigations with brain cells in primary culture have clearly demonstrated that this peptidase is a true ectoenzyme, i.e. it is localized on the plasma membrane with its active site oriented towards the extracellular space. Interestingly, this enzyme is found on the surface of neuronal cells but not on glial cells, and with dispersed pituitary cells it was found to be associated preferentially with lactotrophic cells (Bauer et al., 1990).

The most striking property of the TRH-degrading ectoenzymes is the unusually high degree of substrate specificity (Bauer et al., 1984; O'Conner and O'Cuinn, 1985; Wilk and Wilk, 1989; Elmore et al., 1990) that is unparalleled within the family of neuropeptide hydrolases. Until now, no other naturally occurring substance has been found that is taken as substrate. Even LH-RH, a decapeptide amide with the sequence pyroGlu-His-... is not degraded by this enzyme and extensive studies on the degradation of TRH-analogues suggest that the emzyme recognizes the entire entity of the TRH-molecule. The TRH-degrading ectoenzyme also exhibits an unusual tissue distribution. Highest activities are found in brain, followed by lung, pituitary, retina and liver while other tissues, and surprisingly even the kidneys, are devoid of this peptidasic activity. In serum the degradation of TRH is catalyzed by an enzyme that exhibits identical chemical characteristics including the high degree of substrate specificity. The biochemical data presently available strongly indicate that the TRH-degrading ectoenzyme and the TRH-degrading serum enzyme derive from the same gene. It remains the subject of further investigation to analyze whether the two enzymes result from differential splicing or differential post-translational processing.

Purification, cDNA-cloning and primary structure

Initially, various detergents were tested to solubilize the TRH-degrading ectoenzyme. Unfortunately, however, all attempts failed since the enzymatic activity could not be stabilized. In contrast, release of the enzymatically active domain from guinea pig synaptosomes by papain and from rabbit brain by trypsin yielded preparations that retained full activity for long periods of time (O'Conner and O'Cuinn, 1985; Wilk and Wilk, 1989). Final purification to electrophoretic homogeneity (200 000 fold) was accomplished from rat and pig brain after very mild trypsin treatment (Bauer, 1994). Enzymatic and chemical fragmentation of purified enzyme yielded several peptides that could be isolated and sequenced. This information was succesfully used for cloning of the corresponding cDNA from rat brain and pituitary (Schander et al., 1994). The deduced primary structure of the TRH-degrading ectoenzyme predicts a polypeptide core of 1025 amino acids including a transmembrane spanning domain near the amino terminus. Part of the enzymatically active extracellular domain reveals high homology to aminopeptidases N and A (Fig. 1). In view of the unique substrate specificity, the restricted tissue distribution and the chemical characteristics we did not anticipate that the TRH-degrading ectoenzyme and the aminopeptidases share common ancestry. Instead of inventing a totally new protein, mother nature apparently preferred to modify the backbone of Zn-dependent aminopeptidases.

```
rAmPN     341   DQIALPDFNAGAMENWGLVTYRESALVFDPQSSSISNKERVV

hAmPA     347   DKIAIPDFGTGAMENWGLITYRETNLLYDPKESASSNQQRVA

rTRHase   395   DLLAVPKHPYAAMENWGLSIFVEQRILLDPSVSSISYLLDVT

rAmPN     383   TVIAHELAHQWFGNLVTVDWWNDLWLNEGFASYVEFLGADY

hAmPA     389   TVVAHELVHQWFGNIVTMDWWEDLWLNEGFASFFEFLGVNH

rTRHase   437   MVIVHEICHQWFGDLVTPVWWEDVWLKEGFAHYFEFVGTDY
```

Fig. 1. Alignment of selected amino acid sequences from the extracellular domains of rat aminopeptidase N (rAmPN), human aminopeptidase A (hAmPA) and rat TRH-degrading ectoenzyme (rTRHase). The residue numbers are given, identical amino acids are underlined and the consensus sequence of Zn-dependent metallopeptidases is indicated by bold letters.

Endocrine regulation

The activity of the adenohypophyseal TRH-degrading ectoenzyme is tightly regulated by thyroid hormones (Bauer, 1987; Ponce et al., 1988; Suen and Wilk, 1989). When euthyroid rats received a single injection of tri-iodothyronine (T3), the enzyme activity increased severalfold and conversely declined after rendering the animals hypothyroid by treatment with the goitrogenic agent propylthiouracil.

This feedback regulation is even more pronounced at the mRNA-level (Schomburg and Bauer, 1995). The direct comparison with transcript levels of TRH-receptor and TSH documented the dynamics and the extraordinary extent of the T3-effects on the mRNA-levels of the ectoenzyme (Fig. 2).

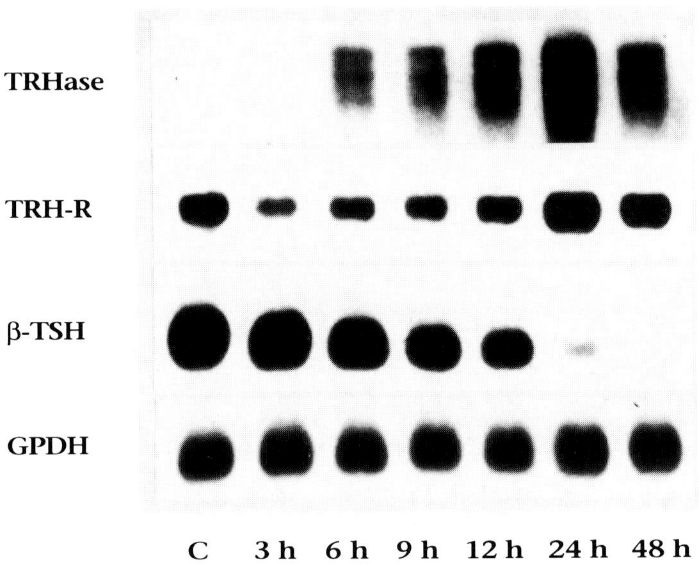

Fig. 2. Time course of the effect of a single injection of T3 on hypophyseal mRNA-levels. Female Sprague-Dawley rats received a single injection of 30 g T3/100 g BW. After the indicated time periods, the pituitaries were removed. mRNA was prepared from the anterior lobes and analyzed by Northern blot. C, Control; TRHase, TRH-degrading ectoenzyme; TRH-R, TRH-receptor; β-TSH, TSH-specific β-subunit; GPDH, glyceraldehyde-3-phosphate dehydrogenase.

Comparably tight regulation of the TRH-degrading activity was also observed with estradiol (Bauer, 1988). Apparently, the expression of the adenohypophyseal TRH-degrading ectoenzyme is dually controlled by these peripheral hormones. Interestingly enough, these hormonal effects are restricted to the anterior pituitary. Neither the activity nor the mRNA-levels of the brain and hypothalamic TRH-degrading ectoenzyme are influenced by thyroid hormones (Bauer, 1987; Schomburg and Bauer, 1995) and estradiol (Bauer, 1988).

Conclusions and perspectives

Compared to other neuropeptides, TRH is very special in that its extracellular inactivation is catalyzed by only one, highly specific enzyme. The characterization of the TRH-inactivating ectoenzyme should therefore contribute further to our understanding of TRH biology in general. The challenging task for the future will be to synthesize specific inhibitors and to generate antibodies that can be used to define the precise localization of the enzyme with respect to the TRH-receptor and the TRH-synthesizing neuron.

References

Bauer, K., Nowak, P. and Kleinkauf, H. (1981) Specificity of a serum peptidase capable of hydrolyzing thyroliberin at the pyroglutamyl-histidine bond. *Eur. J. Biochem.* 118:173-176.

Bauer, K. (1987) Adenohypophyseal degradation of thyrotropin releasing hormone regulated by thyroid hormones. *Nature* 330:375-377.

Bauer, K. (1988) Degradation and biological inactivation of thyrotropin releasing hormone (TRH): regulation of the membrane-bound TRH-degrading enzyme from rat anterior pituitary by estrogens and thyroid hormones. *Biochimie* 70:69-74.

Bauer, K., Carmelit, P., Schulz, M., Baes, M. and Denef, C. (1990) Regulation and cellular localization of the membrane-bound thyrotropin-releasing hormone-degrading enzyme in primary cultures of neuronal, glial and adenohypophyseal cells. *Endocrinology* 127:1224-1233.

Bauer, K. (1994) Purification and characterization of the thyrotropin-releasing hormone degrading ectoenzyme. *Eur. J. Biochem.* 224:387-96.

Elmore, M.A., Griffiths, E.C., O'Connor, B. and O'Cuinn, G. (1990) Further characterization of the substrate specificity of a TRH hydrolyzing pyroglutamate aminopeptidase from guinea-pig brain. *Neuropeptides* 15:31-36.

Guillemin, R. (1978) Peptides in the brain: the new endocrinology of the neuron. *Science* 202:390-402.

O'Connor, B. and O'Cuinn, G. (1985) Purification of and kinetic studies on a narrow specificity synaptosomal membrane pyroglutamate aminopeptidase from guinea-pig brain. *Eur. J. Biochem.* 150:47-52.

Ponce, G., Charli, J.L., Pasten, J.A., Aceves, C. and Joseph, B.P. (1988) Tissue-specific regulation of pyroglutamate aminopeptidase II activity by thyroid hormones. *Neuroendocrinology* 48:211-213.

Schally, A.V. (1978) Aspects of hypothalamic regulation of the pituitary gland. *Science* 202:18-28.

Schauder, B., Schomburg, L., Köhrle, J. and Bauer, K. (1994) Cloning of a cDNA encoding an ectoenzyme that degrades thyrotropin-releasing hormone. *Proc. Natl. Acad. Sci. USA* 91:9534-9538.

Schomburg, L. and Bauer, K. (1995) Thyroid hormones rapidly and stringently regulate the messenger RNA levels of the thyrotropin-releasing hormone (TRH) receptor and the TRH-degrading ectoenzyme. *Endocrinology* 136:3480-3485.

Suen, C.S., Wilk and S. (1989) Regulation of thyrotropin releasing hormone degrading enzymes in rat brain and pituitary by L-3,5,3'-triiodothyronine. *J. Neurochem.* 52:884-888.

Wilk, S. and Wilk, E.K. (1989) Pyroglutamyl peptidase II, a thyrotropin releasing hormone degrading enzyme: purification and specificity studies of the rabbit brain enzyme. *Neurochem. Int.* 15:81-90.

The function of glial cells in the inactivation of neuropeptides

R. Mentlein, P. Dahms, R. Lucius and D. Plogmann

Universität Kiel, Anatomisches Institut, Olshausenstrasse 40, D-24108 Kiel, Germany

Summary. As compared to cultivated rat cortical neurons, cultivated glial cells have a much higher potential to degrade neuropeptides by proteolytic cleavage. Rat microglial cells in culture cleave neuropeptides by action of plasma membrane-bound aminopeptidase N and plasmin/plasminogen activators released into the medium. Cultivated astrocytes catabolize preferentially medium chain length peptides like neurotensin, bradykinin, substance P or somatostatin. By comparison of substrate specificity, influence of inhibitors and immunostaining, endopeptidases 24.15 and 24.16 were identified to be responsible for this cleavage. Since astrocytes embrace neurons and their synapses, these astrocytic proteases appear to be relevant to inactivate neuropeptides after their release in the central nervous system.

Introduction

Glial cells actively participate in the metabolism of interneuronal signal molecules. The role of astrocytes in uptake and metabolism of amino acid and catecholamine transmitters is well established (Hansson, 1988). By analogy, glial cells might also be involved in the inactivation of neuropeptides, especially because they are themselves targets for neuropeptides (Krisch and Mentlein, 1994). Therefore, we evaluated the potency of glial cells in culture to metabolize neuropeptides and identified some glial proteases responsible for peptide degradation.

Materials and methods

Cell cultures. Primary glial cell cultures were prepared from dissociated cerebral cortices of 2-day postnatal Wistar rats (strain Han:WIST). Astrocytes were separated from microglia and oligodendrocytes by mechanical shaking as described by McCarthy and de Vellis (1980). Astrocytes were subcultivated in Dulbecco's modified Eagle medium (DMEM) supplemented with 10% fetal calf serum, oligodendrocytes in DMEM with 2% serum. Microglia were collected as free floating cells in primary glial cultures (Gebicke-Haerter, 1989; Lucius et al., 1995) and cultivated in astrocyte-conditioned medium. Neurons were prepared from dissociated cortices of 17-day embryonic rats and cultivated in a sandwich system in serum-free medium (Lucius and Mentlein, 1991, 1995). Purity of cultures was verified by immunostaining with cell specific markers and was generally more than 90% (Mentlein et al., 1990; Lucius et al., 1995).

Degradation experiments. Cells ($0.05-4 \times 10^6$) were carefully washed (2×15 min) with 37°C-thermostated incubation medium that consisted of 145 mM NaCl, 5.4 mM KCl, 1.8 mM $CaCl_2$, 1.0 mM $MgCl_2$, 20 mM glucose, and 20 mM Hepes, pH 7.4, 330 mosmol/l (Horsthemke et al., 1984). After addition of fresh medium, peptides (and inhibitors) were added as 1 mM (or 10 mM) stock solutions in water, and cultures were incubated for up to 5 h. After 0.5, 1, 2, 4 and 5 h aliquots of 200 µl were withdrawn, acidified with 10 µl 10% trifluoroacetic acid, centrifuged and applied to a 250×4.6-mm reversed-phase HPLC column (Vydac C_{18}, 5 µm particles, 300 Å pores). The column was eluted with a linear gradient of 0-50% acetonitrile in 0.1% trifluoroacetic acid formed over 30 min at a flow rate of 1 ml/min, and peptides or their fragments in the eluate detected and quantified by measuring their absorbance at 220 (peptide bonds) or 280 (aromatic amino acids) nm.

Protein and peptide chemistry. Protein was measured by a micromodification of the Coomassie Brilliant Blue binding assay for membrane-bound proteins, and peptides were analyzed by amino acid analysis (Lucius and Mentlein, 1991). Endopeptidases-24.15 and -24.16 were purified from rat brain as described (Dahms and Mentlein, 1992).

Antibody and immunocytochemistry. Antibodies against purified endopeptidase-24.15 were raised in White New Zealand rabbits by subcutaneous injections of emulsions from 0.1 mg purified protein in 0.5 ml phosphate-buffered saline (PBS), pH 7.4, and 0.5 ml Freund's complete adjuvant (Difco Laboratories, Detroit, Mich.). Booster injections with enzyme emulsions in incomplete adjuvant followed at intervals of 4 weeks. Sera were collected 2 weeks after booster injections. The titers were controlled by dot-blot assay (Krisch and

Mentlein, 1989). Crude antisera were purified by immuno-affinity chromatography. 0.3 mg of purified protease was dialysed versus 0.1 M NaHCO$_3$ buffer, pH 8.0, and incubated for 16 h at 4°C with 6 g CNBr-activated Sepharose 4B (Pharmacia, Freiburg, Germany) previously swollen in ice-cold 1 mM HCl, washed and suspended in 15 ml conjugation buffer. Excess binding sites were blocked (2 h 20°C) with 0.1 M Tris-HCl buffer, pH 8.0, the gel washed extensively (repeated cycles of 0.5 M NaCl either in acetate buffer, pH 4.0, or Tris-HCl buffer, pH 8.0) and equilibrated with 0.14 M NaCl in 20 mM Hepes buffer, pH 7.4. Crude antiserum (1-3 ml) was passed slowly through the gel (20°C), the column washed with 30 ml buffer, and antibodies eluted with 0.2 M acetic acid. When the pH of the eluate dropped below pH 7, fractions of 0.5 ml were collected, immediately neutralized with 0.2 M potassium carbonate and assayed by dot-blot for immunoreactivity. Active fractions (usually 3-5 ml) were combined, supplemented with bovine serum albumin (0.01%) and concentrated by ultrafiltration (YM 10 membrane from Amicon, Witten, Germany).

For light-microscopic immunocytochemistry, cells were fixed for 30 min at 20°C with 4% paraformaldehyde 3% sucrose in PBS, etched with 0.5% Triton X-100 in PBS (2 x 15 min) and washed with PBS (10 min). After blocking with 1% normal goat serum in PBS (10 min), samples were incubated with the antibody (1:50 in PBS) for 24 h at 4°C in a humid chamber, washed, and bound antibody was visualized with biotinylated goat anti-rabbit IgG followed by avidin-conjugated peroxidase and diaminobenzidine/H$_2$O$_2$-reaction (universal anti-rabbit kit SIH 918 from Sigma, Munich, Germany). Nuclei were counterstained with Mayer's haematoxylin. Controls were done by omitting primary antibody and by preincubation of antibody with purified enzyme. For electron microscopy, cells were fixed with 2.5% glutaraldehyde 2% paraformaldehyde in 0.1 M phosphate buffer, pH 7.4 (1 h 20°C), etched and incubated with antibody as above. After washing bound IgG was reacted with gold-conjugated goat anti-rabbit IgG, samples osmicated, embedded in Araldite, cut and counterstained with lead citrate and uranyl acetate.

R. Mentlein et al.

Results and discussion

Among cultivated brain cells, astrocytes and microglial cells had the highest potential to catabolize neuropeptides (Fig. 1). Neuronal (N1E-115) and astroglial (C6) cell lines exhibited a comparatively low proteolytic capacity and could therefore not serve as models to replace primary or secondary cultures.

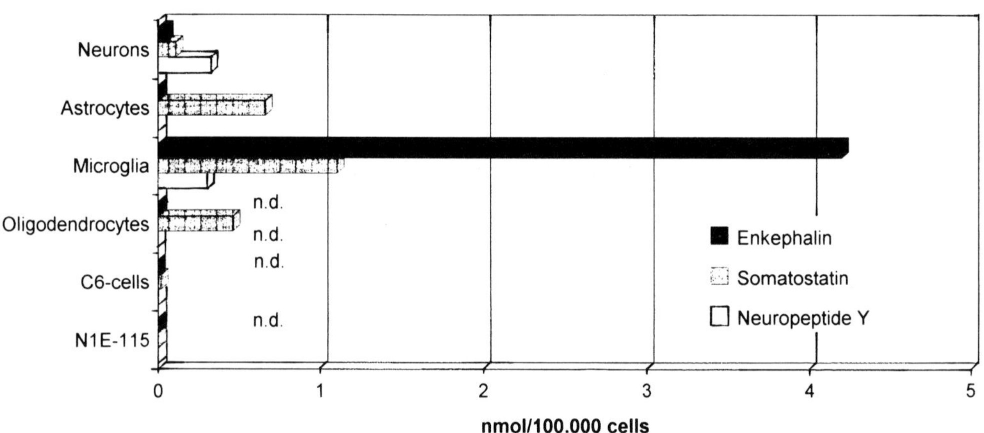

Fig. 1. Comparison of the degradation rates for 50 µM Leu-enkephalin, somatostatin and neuropeptide Y by cultivated brain cells. The peptides were selected to cover a range of different amino acid chain lengths (5, 14 and 36 residues). Neurons and microglia were primary, astrocytes and oligodendrocytes subcultures obtained from embryonic (E 17, neurons) or postnatal (P 2) rat brain. C6 is a rat tumourigenic astrocyte cell line, N1E-115 a mouse neuroblastoma cell line. Purity of all cultures was more than 95% as tested by immunocytochemistry with cell-specific markers. It should be noted that primary glia (astrocyte) cultures often contain significant impurities of microglial cells. n. d., not determined.

The significant high cleavage rate for enkephalin by microglial cells is produced by high levels of aminopeptidase N/M (EC 3.4.11.2, CD 13) at their surface (Lucius et al., 1995). Aminopeptidase N is an integral membrane glycoprotein anchored by a N-terminal 24 residue domain as an ectoenzyme in the plasma membranes. Microglia cells and neurons cleave somatostatin and neuropeptide Y by plasmin and plasminogen activators secreted by the cultured cells (Ludwig et al., 1996). Low levels of aminopeptidase N could also be immunostained and measured in neuronal cultures.

Apart from microglia, astrocytes exhibit a relatively high cleaving rate for somatostatin and other neuropeptides of chain lengths between 8 and 14 residues. Shorter peptides like enkephalin (5 residues) and longer peptides like neuropeptide Y, calcitonin gene-related peptide and others (36 and more residues) are virtually not attacked by cultivated astrocytes (Fig. 2). Moreover, the cyclic peptides oxytocin and vasopressin are not cleaved by the astrocytic proteases.

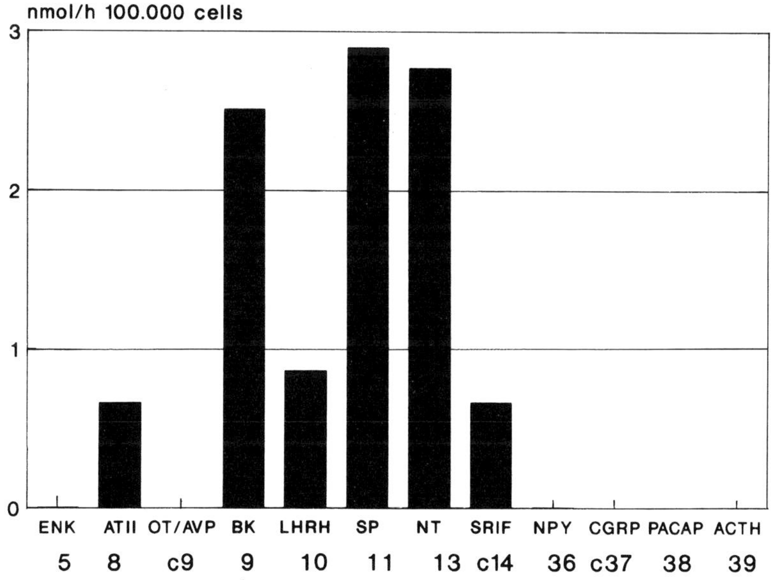

Fig. 2. Degradation rates of neuropeptides with varying chain lengths by astrocytes in culture. The numbers of amino acid residues are given below the peptides; c, cyclic peptide; ENK, Leu-enkephalin; AT II, angiotensin II; OT, oxytocin, AVP, Arg-vasopressin; BK, bradykinin; LHRH, gonadoliberin; SP, substance P; NT, neurotensin; SRIF, somatostatin; NPY, neuropeptide Y; CGRP, calcitonin gene-related peptide; PACAP, pituitary adenylate-cyclase activating peptide; ACTH, adrenocorticotropin.

 R. Mentlein et al.

Lucius, R., Sievers, J. and Mentlein, R. (1995) Enkephalin metabolism by microglia aminopeptidase N (CD13). *J. Neurochem.* 64: 1841-1847.

Ludwig, R., Feindt, J., Lucius, R., Petersen, A. and Mentlein, R. (1996) Metabolism of neuropeptide Y and calcitonin gene-related peptide by cultivated neurons and glial cells. *Mol. Brain Res.; in press.*

McCarthy, K. D. and de Vellis, J. (1980) Preparation of separate astroglial and oligodendroglial cultures from rat cerebral tissue. *J. Cell Biol.* 85: 890-902.

Mentlein, R., von Kolszynski, M., Sprang, R. and Lucius, R. (1990b) Proline-specific proteases in cultivated neuronal and glial cells. *Brain Res.* 527: 159-162.

III. Neuropeptide receptors

Molecular biology of peptide receptors

G. Liapakis and T. Reisine

Department of Pharmacology, University of Pennsylvania School of Medicine, Room 103 John Morgan Building, 36th Street and Hamilton Walk, Philadelphia Pennsylvania, 19104, USA

Summary. Peptides are a large class of endogenous molecules involved in neurotransmission. Most peptide receptors have a typical seven transmembrane spanning-like structure. Structure-function analysis of cloned peptide receptors has revealed important information on the ligand binding domains and regions of the receptors involved in coupling to G proteins and cellular effector systems. Somatostatin receptors consist of a family of five receptor subtypes which have approximately 50% amino acid sequence identity. Selective ligands have been identified at 3 of the 5 receptors and have been useful in revealing distinct functions of those subtypes. The subtype SSTR2 mediates important physiological actions of somatostatin including regulation of growth hormone release and is a target of anticancer agents. Ligand binding domains of this receptor have been identified using site-directed mutagenesis approaches. A region of four amino acids at the juncture of the third extracellular loop and transmembrane seven is involved in binding of synthetic hexa- and octapeptide analogs of somatostatin. A phenylalanine within this region is especially critical for binding octapeptide analogs such as Sandostatin. The third intracellular loop of this receptor may be particularly important in coupling the receptor to G proteins and appears to contain sites that may be involved in the desensitization of the receptor. In fact, this receptor becomes phosphorylated during desensitization. Phosphorylation of sites within the third intracellular loop of the receptor may be responsible for uncoupling the receptor from G proteins and cellular effector systems. Biochemical studies have revealed which G proteins associate with this receptor and have indicated that different G proteins link the receptor to distinct cellular effector systems. Because of the important physiological roles of this receptor and its involvement in mediating therapeutic actions of somatostatin analogs, non-peptide SSTR2 drugs may have a number of clinical uses. Structure-function analysis of this receptor may facilitate the development of these drugs.

Introduction

Somatostatin (SRIF) is a neuropeptide that is an important regulator of hormone secretion from endocrine organs and is a neurotransmitter in brain (Brazeau et al., 1972; Reisine and Bell, 1995a,b; Reichlin, 1983). It induces its biological actions by interacting with a family of receptors, of which five have recently been cloned (Bell and Reisine, 1993; Bruno et al., 1992; Hoyer et al., 1994; Kluxen et al., 1992; O'Carroll et al., 1992, 1994; Reisine and

Bell, 1995a,b; Rohrer et al., 1993; Xu et al., 1993 Yamada et al., 1992, 1993; Yasuda et al., 1992). The five receptors have the typical structure of a seven transmembrane spanning receptor (Reisine and Bell, 1995a,b). They share approximately 50% amino acid sequence identity. Regions of greatest diversity are the N- and C-termini. Regions of highest similarity are the transmembrane spanning regions.

The somatostatin receptors have no similarity to any other receptors, except the opiate receptors, for which they share approximately 40% amino acid identity (Reisine and Bell, 1993; Yasuda et al., 1993). The opiate and somatostatin receptors appear to also share some similarities in ligand binding since somatostatin analogs such as SMS 201-995 (Sandostatin) bind potently to μ opiate receptors (Maurer et al., 1982). This somatostatin analog served as the basis for the development of CTOP (DPhe-Cys-Tyr-DTrp-Orn-Thr-Pen-Thr-NH$_2$), which is a highly selective μ receptor antagonist (Pelton et al., 1985).

In an effort to localize the expression of the different somatostatin receptors (SSTR), mRNA encoding the receptors has been identified in various regions of the body by either RNA blotting, polymerase chain reaction analysis or *in situ* hybridization (Breder et al., 1992; Bruno et al., 1993; Kong et al., 1994a; Perez et al., 1994; O'Carroll et al., 1992, 1994; Wulfsen et al., 1993). SSTR1-4 mRNA are detected in brain using all three methodologies. SSTR5 mRNA is very low in brain and is only detected with the most sensitive approaches. SSTR5 mRNA is found predominantly in the anterior pituitary, although in human it is also found in the heart and gut (O'Carroll et al., 1994). SSTR3 mRNA is highly expressed in the pancreatic islets, although a role for this receptor in control of insulin and glucagon secretion is not established (Yamada et al., 1993). SSTR2 mRNA is highly expressed in pituitary and this receptor has been suggested to have a primary role in regulating growth hormone secretion (Kong et al., 1994a). SSTR2 mRNA is also found in low to moderate levels in pancreas and this subtype has been suggested to be involved in mediating the inhibition of glucagon secretion by somatostatin.

SSTR1 and SSTR2 mRNA are highly expressed in a variety of tumours (Greenman and Melmed, 1994; Kubota et al., 1994; Reubi et al., 1994; Taylor et al., 1994). The somatostatin analog SMS-201-995, which interacts potently with SSTR2, is used to treat pituitary tumours and VIPomas (Lamberts et al., 1991; Kubota et al., 1994). Both the cloned SSTR1 and SSTR2 mediate antiproliferative effects of somatostatin (Buscail et al., 1994, 1995). Agents that continuously activate these receptors may therefore be particularly useful in suppressing tumour growth.

While studies examining the distribution of somatostatin receptor mRNAs have been useful in identify potential functional roles of the receptor subtypes, these studies do not actually identify where the receptor is expressed. It is not clearly established that there is a direct link between somatostatin receptor mRNA expression and somatostatin receptor expression. For example, the rat cerebellum highly expresses SSTR3 mRNA (Kong et al., 1994a). However, this brain region has very little if any somatostatin receptor binding sites. Thus, to gain a better insight into the localization of the receptor subtypes, subtype selective ligands are needed that could be used to detect the receptor expression.

By screening a number of somatostatin analogs for binding to the cloned somatostatin receptors, it was possible to identify selective agonists at some of the receptors (Raynor et al., 1993a,b). The analogs BIM 23027, NC8-12 and NC4-28B are selective SSTR2 agonists. Their potencies at binding to SSTR2 are at least 100-fold greater than binding to the other cloned somatostatin receptors. BIM 23052 is a selective agonist at the rodent SSTR5 and L362,855 has partial agonist/antagonistic properties at SSTR5. BIM 23052 is not selective at binding to human SSTR5 and therefore is only useful in distinguishing the functional role of this receptor in rodents (O'Carroll et al., 1994).

Using subtype selective ligands, specific functions of the receptor subtypes have been identified (Raynor et al., 1993a,b; Rossowski and Coy, 1993, 1994). SSTR2 appears to be primarily involved in the control of growth hormone and glucagon secretion whereas SSTR5 is involved in the regulation of insulin secretion. SSTR2 may be a primary mediator of the anti-cancer actions of SMS-201-995 in humans. Stimulation of SSTR2 has been reported to inhibit tumour cell growth (Buscail et al., 1994, 1995). This inhibition has been related to the stimulation of a tyrosine phosphatase. Activation of SSTR1 also mediates inhibitory effects of somatostatin on cell growth and like SSTR2 is coupled to a tyrosine phosphatase (Buscail et al., 1994). Activation of SSTR5 also inhibits tumour cell growth but this receptor is not coupled to the tyrosine phosphatase and may inhibit cell growth through modulation of calcium mobilization (Buscail et al., 1995). Agents that selectively activate SSTR1, SSTR2 and SSTR5 may have clinical uses in treating cancer.

Structure-function analysis

Availability of the cloned receptors has allowed studies to identify ligand binding domains, G protein coupling domains, the G proteins associated with the receptors and the investigation of the mechanisms involved in the regulation or desensitization of the receptors. The first mutagenesis studies done on the somatostatin receptors were to test the role of a conserved aspartate in the second transmembrane spanning region of SSTR2 in mediating Na^+ regulation of agonist binding (Kong et al., 1993a). Sodium ions have been shown to regulate agonist binding to a number of receptors including adrenergic and opiate receptors by directly interacting with the receptors (Horstman et al., 1990; Kong et al., 1993b). The interaction of Na^+ with the receptor results in an uncoupling of the receptor from G proteins, thereby converting the receptor into a low affinity state for agonists. Studies by Horstman et al. (1990) showed that the conserved aspartate in the second transmembrane region of the α_2-adrenergic receptor was necessary for Na^+ to regulate agonist binding. Similarly, a conserved aspartate in the second transmembrane spanning region of the δ opiate receptor is also involved in mediating Na^+ regulation of agonist binding (Kong et al., 1993b). Mutation of aspartate 89 in SSTR2 to an asparagine abolished Na^+ regulation of agonist binding (Kong et al., 1993a). The mutant receptor remained associated with G proteins since GTP analogs were still able to reduce agonist binding to the mutant receptor. However the ability of Na^+ to uncouple this mutant receptor from G proteins was lost.

Several recent studies have employed site-directed mutagenesis to investigate regions in somatostatin receptors involved in ligand binding. Fitzpatrick and Vandlen (1994) reported that the second and third extracellular loops of SSTR2 were critical for the binding of the hexapeptide analog MK 678. For their studies, they generated chimeric receptors between SSTR1 and SSTR2. Both of these receptors bind somatostatin28 with high affinity.

However, only SSTR2 binds MK 678 with high affinity. By shifting different parts of SSTR1 and SSTR2 amongst each receptor, these investigators were able to come to the conclusion that these extracellular loops were particularly important for the binding of the small synthetic peptide.

Kaupmann et al. (1994) reported that a phenylalanine at the border of the third extracellular loop and transmembrane seven was critical for the binding of the octapeptide SMS 201-995 to the receptor. SMS 201-995 has very low affinity for SSTR1. These authors showed that conversion of a serine residue in transmembrane seven of SSTR1 to the corresponding phenylalanine of SSTR2 created an SSTR1 mutant with higher affinity for SMS 201-995. These findings suggest that critical recognition sites for binding of constrained analogs of somatostatin to SSTR2 are within rather hydrophillic, extracellular loops of the receptors.

To further investigate regions involved in ligand binding to SSTR2, Liapakis et al. (1995) have found that a small region of SSTR2, consisting of four amino acids (Phe-Asp-Phe-Val), at the junction the third extracellular loop and transmembrane seven is essential for the high affinity binding of hexa- and octapeptide analogs. Mutation of these amino acids to the corresponding residues of SSTR1 generated a mutant SSTR2 with high affinity for native somatostatins but very low affinity for SMS 201-995 and MK 678. A phenylalanine within this region is essential for octapeptide binding since mutation of a corresponding serine in SSTR1 to a phenylalanine conferred onto SSTR1 high affinity binding for octapeptide analogs of somatostatin. These findings are similar to those reported by Kaupmann et al. (1995). However, this amino acid does not appear to be critical for the binding of hexapeptides such as MK 678. This single amino acid can distinguish between hexa- and octapeptide binding to SSTR2 and hexa- and octapeptides appear to have different determinants for binding to this receptor subtype.

These mutagenesis studies revealed a restricted region of SSTR2 involved in the binding of synthetic peptides. These sites are at hydrophil, extracellular domains of the receptor. Similar results have recently been reported for other peptide receptors including opiate receptors (Kong et al., 1994b). In contrast, adrenergic receptors appear to have ligand binding domains more closely associated with hydrophobic, transmembrane spanning regions of the receptors (Strader et al., 1987). Because peptides are so much larger and contain considerably more charged regions than classical transmitters such as noradrenaline, it is understandable that the ligand binding domains of the peptide receptors may differ considerably from the non-peptide, G protein linked receptors.

Identification of the selective ligand binding domains of SSTR2 may facilitate development of agonists that are better drugs. In particular, such information, in conjunction with structural analysis of the somatostatin peptides (Huany et al., 1992) may facilitate rational design of non-peptide SSTR2 agonists as well as antagonists.

Coupling of somatostatin receptors to G proteins and effector systems

Two forms of SSTR2 are generated by differential splicing (Vanetti et al., 1992). The variations in amino acid sequences of the splice variants occur in the C-terminal region of the receptor. SSTR2B has a shorter C-terminus than SSTR2A and some of the C-terminal amino acids differ between the receptors. SSTR2B is much more effective in coupling to adenylyl cyclase than SSTR2A, suggesting that differences in the C-terminal sequence may hinder coupling (Vanetti et al., 1993; Reisine et al., 1993). In fact, truncation of the C-terminus of SSTR2 to remove most of the C-terminal tail results in a receptor that effectively mediates the inhibition of cAMP accumulation by somatostatin (Woulfe et al., 1994; Law et al., 1995). These findings indicate that the C-terminus of SSTR2 is not required for coupling to adenylyl cyclase. It is more likely that intracellular loops, such as the third intracellular loop, may be more essential in coupling the receptor to adenylyl cyclase, as proposed for other G protein linked receptors.

G proteins link somatostatin receptors to its various cellular effector systems. In addition to adenylyl cyclase, somatostatin receptors are coupled to Ca^{++} channels (Wang et al., 1990a; Raynor et al., 1991) and K^+ channels (Wang et al., 1989, 1990b; Raynor et al., 1991). A subfamily of G proteins sensitive to pertussis toxin appear to be primarily involved in coupling somatostatin receptors to adenylyl cyclase and ionic conductance channels since most if not all of the cellular effects mediated by SSTR2 are blocked by pertussis toxin treatment.

To investigate which G proteins associate with somatostatin receptors, Law et al. (1991, 1992) developed a technique to solubilize somatostatin receptors with a mild detergent, and to immunoprecipitate somatostatin receptor/G protein complexes with antibodies against different alpha or beta subunits of G proteins. Using this approach, it was reported that rat brain and AtT-20 cells contain somatostatin receptors with high affinity for SSTR2 selective radioligands associated with $G_{i\alpha1}$, $G_{i\alpha3}$ and $G_{o\alpha}$.

These different G proteins have been proposed to link somatostatin receptors to distinct cellular effector systems. Tallent and Reisine (1992) reported that $G_{i\alpha1}$ couples AtT-20 cell somatostatin receptors to adenylyl cyclase. Kleuss et al. (1993) showed that $G_{o\alpha}$ couples GH_3 cell somatostatin receptors to voltage sensitive Ca^{++} channels and Yatani et al. (1987) suggested that $G_{i\alpha3}$ couples somatostatin receptors to K^+ channels. Therefore, G proteins contribute to the functional diversity of somatostatin receptors.

SSTR2A physically associates with $G_{i\alpha3}$ and $G_{o\alpha}$ (Law et al., 1993). The lack of effective association of SSTR2A with $G_{i\alpha1}$ may explain its inefficient coupling to adenylyl cyclase. SSTR2B may associate with these G proteins as well as $G_{i\alpha1}$ which may link the receptor to adenylyl cyclase.

SSTR3, like SSTR2 couples to adenylyl cyclase and mediates agonist inhibition of cAMP accumulation (Yasuda et al., 1992). The coupling to adenylyl cyclase is mediated by $G_{i\alpha1}$ (Law et al., 1994). SSTR3 expressed in a CHO cell line lacking detectable levels of $G_{i\alpha1}$ and $G_{i\alpha2}$ did not couple to adenylyl cyclase. When $G_{i\alpha2}$ was coexpressed in the cells with SSTR3 no coupling was observed. However, when $G_{i\alpha1}$ was coexpressed, SSTR3 mediated inhibition of cAMP accumulation by somatostatin. Through the generation of chimeric α subunits, it was shown that the C-terminal region of $G_{i\alpha1}$ is essential for coupling SSTR3 to adenylyl cyclase.

In addition to adenylyl cyclase, SSTR2 couples to voltage sensitive Ca^{++} channels. Inhibition of Ca^{++} conductance and influx into secretory cells may be a major mechanism by which somatostatin inhibits hormone and neurotransmitter release (Reisine et al., 1995; Raynor et al., 1991). The cloned SSTR2 expressed in RIN cells was reported to couple to a voltage sensitive Ca^{++} channel (Fujii et al., 1994) and SSTR2 endogenously expressed in the cell line AtT-20 couples to an L-type Ca^{++} channel (Reisine et al., 1995; Tallent et al., submitted). The coupling to the L-type Ca^{++} channel is blocked by pertussis toxin treatment but is resistant to agonist induced desensitization. This finding is of interest since the inhibitory effects of SSTR2-agonists on adenylyl cyclase desensitizes (Reisine and Axelrod, 1983) and the cloned SSTR2B coupling to adenylyl cyclase desensitizes suggesting that SSTR2 coupling to various cellular effector systems is differentially sensitive to prolonged agonist treatments (Reisine et al., 1994).

Desensitization of SSTR2 may involve the phosphorylation and uncoupling of the receptor from G proteins linking it to adenylyl cyclase and other cellular effector systems. Using a peptide directed antisera against SSTR2 to immunoprecipitate and identify the receptor, it has been reported that prolonged stimulation of SSTR2 with somatostatin analogues induces the phosphorylation of the receptor (Hines et al., 1993; Reisine, 1995). An enzyme that may phosphorylate the receptor is beta-adrenergic receptor kinase (BARK). Prolonged stimulation of S49 lymphoma cells with somatostatin has been reported to desensitize somatostatin receptors and activate BARK (Mayor et al., 1987). In preliminary studies a BARK dominant negative mutant blocked the desensitization of SSTR2 consistant with a role of BARK in the rapid inactivation of SSTR2. Since the third intracellular loop of SSTR2 may have contact sites with G proteins linking the receptor to diverse cellular effector systems, it is conceivable that this region may also be phosphorylated during desensitization to cause receptor uncoupling from G proteins. The third intracellular loop of SSTR2 has multiple phosphorylation acceptor sites for BARK and other protein kinases which may be loci for the molecular events involved in somatostatin desensitization.

In addition to SSTR2 and SSTR3, SSTR5 effectively couples to adenylyl cyclase (O'Carroll et al., 1992, 1994). SSTR5 also mediates antiproliferative effects of somatostatin possibly via an inhibition of Ca^{++} mobilization (Buscail et al., 1995). Furthermore, electrophysiological studies have shown that SSTR5, like SSTR2, couples to an L-type Ca^{++} channel and mediates agonist inhibition of Ca^{++} mobilization (Reisine et al., 1995; Tallent et al., submitted). Unlike SSTR2, SSTR5 coupling to an L-type Ca^{++} channel desensitizes, suggesting that its coupling mechanisms to this effector system may differ from that of SSTR2.

SSTR5 has the unique characteristic of having higher affinity for somatostatin-28 than somatostatin-14 (O'Carroll et al., 1992). Structure-function analysis has suggested that a simple hydroxyl group in SSTR5 may be responsible for the ability of this receptor to distinguish these two peptides (Ozenberger and Hadcock, 1994). SSTR1, SSTR2, SSTR3 and SSTR4 have a tyrosine in transmembrane six that is a phenylalanine in SSTR5. Mutation of the phenylalanine to a tyrosine by Ozenberger and Hadcock (1994) created a mutant SSTR5 with similar affinities for somatostatin-28 and somatostatin-14. The mutation in effect increased the affinity of the receptor for somatostatin by 50-fold. The only difference between phenylalanine and tyrosine is a hydroxyl group, which appears to be critical for high affinity binding of somatostatin.

A major limitation to identifying the functional roles of somatostatin receptors has been the lack of antagonists. Recent studies have identified an antagonist at SSTR5 (Reisine et al., 1995; Tallent et al., submitted). The peptide L-362,855 binds with high affinity to SSTR5. It had minimal effects in inhibiting cAMP formation in cells expressing rat or human SSTR5 and at concentrations below 100 nM did not reduce Ca^{++} conductance in AtT-20 cells. In contrast, the SSTR5 selective agonist BIM23052 reduced Ca^{++} conductance in AtT-20 cells and inhibited cAMP accumulation in cells expressing the cloned SSTR5. L362,855 blocked the inhibitory effects of somatostatin agonists on Ca^{++} conductance and cAMP accumulation. L362,855 is a partial agonist/antagonist because at concentrations above 100 nM it is able to reduce Ca^{++} conductance and cAMP accumulation.

L362,855 has two phenylalanine groups in its structure. Conversion of one of these residues to a tyrosine generated a pure agonist at SSTR5. This result indicates that a simple hydroxyl group is critical for the intrinsic activity of L-362,855. Such information may be useful to molecular modelers to design non-peptide agonists and antagonists at the somatostatin receptors.

A unique somatostatin receptor

A major physiological response of somatostatin is the potentiation of K$^+$ conductance. The increase in K$^+$ currents may be critical in somatostatin's ability to hyperpolarize neurons and secretory cells to inhibit firing activity and to reduce transmitter and hormone release. The receptor mediating somatostatin's potentiation of K$^+$ currents has not been identified. Studies on the cell line AtT-20 suggest that a unique somatostatin receptor may be coupled to K$^+$ channels (Tallent et al., 1995; Reisine et al., 1994). In addition to somatostatin, hexapeptides such as MK 678 and BIM 23027 stimulated the K$^+$ current. Both agonists potently interact with SSTR2. However, octapeptides such as SMS-201-995 and NC8-12, which bind potently to SSTR2 did not increase K$^+$ currents in AtT-20 cells, indicating that the ligand selectivities of this receptor are unique.

This was further established by the ability of the peptide c[Aha-Phe-DTrp-Lys-Thr(Bzl)] (SA) to block the ability of MK 678 to potentiate this K$^+$ current. SA had little effect of its own. This peptide does not bind to SSTR2 and has very low affinity for the other cloned somatostatin receptors. Its ability to block MK 678 effects further suggests that the receptor coupled to the K$^+$ channel is a unique somatostatin receptor subtype.

The receptor mediating the effects of somatostatin on the K$^+$ current is associated with G proteins, since pertussis toxin blocked somatostatin evoked K$^+$ currents. Furthermore, the potentiation of the K$^+$ currents desensitized following continuous agonist stimulation. The desensitization distinguishes this response from somatostatin's inhibition of Ca^{++} conductance in AtT-20 cells, which is resistant to desensitization.

Previous biochemical studies have suggested that AtT-20 cells may express a unique somatostatin receptor subtype. Theveniau et al. (1992) generated antisera against a rat brain somatostatin receptor that reacted with a 60 kDa somatostatin from AtT-20 cells. The size of the somatostatin receptor detected by this antisera is different than SSTR2 (Theveniau et al., 1994) and this antisera did not cross-react with SSTR2 or SSTR1 and SSTR3. The receptor detected by this antisera has high affinity for MK 678 suggesting that this receptor may be similar to the unique receptor coupled to K$^+$ channels. Isolation and cloning of this unique receptor may reveal the identity of this potentially important somatostatin receptor.

Future directions

Somatostatin is known to have a number of important endocrine functions including inhibition of growth hormone, insulin and glucagon secretion (Hellman and Lernmark, 1969; Brazeau et al., 1972; Brown et al., 1977; Mandarino et al., 1981; Reichlin, 1983). Furthermore, it is a neurotransmitter involved in cognitive functions and locomotor activity (Haroutunian et al., 1987; DeNoble et al., 1989; Raynor and Reisine, 1992; Raynor et al., 1993c). Somatostatin analogs are presently used to treat pituitary tumours and VIPomas (Lamberts et al., 1991). The potential therapeutic uses of somatostatin analogs are widespread both in the control of hyperinsulin secretion and as anticancer agents. They may also be useful in treating central nervous system disorders involving imbalances in somatostatin transmission such as epilepsy and Alzheimer's disease. The cloning of the somatostatin receptors has now provided the means to develop new somatostatin analogs, in particular non-peptide analogs, that could be useful in treating these diseases of the nervous systems and endocrine systems.

Acknowledgments
This work was supported by NIH grants MH45533 and MH48518.

References

Bell, G.I. and Reisine, T. (1993) Molecular biology of SRIF receptors. *Trends Neurosci.* 16:34-38.
Brazeau, P., Vale, W., Burgus, R., Ling, N., Rivier, J. and Guillemin, R (1972) Hypothalamic polypeptide that inhibits the secretion of immunoreactive pituitary growth hormone. *Science* 129:77-79.
Breder, C.D., Yamada, Y., Yasuda, K., Seino, S., Saper, C.B. and Bell, G.I. (1992) Differential expression of SRIF receptor subtypes in brain. *J. Neurosci.* 12:3920-3934.
Brown, M., Rivier, J. and Vale, W. (1977) SRIF analogs with selected biological activities. *Science* 196:1467-1468.
Bruno, J.-F., Xu, Y., Song, J. and Berelowitz, M. (1993) Tissue distribution of SRIF receptor subtype messenger ribonucleic acid in the rat. *Endocrinology.* 133:2561-2567.
Bruno, J.F., Xu, Y., Song, J. and Berelowitz, M. (1992) Molecular cloning and functional expression of a novel brain specific SRIF receptor. *Proc. Natl. Acad. Sci. USA* 89:11151-11155.
Buscail, L., Delesque, N., Esteve, J.-P., Saint-Laurent, N., Prats, H., Clerc, P., Robberecht, D., Bell, G.I., Liebow, C., Schally, A.V., Vaysse, N. and Susini, C. (1994) Stimulation of tyrosine phosphatase and inhibition of cell proliferation by SRIF analogues: mediation by human SRIF receptor subtypes SSTR1 and SSTR2. *Proc. Natl. Acad. Sci. USA* 91:2315-2319.

Buscail, L., Esteve, J.P., Saint-Laurent, N., Bertrand, V., Reisine, T., O'Carroll, A.M., Bell, G.I., Schally A., Vaysse, N. and Susini, C. (1995) Inhibition of cell proliferation by the somatostatin analogue RC-160 is mediated by SSTR2 and SSTR5 somatostatin receptor subtypes through different mechanisms. *Proc. Natl. Acad. Sci.* 92:1580-1584.

DeNoble, V., Hepler, D. and Barto, R. (1989) Cysteamine-induced depletion of SRIF produces differential cognitive deficits in rats. *Brain Res.* 482:42-48.

Fitzpatrick, V. and Vandlen, R. (1994) Agonist selectivity determinants in SRIF receptor subtypes I and II. *J. Biol. Chem.* 269:24621-24626.

Fujii, Y., Gonoi, T., Yamada, Y., Chihara, K., Inagaki, N. and Seino S. (1994) Somatostatin receptor subtype SSTR2 mediates inhibition of high-voltage activated calcium channels by somatostatin and its analogue SMS 201-995. *FEBS Lett.* 355:117-120.

Greenman, Y. and Melmed, S. (1994) Heterogeneous expression of two SRIF receptor subtypes in pituitary tumours. *J. Clin. Endocrinol. Metab.* 78:398-403.

Haroutunian, V., Mantin, G., Campell, G., Tsuboyama, G. and Davis, K. (1987) Cysteamine-induced depletion of central SRIF-like immunoreactivity: effects on behaviour, learning, memory and brain neurochemistry. *Brain Res.* 403:234-242.

Hellman, B. and Lernmark, A. (1969) Inhibition of the in vitro secretion of insulin by an extract of pancreatic alpha$_1$ cells. *Endocrinology.* 84:1484-1488.

Hines, J., Theveniau, M., Benovic, J. and Reisine, T. (1993) Desensitization of the SRIF receptor SSTR2 involves beta-adrenergic receptor kinase. *Soc. Neurosci. Abstr.* 19:1541.

Horstman, D., Brandon, S., Wilson, A., Guyer, C., Cragoe, E. and Limbird, L. (1990) An aspartate conserved among G-protein linked receptors confers allosteric regulation of alpha$_2$-adrenergic receptors by sodium. *J. Biol. Chem.* 265:21590-21595.

Hoyer, D., Lübbert, H. and Bruns, C. (1994) Molecular Pharmacology of somatostatin receptors. *Naunyn-Schmiedeberg's Arch. Pharmacol.* 350:441-453.

Huany, Z., He, Y., Raynor, K., Tallent, M., Reisine, T. and Goodman, M. (1992) Side chain chiral methylated SRIF analog synthesis and conformational analysis. *J. Am. Chem. Soc.* 114:9390-9401.

Kaupmann, K., Bruns, C., Raulf, R., Weber, H., Mattes, H. and Lübbert, H. (1995) Two amino acids located in transmembrane domains VI and VII, determine the selectivity of the peptide agonist SMS 201-995 for the SSTR2 somatostatin receptor. *EMBO J.* 14:727-735.

Kleuss, C., Scherubl, H., Hescheler, J., Schultz, G. and Wittig, B. (1993) Selectivity in signal transduction determined by gamma subunits of heterotrimeric G proteins. *Science* 259:832-834

Kluxen, F.-W., Bruns, C. and Lübbert, H. (1992) Expression cloning of a rat brain SRIF receptor cDNA. *Proc. Natl. Acad. Sci. USA* 89:4618-4622.

Kong, H., Raynor, K., Yasuda, K., Bell, G.I. and Reisine, T. (1993a) Mutation of an aspartate at residue 79 in the SRIF receptor subtype SSTR2 prevents Na$^+$ regulation of agonist binding but does not affect apparent receptor/G protein association. *Mol. Pharmacol.* 44:380-384.

Kong, H., Raynor, K., Yasuda, K., Moe, S., Portoghese, P., Bell, G.I. and Reisine, T. (1993b) A single residue, aspartic acid 95, in the delta opioid receptor specifies selective high affinity agonist binding. *J. Biol. Chem.* 268:23055-23058.

Kong, H., DePaoli, A.M., Breder, C.D., Yasuda, K., Bell, G.I. and Reisine, T. (1994a) Differential expression of SRIF receptor subtypes SSTR1, SSTR2 and SSTR3 in adult rat brain, pituitary and adrenal gland. Analysis by RNA blotting and in situ hybridization. *Neuroscience.* 59:175-184.

Kong, H., Raynor, K., Yano, H., Takeda, J., Bell, G.I. and Reisine, T. (1994b) Agonists and antagonists bind to different domains of the cloned kappa opioid receptor. *Proc. Natl. Acad. Sci.* 91:8042-8046.

Kubota, A., Yamada, Y., Kagimoto, S., Shimatsu, A., Imamura, M., Tsuda, K., Imura, H., Seino, S. and Seino, Y. (1994) Identification of SRIF receptor subtypes and an implication for the efficacy of SRIF analog SMS 201-955 in treatment of human endocrine tumours. *J. Clin. Invest.* 93:1321-1325.

Lamberts, S.W., Krenning, E. and Reubi, J.-C. (1991) The role of SRIF and its analogs in the diagnosis and treatment of tumours. *Endocrine Rev.* 12:450-482.

Law, S. and Reisine, T. (1992) Agonist binding to rat brain SRIF receptors alters the interaction of the receptor with guanine nucleotide binding regulatory proteins. *Mol. Pharmacol.* 42:398-402.

Law, S., Woulfe, D. and Reisine, T. (1995) SRIF receptor activation of cellular effector systems. Minireview. *Cellular Signalling* 7:1-8.

Law, S., Manning, D. and Reisine, T. (1991) Identification of the subunits of GTP binding proteins coupled to SRIF receptors. *J. Biol. Chem.* 266:17885-17897.

Law, S., Yasuda, K., Bell, G.I. and Reisine T (1993) $G_{i\alpha3}$ and $G_{o\alpha}$ selectively associate with the cloned SRIF receptor subtype SSTR2. *J. Biol. Chem.* 268:10721-10727.

Law, S., Zaina, S., Sweet, R., Yasuda, K., Bell, G.I., Stadel, J. and Reisine, T. (1994) $G_{i\alpha1}$ selectively couples the SRIF receptor subtype SSTR3 to adenylyl cyclase: Identification of the functional domains of this α subunit necessary for mediating SRIF's inhibition of cAMP formation. *Mol. Pharmacol.* 45:587-590.

Liapakis, G., Fitzpatrick, D., Codispoti, C., Vandlen, R. and Reisine, T. (1995) Ligand binding domains of the somatostatin receptor SSTR2. *Soc. Neurosci. Abstr.* 21 (in press).

Mandarino, L., Stenner, D., Blanchard, W., Nissen, S., Gerich, J., Ling, N., Brazeau, P., Bohlen, P., Esch, F. and Guillemin, R. (1981) Selective effects of SRIF-14, -25, and -28 on in vitro insulin and glucagon secretion. *Nature* 291:76-77.

Maurer, R., Gähwiler, B., Buescher, H., Hill, R. and Roemer, D. (1982) Opiate antagonistic properties of an octapeptide somatostatin analog. *Proc. Natl. Acad. Sci.* 79:4815-4817.

Mayor, F., Benovic, J., Caron, M.G. and Lefkowitz, R.J. (1987) SRIF induces translocation of the beta-adrenergic receptor kinase and desensitizes SRIF receptors in S49 lymphoma cells. *J. Biol. Chem.* 262:6468-6471.

O'Carroll, A.-M., Lolait, S.J., Konig, M. and Mahan, L.C. (1992) Molecular cloning and expression of a pituitary SRIF receptor with preferential affinity for SRIF-28. *Mol. Pharmacol.* 42:939-946.

O'Carroll, A.-M., Raynor, K., Lolait, S.J. and Reisine, T. (1994) Characterization of cloned human SRIF receptor SSTR5. *Mol. Pharmacol.* 48:291-298.

Ozenberger, B. and Hadcock, J. (1995) A single amino acid substitution in the somatostatin receptor subtype 5 increases affinity for SRIF. *Mol. Pharmacol.* 47:82-87.

Pelton, J., Gulya, K., Hruby, V., Duckles, S. and Yamamura, H.I. (1985) Conformationally restricted analogs of somatostatin with high mu-opiate receptor specificity. *Proc. Natl. Acad. Sci.* 82:236-239.

Perez, J., Rigo, M., Kaupmann, C., Bruns, C., Yasuda, K., Bell, G.I., Lübbert, H. and Hoyer, D. (1994) Localization of SRIF (SRIF) SSTR-1, SSTR-2 and SSTR-3 receptor mRNA in rat brain by in situ hybridization. *Naunyn-Schmiedeberg's Arch. Pharmacol.* 349:145-160.

Raynor, K. and Reisine, T. (1992) SRIF receptors. *Crit. Rev. Neurobiol.* 16:273-289.

Raynor, K., Wang, H., Dichter, M. and Reisine, T. (1991) Subtypes of brain SRIF receptors couple to multiple cellular effector systems. *Mol. Pharmacol.* 40:248-253.

Raynor, K., Murphy, W., Coy, D., Taylor, J., Moreau, J.-P., Yasuda, K., Bell, G.I. and Reisine, T. (1993a) Cloned SRIF receptors: Identification of subtype selective peptides and demonstration of high affinity binding of linear peptides. *Mol. Pharmacol.* 43:838-844.

Raynor, K., O'Carroll, A.-M., Kong, H., Yasuda, K., Mahan, L., Bell, G.I. and Reisine, T. (1993b) Characterization of cloned SRIF receptors SSTR4 and SSTR5. *Mol. Pharmacol.* 44:385-392.

Raynor, K., Lucki, I. and Reisine, T. (1993c) $SRIF_1$ receptors in nucleus accumbens selectively mediate the stimulatory effect of SRIF on locomotor activity in rats. *J. Pharmacol. Expt. Therap.* 265:67-73.

Reichlin, S. (1983) Somatostatin *New Engl. J. Med.* 309:1495-1563.

Reisine, T. (1995) Somatostatin receptors. *Am. J. Physiol.* 32: G813 - G820.

Reisine, T. and Axelrod, J. (1983) Prolonged SRIF pretreatment desensitizes SRIF inhibition of receptor-mediated release of adrenocorticotropin and sensitizes adenylyl cyclase. *Endocrinology.* 113:811-813.

Reisine, T. and Bell, G.I. (1993) Molecular biology of opioid receptors. *Trends Neurosci.* 16:506-510.

Reisine, T. and Bell, G.I. (1995a) Molecular properties of somatostatin receptors. *Neuroscience* 67:777-790.

Reisine, T. and Bell, G.I. (1995b) Molecular biology of somatostatin receptors. *Endocrinology* 16:427-442.

Reisine, T., Kong, H., Raynor, K., Yano, H., Takeda, J., Yasuda, K. and Bell, G.I. (1993) Splice variant of the SRIF receptor 2 subtype, SSTR2B, couples to adenylyl cyclase. *Mol. Pharmacol.* 44:1008-1015.

Reisine, T., Tallent, M. and Dichter, M. (1994) SRIF receptor subtypes endogenously expressed in AtT-20 cells couple to three different ionic currents. *Soc. Neurosci. Abstr.* 20:376.19135.

Reisine, T., Tallent, M., Liapakis, G., O'Carroll, A.-M. and Dichter, M. (1995) An antagonist at the somatostatin receptor SSTR5. *Soc. Neurosci. Abstr.* 21; *in press.*

Reubi, J.C., Schaer, J., Wagner, D. and Mengod, G. (1994) Expression and localization of SRIF receptor SSTR1, SSTR2 and SSTR3 mRNA in primary human tumours using in situ hybridization. *Cancer Res.* 54:3455-3459.

Rohrer, L., Raulf, F., Bruns, C., Buettner, R., Hofstaedter, F. and Schule, R. (1993) Cloning and characterization of a fourth human SRIF receptor. *Proc. Natl. Acad. Sci. USA* 90:4196-4200.

Rossowski, W. and Coy, D. (1993) Potent inhibitory effects of a type four receptor selective SRIF analog on rat insulin release. *Biochem. Biophys. Res. Commun.* 197:366-371.

Rossowski, W. and Coy, D. (1994) Specific inhibition of rat pancreatic insulin and glucagon release by receptor-selective somatostatin analogs. *Biochem. Biophys. Res. Commun.* 205:341-346.

Strader, C., Sigal, I., Register, R., Candelore, M., Rands, E. and Dixon, R. (1987) Identification of residues required for ligand binding to the beta-adrenergic receptor. *Proc. Natl. Acad. Sci.* 84L4384-4388.

Tallent, M. and Reisine, T. (1992) $G_{i\alpha 1}$ selectively couples SRIF receptor to adenylyl cyclase in the pituitary cell line AtT-20. *Mol. Pharmacol.* 41:452-455.

Tallent, M., Dichter, M. and Reisine, T. (1995) Coupling of the cloned kappa and mu opioid receptors to the inward rectifier potassium current is differentially regulated. *Soc. Neurosci. Abst.* 21; *in press.*

Tallent, M., Liapakis, G., O'Carroll, A.-M., Lolait, S., Dichter, M. and Reisine, T. Somatostatin receptor subtypes SSTR2 and SSTR5 couple to an L-type Ca^{++} channel in the pituitary cell line AtT-20 (submitted).

Taylor, J., Theveniau, M., Bashirzdeh, R., Reisine, T. and Eden, P. (1994) Detection of SRIF receptor subtype 2 (SSTR2) in established tumours and tumour cell lines: Evidence for SSTR2 heterogeneity. *Peptides* 15:1229-1236.

Theveniau, M., Yasuda, K., Bell, G.I. and Reisine, T. (1994) Immunological detection of isoforms of the SRIF receptor subtype, SSTR2. *J. Neurochem.* 63:447-455.

Theveniau, M., Rens-Domiano, S., Law, S., Rougon, G. and Reisine, T. (1992) Development of antisera against the rat brain SRIF receptor. *Proc. Natl. Acad. Sci.* 89:4314-4318.

Vanetti, M., Kouba, M., Wang, X., Vogt, G. and Höllt, V. (1992) Cloning and expression of a novel mouse SRIF receptor. *FEBS Lett.* 311:290-294.

Vanetti, M., Vogt, G. and Höllt, V. (1993) The two isoforms of the mouse SRIF receptor (mSSTR2A and mSSTR2B) differ in coupling efficiency to adenylate cyclase and in agonist-induced receptor desensitization. *FEBS Lett.* 331:260-266.

Wang, H., Bogen, C., Reisine, T. and Dichter, M. (1989) SRIF-14 and SRIF-28 induce opposite effects on potassium currents in rat neocortical neurons. *Proc. Natl. Acad. Sci. USA* 86:9616-9620.

Wang, H., Reisine, T. and Dichter, M. (1990a) SRIF-14 and SRIF-28 inhibit calcium currents in rat neocortical neurons. *Neuroscience* 38:335-342.

Wang, H., Dichter, M. and Reisine, T. (1990b) Lack of cross-desensitization of SRIF-14 and SRIF-28 receptors coupled to potassium channels in rat neocortical neurons. *Mol. Pharmacol.* 38:357-361.

Woulfe, D. and Reisine, T. (1994) Splice variants of SSTR2 differentially couple to adenylyl cyclase. *Soc. Neurosci Abst.* 20:907.

Wulfsen, I., Meyerhof, W., Fehr, S. and Richter, D. (1993) Expression patterns of rat SRIF receptor genes in pre- and postnatal brain and pituitary. *J. Neurochem.* 61:1549-1552.

Xu, Y., Song, H., Bruno, J.F. and Berelowitz, M. (1993) Molecular cloning and sequencing of a human SRIF receptor, hSSTR4. *Biochem. Biophys. Res. Commun.* 193:648-652.

Yamada, Y., Post, S.R., Wang, K., Tager, H.S., Bell, G.I. and Seino, S. (1992) Cloning and functional characterization of a family of human and mouse SRIF receptors expressed in brain, gastrointestinal tract, and kidney. *Proc. Natl. Acad. Sci. USA* 89:251-255.

Yamada, Y., Reisine, T., Law, S.F., Ihara, Y., Kubota, A., Kagimoto, S., Seino, M., Seino, Y., Bell, G.I. and Seino, S. (1993) SRIF receptors, an expanding gene family: cloning and functional characterization of human SSTR3, a protein coupled to adenylyl cyclase. *Mol. Endocrinol.* 6:2136-2142.

Yasuda, K., Rens-Domiano, S., Breder, C.D., Law, S.F., Saper, C.B., Reisine, T. and Bell, G.I. (1992) Cloning of a novel SRIF receptor, SSTR3, that is coupled to adenylyl cyclase. *J. Biol. Chem.* 267:20422-20428.

Yasuda, K., Raynor, K., Kong, H., Breder, C.D., Takeda, J., Reisine, T. and Bell, G.I. (1993) Cloning and functional comparison of kappa and delta opioid receptors from mouse brain. *Proc. Natl. Acad. Sci. USA* 90:6736-6740.

Yatani, A., Codina, J., Sekura, R., Birnbaumer, L. and Brown, A. (1987) Reconstitution of SRIF and muscarinic receptor mediated stimulation of K^+ channels by G_k protein in clonal rat anterior pituitary cell membrane. *Mol. Endocrinol.* 1:283-293.

Reduction of somatostatin-14 binding to the rat somatostatin receptor subtype 3 by Na⁺ is enhanced by mutation of the glutamate residue 92 in the transmembrane domain II

R.B. Nehring, W. Meyerhof[1] and D. Richter

Institut für Zellbiochemie und klinische Neurobiologie, UKE, Martinistraße 52, D-20246 Hamburg, Germany
[1]Abteilung für Molekulare Genetik, Deutsches Institut für Ernährungsforschung, Universität Potsdam, Arthur-Scheunert-Allee 114-116, D-14558 Potsdam-Rehbrücke, Germany

Summary. In order to elucidate the amino acid residues that are involved in binding of somatostatin-14 (SST-14), mutations were introduced into the cDNA encoding somatostatin receptor subtype 3 (SSTR 3) using a polymerase chain reaction(PCR-)based approach. A glutamate residue (E92) in the transmembrane domain II of the SSTR3 sequence was subsequently mutated into a valine residue (E92V) referred to the sequence of SSTR5, which is known to preferentially bind SST-28. The mutation showed no significant differences in agonist binding, but caused enhancement of the sodium sensitivity of the receptor-agonist interaction.

Introduction

Somatostatin-14 (SST-14) and its amino-terminal extended form somatostatin-28 (SST-28) are both physiologically active and derived from a single prohormone. They are widely distributed throughout the central nervous system (Epelbaum, 1986) and peripheral tissues (Reichlin, 1983a,b). These peptides inhibit the release of several other peptide hormones, including growth hormone, gastrin, insulin and glucagon. In the nervous system somatostatin acts as a neurotransmitter or modulator (Epelbaum, 1986; Reichlin, 1983a,b). The signal transduction pathways triggered by somatostatin are mediated by at least five different G-protein coupled receptor subtypes (SSTR1-5) that differ in their pharmacological properties (reviewed in Hoyer et al., 1994). SSTR5 displays much higher affinity for SST-28 than for SST-14 (O'Carroll et al., 1992), while SSTR1-4 show about comparable affinities to both ligands. In transmembrane domain II (TM II) of SSTR1-4, but not in TM II of SSTR5 a glutamic acid residue is present that is located directly adjacent to an aspartic acid residue that is highly conserved among many G-protein coupled receptors (Probst et al., 1992).

We have mutated the glutamic acid residue in SSTR3 to glutamine, leucine or valine and analysed the mutants with respect to binding affinity of SST-14, agonist selectivity between SST-14 and SST-28 and modulation of agonist binding by sodium ions.

Materials and methods

Materials. Vent$_R$ DNA polymerase was purchased from New England Biolabs, Schwalbach, Germany; The vector pcDNAI/Amp from Invitrogen, Leek, The Netherlands; ^{125}I-Tyr[11]-somatostatin-14 from Amersham, Braunschweig, Germany, and somatostatin-14 from Saxon Biochemicals, Hannover, Germany.

Site-directed mutagenesis. All mutations were introduced into a *Apa* I fragment of the SSTR3 cDNA (Meyerhof et al., 1992) by polymerase chain reaction-mediated site-directed mutagenesis (Higuchi et al., 1988). Amplified, mutated fragments generated by the Vent$_R$ DNA polymerase were sequenced and the complete cDNA was subcloned into the expression vector pcDNAI/Amp as described recently (Nehring et al., 1995).

Binding assays. Constructs containing the cDNAs of the wild type SSTR3, mutant SSTR3, or vector alone were transfected into COS-7 cells by the calcium phosphate/glycerol method (Ausubel et al., 1987). Membrane preparations and radioligand binding assays were carried out as reported (Nehring et al., 1995). Binding assays and analysis of sodium sensitivity were carried out in the presence of 100 pM ^{125}I-Tyr[11]-SST-14 (specific activity 2000 Ci/mmol).

Results and discussion

Amino acid sequence comparison of SSTR5 with those of the other SSTR subtypes revealed, that SSTR 1-4 possess a conserved glutamate (E92) in TM II, while SSTR5 displays a valine residue (Fig. 1). In the SSTR3 cDNA the glutamate residue was mutated to glutamine (E92Q), leucine (E92L) or valine (E92V). Scatchard analysis of the saturation binding curves revealed only slight alterations of ^{125}I-Tyr^{11}-SST-14 binding to mutant receptors when compared to wild type SSTR3 (wild type, K_D 44 +/- 1 pM; E92L, K_D 49 +/- 2 pM; E92Q, K_D 67 +/- 3 pM; E92V, K_D 75 +/- 2 pM). In addition, the relative affinities of SST-14 and SST-28 were not altered in the mutant receptors (Fig. 2A). These results suggest that E92 is not involved in selective binding of SST-28.

TM II

```
          *   *    **** ** * ** **                    ****
rSSTR1  ...TNIYI │LNLAIADELLMLSVPFLVTSTL│ LRHWPFG...
rSSTR4  ...TNIYL │LNLAVADELFMLSVPFVASAAA│ LRHWPFG...
rSSTR2  ...TNIYI │LNLAIADELFMLGLPFLAMQVA│ LVHWPFG...
rSSTR5  ...TNVYI │LNLAVAD[V]LFMLGLPFLATQNA│ VSYWPFG...
rSSTR3  ...TSVYI │LNLALADELFMLGLPFLAAQNA│ LSYWPFG...
```

Fig. 1. Sequence alignment of TM II (boxed) of the five rat somatostatin receptor subtypes. Amino acid residues that are conserved in all five somatostatin receptors are marked by asterisks. The unique valine residue of SSTR5 is highlighted.

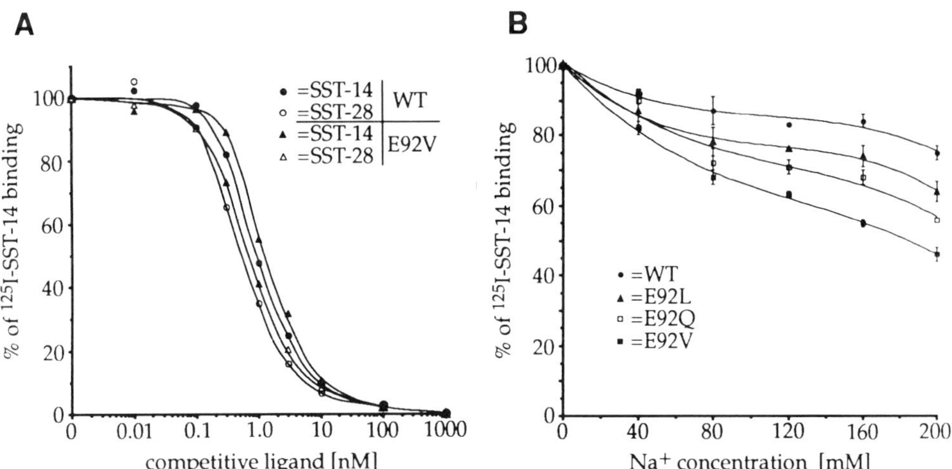

Fig. 2. (A) Displacement of ^{125}I-SST-14 binding to COS-7 cell membranes expressing wild type or E92V mutant of SSTR3 by SST-14 or SST-28. Membranes were incubated in radioligand binding buffer in the presence of unlabeled SST-14 or SST-28 at the indicated concentrations. Results are expressed as percentage of maximal specific binding observed in the absence of competitor. (B) Effect of sodium ions on ^{125}I-SST-14 binding. Membrane preparations of COS-7 cells expressing wild type or mutant receptors were incubated in binding buffer containing the indicated concentrations of sodium chloride. Alterations in radioligand binding due to changes in ionic strength were corrected using corresponding concentrations of N-methyl-D-glucamine.

In the mouse a functional effect on sodium regulation of agonist binding (Kong et al., 1993) has been found for the SSTR2 aspartate 89 residue in TM II (aspartate91 corresponding residue in the rSSTR3 sequence). Mutation of this aspartate to an asparagine residue results in an almost complete loss of Na^+ regulation. This aspartate residue, highly conserved among many G-protein coupled receptors, is located directly adjacent to E92. When the rat SSTR3 mutants were analysed for inhibition of ^{125}I-Tyr^{11}-SST-14 binding by Na^+, we found that the inhibitory effect of the monovalent ion was enhanced in the order of wild type<E92L<E92Q<E92V (Fig. 2B). Thus, the removal of the negative charge at aspartate89 in TM II of the SSTR2 strongly reduces the Na^+ regulation of binding, while the removal of the negative charge at E92 in rSSTR3 significantly increases the regulation.

Therefore, the exact orientation of the amino acid side chains in the second transmembrane domain seems to play a significant role in the regulation of agonist binding by Na^+ ions. Structural changes after the loss of the second charged amino acid may enable Na^+ to get a better access to the binding pocket enhancing the effect of Na^+ ions by stronger ionic interactions. Alternatively, a weaker binding of the G-protein to the receptor may be caused by conformational changes of TM II and therefore a given concentration of sodium ions causes a stronger effect on agonist binding. The biological relevance of Na^+ regulation of somatostatin receptors is still unclear. It is possible that increasing concentrations of Na^+ ions are able to hinder the signal transduction of somatostatin by uncoupling the receptors from G-proteins.

References

Ausubel, F.M., Brent, R., Kingston, R.E., Moore, D.D., Smith, J.A., Seidman, J.G. and Struhl, K. (1987) Introduction of DNA into mammalian cells. In: *Current Protocolls in Molecular Biology,* Greene Pub. Ass., Brooklyn, N.Y., USA, Chapter 9, 9.0.1-9.5.6.

Epelbaum, J. (1986) Somatostatin in the central nervous system: physiology and pathological modifications. *Prog. Neurobiol.* 27: 63-100.

Higuchi, R., Krummel, B. and Saiki, R.K. (1988) A general method of *in vitro* preparation and specific mutagenesis of DNA fragments: study of protein and DNA interactions. *Nucleic. Acids Res.* 16: 7351-7367.

Hoyer, D., Luebbert, H. and Bruns, C. (1994) Molecular pharmacology of somatostatin receptors. *Nauyn-Schmiedeberg's Arch. Pharmacol.* 350: 441-453.

Kong, H., Raynor, K., Yasuda, K., Bell, G.I. and Reisine, T. (1993) Mutation of an aspartate at residue 89 in somatostatin receptor subtype 2 prevents Na⁺ regulation of agonist binding but does not alter receptor-G protein association. *Mol. Pharmacol.* 44: 380-384.

Meyerhof, W., Wulfsen, I., Schoenrock, C., Fehr, S. and Richter, D. (1992) Molecular cloning of a somatostatin-28 receptor and comparison of its expression pattern with that of a somatostatin-14 receptor in rat brain. *Proc. Natl. Acad. Sci. USA* 89: 10267-10271.

Nehring, R.B., Meyerhof, W. and Richter D. (1995) Aspartic acid residue 124 in the third transmembrane domain of the somatostatin receptor subtype 3 is essential for somatostain-14 binding. *DNA Cell Biol.* 14: 939- 944.

O′Carroll, A.-M., Lolait, S.J., König, M. and Mahan, L.C. (1992) Molecular cloning and expression of a pituitary somatostatin receptor with preferential affinity for somatostatin-28. *Mol. Pharmacol.* 42: 939-946.

Probst, W.C., Snyder, L.A., Schuster, D.I., Brosius, J. and Sealfon, S.C. (1992) Sequence alignment of the G-protein coupled receptor superfamily. *DNA Cell Biol.* 11: 1-20.

Reichlin, S. (1983a) Somatostatin (first part). *N. Engl. J. Med.* 309: 1495-1505.

Reichlin, S. (1983b) Somatostatin (second part). *N. Engl. J. Med.* 309: 1556-1563.

The Peptidergic Neuron
B. Krisch and R. Mentlein (eds)
© 1996 Birkhäuser Verlag Basel/Switzerland

Somatostatin receptor subtypes in human astrocytes and gliomas: Influence of cultivation process

J. Feindt, H.-H. Hugo[1], R. Mentlein and B. Krisch

Universität Kiel, Anatomisches Institut, Olshausenstrasse 40, D-24098 Kiel, Germany
[1] *Universität Kiel, Klinik für Neurochirugie, Weinmarer Str. 8, D-24106 Kiel, Germany*

Summary. Expression of somatostatin receptors was investigated on normal human astrocytes and human glial tumours. All cultivated glial cells and gliomas, directly embedded in paraffin were immunopositive for the astrocytic marker glial fibrillary acidic protein. Moreover, somatostatin-binding sites could be visualized on all cell types by affinity labelling with a somatostatin-gold conjugate. Thereby, the normal astrocytes showed a fine, stippled pattern of the conjugate all over the cell surface wheras the tumourous cells had a more thread-like pattern preferentially on the cell processes. The transcripts of the different somatostatin receptor subtypes were detected by reverse transcription - poymerase chain reaction (RT-PCR) with oligonucleotides specific for five human somatostatin receptor subtypes (SSTRs). Normal astrocytes expressed SSTR-1, (SSTR-2) and SSTR-4 specific transcripts, glioma cells showed an overexpression of SSTR-2 compared to normal astrocytes (relative to equal intensities for β-actin amplificates). This overexpression of SSTR-2 transcript could be detected in cultivated tumour glial cells as well as in solid gliomas. Thus the cultivation process had no influence on the individual SSTR-2 expression on normal and tumourous cells.

Introduction

The neuropeptide somatostatin - first isolated from ovine hypothalamus (Brazeau et al., 1973) - is widely distributed in the central nervous system and in the periphery, including pancreas, gut and pituitary. In the brain somatostatin is believed to function as neurotransmitter and neuromodulator (Epelbaum, 1986). The effects of somatostatin are mediated by seven transmembrane domain G-protein coupled receptors. Up to now, five different somatostatin receptor subtypes are known (Bruno et al., 1992; Yamada et al., 1992a; Yamada et al., 1992b; Yamada et al., 1993), which differ in their interaction with an extended form of the neuropeptide (somatostatin-28) or synthetic derivates (Patel and Srikant, 1994) and their tissue distribution (Bell and Reisine, 1993). Somatostatin-binding

                    J. Feindt et al.

sites were detected on glial cells *in situ* (Krisch, 1994) and on glial tumours *in situ* and *in vivo* (Luyken et al., 1994). However, a comparative study on the somatostatin receptor subtypes expressed on normal glial cells and glial tumours has not yet been published. Therefore, we wanted to know, which somatostatin receptor subtypes are expressed on normal and tumourous glial cells and which morphological distribution of receptors is found on their cell surface. Moreover, we wanted to elucidate the influence of the cultivation process on somatostatin receptor subtype expression.

Materials and methods

Cell cultures. Surgical human gliomas were freed from blood vessels, connective tissue and meningeal cells as far as possible, dissociated mechanically in Dulbeccos´s modified Eagle´s medium (DMEM) plus 5 % DNase plus 3 % trypsin, plated into culture dishes, cultivated in DMEM plus 10 % fetal calf serum, and subcultivated after 2-4 weeks by trypsin treatment. All tumours were classified as glioblastoma WHO grade IV, except for three which were astrocytomas grade I, II or III (see Tables I and II). Human astrocytes were from non-tumourous surgical samples and were cultivated similar to the gliomas. Part of some gliomas was embedded directly in paraffin and analysed by immunhistochemistry parallel to the cultivation process. Another part of the same glioma material was used to isolate RNA.

Immunohistochemistry. To prove the identity of the tissues and cultivated cells, they were analysed by immunohistochemistry for the astroglial marker glial fibrillary acidic protein (GFAP). Tissue was directly embedded in paraffin, cut into 3 μm sections; cultivated cells were fixed with -20 °C acetone, and samples immunostained as described (Mentlein et al., 1990).

Visualization of somatostatin receptors. Binding sites on cultivated cells were visualized by affinity labelling with 1 nM somatostatin-gold conjugate as described earlier (Mentlein et al., 1990; Krisch, 1994). For electron microscopy, cells were osmicated, embedded in Araldite, ultrathin sections were cut with a diamond knife, and the sections were stained in aqueous uranyl acetate and viewed in a Phillips 300 electron microscope.

RNA-isolation; reverse transcription - polymerase chain reaction (RT-PCR). RNA was isolated from cultivated tumour material and human astrocytes by CsCl density centrifugation in 4 M guanidinium isothiocyanate / 0.5 % lauryl sarconisate as described (Chirgwin et al., 1979), from solid human tumours (see Table II) with the commercial "Qiagen Total RNA Isolation Kit" (Qiagen).

Reverse transcription of 3 µg RNA was done in a final volume of 20 µl as described. PCR for cultivated cells (Tables I and II) were performed with a single primer-pair polymerase chain reaction (PCR) with human somatostatin receptor subtype (SSTR) -1 to -5 specific primer pairs as published. The single specific PCR products were of the expected sizes: SSTR-1 (542 bp), SSTR-2 (377 bp), SSTR-3 (376 bp), SSTR-4 (371 bp), SSTR-5 (361 bp). Electrophoresis bands of human β-actin were semiquantitatively compared to those of SSTRs. In contrast to single primer-pair experiments PCR for transcripts from solid human tumours (Table II) were performed simultaneously with 10 pmol of human SSTR-2 specific and human β-actin specific sense / antisense primers in a final volume of 100 µl in a reaction mixture containing 4 µl 25 mM $MgCl_2$ and 1x PCR buffer IV (Biomol) under same PCR conditions. The two different PCR products were of the expected sizes: SSTR-2 (377 bp), β-actin (650 bp). Controls for specificity of PCR products and selectivity of PCR conditions were carried out as described earlier (Feindt et al., 1995).

Results and discussion

After 2-4 subcultures, all cultivated human gliomas and normal human astrocytes (Tables I and II) were immunohistochemically > 95 % positive for the astroglial marker GFAP and were then used for affinity labelling with the somatostatin-gold conjugate and RT-PCR experiments.

144   J. Feindt et al.

Glioma material embedded directly in paraffin (Table II) contained > 70 % GFAP-immunopositive glial cells (Fig. 1, thick arrows), which showed the typical astrocytic morphology. As expected, GFAP-immunonegative endothelial cells and cells from the surrounding neuropile were intermingled with GFAP-positive glioma cells in the solid human brain tissue (Fig. 1, thin arrows).

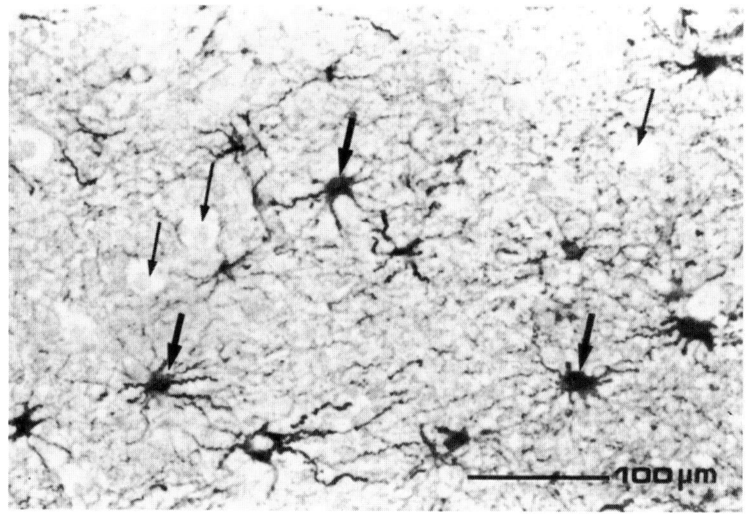

Fig. 1. Light-microscopic visualization of GFAP-immunoreactivity in paraffin-embedded surgical samples from patient H.-P.W., glioma grade WHO IV. More than 70 % of the tumour tissue contained GFAP-immunopositive glial cells (thick arrows), some GFAP- negative cells are indicated by thin arrows. This tumour tissue could be used for direct RNA-isolation and RT-PCR experiments for somatostatin receptor subtypes. Original magnification: x 220.

Although not pure with respect to a single cell type, but retaining the initial transcription pattern of the glioma cells, RNA was isolated directly from parts of surgical samples, analysed by RT-PCR for SSTR subtypes and compared to SSTR subtypes of the same but (sub-)cultivated and immunohistochemically pure tumour material.

After confirming the glial character of the cultivated tumourous cells and normal astrocytes, somatostatin receptors on the cell surface - apart from mRNA transcripts - of the different cell types was proved by affinity labelling with a somatostatin-gold conjugate. On all cultivated human glioma material and human astrocytes, somatostatin receptors could be detected (Table I and II) at the light- and electron-microscopic level. Usually, somatostatin receptors were expressed earliest two days after a subcultivation step. The somatostatin-gold conjugate, non-selective for different somatostatin receptor subtypes (SSTRs), showed a specific and distinct distribution pattern of binding sites on normal astrocytes and glioma cells (Fig. 2 A and B).

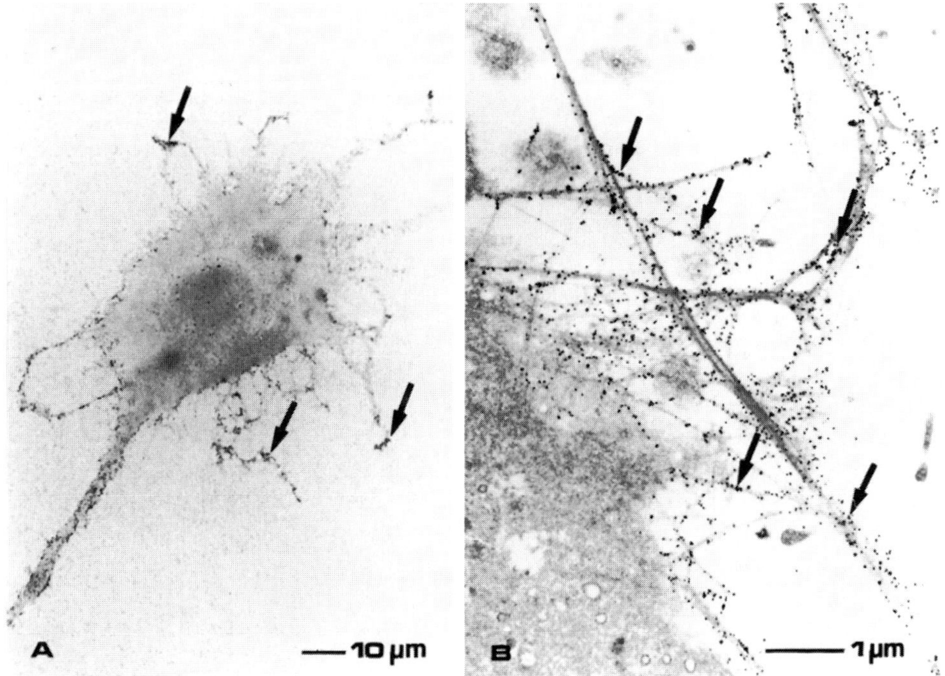

Fig. 2. Light- and electron-microscopic visualization of somatostatin-binding sites on human astrocytes and glioma cells. **A:** Light microscopy of cultivated normal human astrocyte from patient A.R.; somatostatin-gold conjugate decorated cell somata and processes in a fine stippled pattern (arrows). **B:** Electron microscopy of cultivated human glioma cells from patient H.D. glioma grade WHO IV; somatostatin-gold conjugate showed a heavily increased threadlike labelling pattern on processes (arrows). Original magnification **A:** x 690, **B:** x 4,400.

On normal human astrocytes the somatostatin-gold conjugate showed a fine, stippled distribution pattern over cell somata and processes (Fig. 2 A, arrows). Usually, some astrocytes remained unlabelled. In most cases the glioma material was more heavily stained than normal astrocytes. Moreover, the somatostatin-gold conjugate bound preferentially on the cell processes of glioma tumours (Fig. 2 B, arrows). Only one tumour (U.L. WHO II-III) showed a less intense staining, but clearly more heavily than normal astrocytes.

These results correspond to previous reports on somatostatin receptors on astrocytes *in vitro* (see Krisch and Mentlein, 1994) and on subpopulations *in situ* (Mentlein et al., 1990; Krisch et al., 1991). Moreover, somatostatin-binding sites were found *in vivo* and *in situ* on most human glial tumours (Luyken et al., 1994). Obviously, besides neurons, astrocytes and glioma cells are targets for neuropeptides.

RT-PCR experiments with human subtype-specific oligonucleotide sense and antisense primers were performed in addition to the detection of SSTRs by affinity-labelling with the somatostatin-gold conjugate. First cultivated human astrocytes and human glioma tumours (Tables I and II) were investigated by single primer-pair polymerase chain reactions (PCRs). Cultivated cells were taken to ensure the homogeneity and purity of the samples. In each case, the quality of the mRNA was related to the RT-PCR products β-actin used as reference. In addition, the intensity of electrophoresis bands for SSTR products of different samples could be compared to the corresponding PCR experiments of the respective single primer-pair βactin PCR products.

Cultured human astrocytes from patient A.R. expressed mRNA of three subtypes: SSTR-1, SSTR-4, and, detected with a very faint electrophoresis band, SSTR-2 (Table I); for cultivated human astrocytes from patient G.B. only transcripts of SSTR-2 were analysed and detected as a faint electrophoresic signal (Table II). Much higher expression of SSTR-2 as compared to normal cells (relative to equal intensities for β-actin PCR products) was found in all cultivated human glial tumours (Tables I and II) - indicated by a strong electrophoresis band. In addition to SSTR-2 some tumours expressed SSTR-1 and SSTR-4, but clearly with less intensity.

Table I. Expression of somatostatin receptor subtype (SSTR)-mRNA in human astrocytes and gliomas after cell cultivation.

Cultivated cells	SSTR-mRNA [a]					Gold-labelling
	1	2	3	4	5	
Astrocytes A.R.	+	+	-	(+)	-	positive
Human gliomas [b]						
A.K. WHO I	+	++	-	-	-	positive
J.K. WHO II	+	++	-	+	-	positive
U.L. WHO II-III	+	++	-	(+)	-	positive
H.D. WHO IV	+	++	-	(+)	-	positive
N.H. WHO IV	+	++	-	-	-	positive
H.C. WHO IV	-	++	-	-	-	positive

[a] SSTR-mRNA was probed by single primer-pair RT-PCR as described in Materials and methods. ++, strong electrophoresis band after 40 cycles; +, strong band after 40 cycles; (+), faint band after 40 cycles. Semiquantitative comparison is valid between different samples for a single receptor subtype. β-actin served as a reference mRNA. [b] The initials indicate the individual patient, the Roman numerals the tumour rating according to the WHO classification.

The three receptor subtypes SSTR-1, SSTR-2, and SSTR-4 that are expressed in cultured astrocytes have been detected previously in various rat brain regions (Bruno et al., 1993; Kaupmann et al., 1993). The high expression of SSTR-2 in all human glial tumours has also been reported in several non-glioma cell lines: rodent pituitary cell lines - GH3 - (Patel et al., 1993) and several rodent pancreatic cell lines (Eden and Taylor, 1993). Obviously, the expression of SSTR-2 is associated with a shift from normal cells to malignancy and SSTR-2 expression is especially correlated with a high malignancy in glial tumours.

To prove SSTR-2 overexpression on fresh (uncultivated) glial tumours, some tumours (Table II), in parallel to cultivation were prepared for a direct RNA-isolation process and used for double primer-pair PCR experiments with specific SSTR-2 and β-actin sense and antisense oligonucleotides. Double primer-pair PCR experiments with an internal β-actin reference (i.e. PCR products for β-actin and SSTR-2 were performed from one RNA

J. Feindt et al.

sample of a specific cell type), yielded more precise quantitative values in comparison to two separate PCR experiments with single primer-pairs.

Table II. Comparison of the expression of somatostatin receptor subtype (SSTR)-2 mRNA in human astrocytes and gliomas after cell cultivation / direct RNA-isolation.

Cell type		SSTR-mRNA expression after [a]		Gold-labelling after cell cultivation
		direct isolation of mRNA	cell cultivation	
Astrocytes G.B.		+	+	positive
Human gliomas [b]				
E.D.	WHO IV	++	++	positive
M.E.-H.	WHO IV	++	n.d.	positive
H.-P.B.	WHO IV	++	n.d.	positive

[a] SSTR-2 mRNA was probed by double (direct RNA-isolated RNA samples) / single (after cell cultivation isolated RNA samples) primer-pair RT-PCR described in Materials and methods. ++, strong electrophoesis band in comparison to the respective β-actin reference PCR products after 40 cycles; +, band detected in comparison to β-actin standard after 40 cycles; n.d.= not detected. [b] The initials indicate the individual patient, the Roman numerals the tumour rating according to the WHO classification.

Normal human astrocytes from patient G.B. and the directly isolated RNA from the glial tumour H.-P.W. WHO IV, showed a weak SSTR-2 electrophoresis band in comparison to equal intensities of internal β-actin PCR amplificates (Fig. 3 and Table II). Compared to normal astrocytes (with equal intensities of β-actin PCR products) the directly isolated gliomas M.E.-H. WHO IV and E.D. WHO IV exhibited a strong SSTR-2 electrophoresis band (Fig. 3 and Table II).

The expression of SSTR-2 transcripts was comparable in directly isolated RNA from surgical samples or RNA isolated from cultivated, immunocytochemical pure cells (Table II). Obviously, the cultivation process had no influence on SSTR-2 mRNA expression and single / double primer-pair PCR experiments yielded identical results. Most of the glial tumours investigated - uncultivated and cultivated tumour material - showed an overexpression of the SSTR-2. Only one glial tumour (H.-P.W.) expressed an SSTR-2 mRNA amount identical to normal astrocytes.

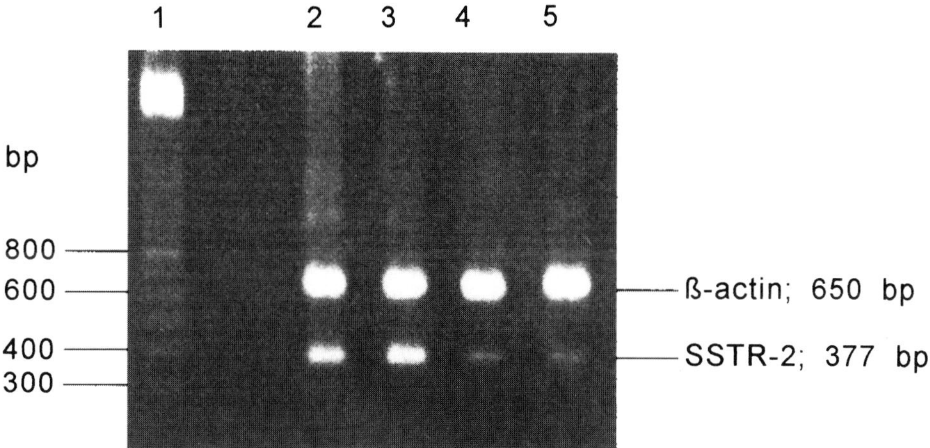

Fig. 3. Example of an ethidium bromide-stained agarose electrophoresis gel after separation double primer-pair RT-PCR products with primers from human β-actin and SSTR-2 from human gliomas M.E.-H. WHO IV (3), E.D. WHO IV (2), H.-P.W. WHO IV (4) and normal astrocytes G.B. (5). The size of the base pairs (bp) of marker fragments (1) is indicated. In comparison to normal astrocytes - apart from H.-P.W. (2) - the human gliomas showed an overexpression of SSTR-2 with equal intensities of the internal β-actin PCR products.

Conclusion

Normal glial cells and glial tumours in culture express somatostatin receptors on their cell surface. For normal astrocytes a fine, stippled pattern of somatostatin binding sites can be detected on the plasma membrane, glial tumours have a more threadlike distribution preferentially on the cell processes. In comparison to normal cells, glial tumours - especially those of high malignancy - show an overexpression of the somatostatin receptor subtype-2 (SSTR-2) transcripts, normal glial cells favour SSTR-1 and SSTR-4. Cultivation has no influence on subtype expression in qualitative and quantitative terms: SSTR-2 overexpression can be verified for mRNA samples from fresh and from cultivated tumour tissue.

Acknowledgements
We thank Dagmar Freier, Martina Burmester, Helga Prien and Doris Weinstein for their expert technical assistance. This work was supported by the Deutsche Forschungsgemeinschaft (grants Kr 569 / 5-7) and the Stiftung Volkswagenwerk (R.M., project "Neuroimmunologie").

References

Bell, G.I. and Reisine, T. (1993) Molecular biology of somatostatin receptors. *Trends Neurosci.* 16: 34-38.

Bruno, J.F., Xu, Y., Song, J. and Berelowitz, M. (1992) Molecular cloning and functional expression of a brain-specific somatostatin receptor. *Proc. Natl. Acad. Sci. USA* 89: 11151-11155.

Bruno, J.F., Xu, Y., Song, J. and Berelowitz, M. (1993) Tissue distribution of somatostatin receptor subtype messenger ribonucleic acid in the rat. *Endocrinology* 133: 2561-2567.

Chirgwin, J.J., Prybla, A.E., MacDonald, R.J. and Rutter, W.J. (1979) Isolation of biologically active ribonucleic acid from sources enriched in ribonuclease. *Biochemistry* 18: 5294-5299.

Eden, P.A. and Taylor, J.E. (1993) Somatostatin receptor subtype expression in human and rodent tumours. *Life Sci.* 53: 85-90.

Epelbaum, J. (1986) Somatostatin in the central nervous system: physiological and pathological modifications. *Progr. Neurobiol.* 27: 63-100.

Feindt, J., Becker, I., Blömer, U., Hugo, H.-H., Mehdorn, H.M., Krisch, B. and Mentlein, R. (1995) Expression of somatostatin receptor subtypes in cultured astrocytes and gliomas. *J. Neurochem.* 65: 1997-2005.

Kaupmann, K., Bruns, C., Hoyer, D., Seuwen, K. and Lübbert, H. (1993) Distribution and second messenger coupling of four somatostatin receptor subtypes expressed in brain. *FEBS Lett.* 331: 53-59.

Krisch, B., Buchholz, C. and Mentlein, R. (1991) Somatostatin-binding sites on rat diencephalic atsrocytes. *Cell Tissue Res.* 263: 253-263.

Krisch B. (1994) Somatostatin binding sites in functional systems of the brain. *Progr. Histochem. Cytochem.* 28: 1-40.

Krisch, B. and Mentlein, R. (1994) Neuropeptide receptors and astrocytes. *Internat. Rev. Cytol.* 148: 119-169.

Luyken, C., Hildebrandt, G., Scheidhauer, K., Krisch, B., Schicha, H. and Klug, N. (1994) [111]Indium (DTPA-Octreotide) scintigraphy in patients with cerebral gliomas. *Acta Neurochir.* 127: 60-64.

Mentlein, R., Buchholz, C. and Krisch, B. (1990) Somatostatin-binding sites on rat telencephalic astrocytes. *Cell Tissue Res.* 262: 431-443.

Patel, Y.C. and Srikant, C.B. (1994) Subtype selectivity of peptide analogs for all five cloned human somatostatin receptors (hsstr 1-5). *Endocrinology* 135: 2814-2817.

Patel, Y.C., Greenwood, M., Kent, G., Panetta, R. and Srikant, C.B. (1993) Multiple gene transcripts of the somatostatin receptor subtype SSTR2: tissue selective distribution and cAMP regulation. *Biochem. Biophys. Res. Commun.* 192: 288-194.

Sambrook, J., Fritsch, E.F. and Maniatis, T. (1989) *Molecular Cloning: A Laboratory Manual*, Second Edition, Cold Spring Harbor Laboratory Press, Cold Spring Harbor, New York.

Yamada, Y., Post, S.R., Wang, K., Tager, H.S., Bell, G.I. and Seino, S. (1992a) Cloning and functional characterization of a family of human and mouse somatostatin receptors expressed in brain, gastrointestinal tract and kidney. *Proc. Natl. Acad. Sci. USA* 89: 251-255.

Yamada, Y., Reisine, T., Law, S.F., Ihara, Y., Kagimotot, S., Seino, Y. Bell, G.I. and Seino, S. (1992b) Somatostatin receptor, an expanding gene family: cloning and functional characterization of human SSTR3, a protein coupled to adenylyl cyclase. *Mol. Endocrinology* 6: 2136-2142.

Yamada, Y., Kagimoto, S., Kubota, A., Yasuda, K., Masuda, K., Someya, Y., Ihara, Y., Li, Q., Imura, H., Seino, S. and Seino, Y. (1993) Cloning, functional expression and pharmacological characterization of a fourth (hSSTR4) and a fifth (hSSTR5) human somatostatin receptor. *Biochem. Biophys. Res. Commun.* 195: 844-852.

Coexistence of angiotensin receptors and angiotensin in hypothalamic neurons of the rat

C. Spengler, J. Pfister, R. Mosimann, M. Raizada[1], D. Felix and H. Imboden

University of Berne, Div. of Neurobiology, Erlachstrasse 9a, 3012 Berne, Switzerland
[1] *Unviversity of Florida, Dept. of Physiology, Gainesville, FL, USA*

Summary. By the use of immunocytochemical methods the colocalization of the octapeptide angiotensin II or fragments of it and the AT_1 receptor subtype is shown in individual neurons of the paraventricular, the accessory and the supraoptic nucleus of the hypothalamus. Similar results were obtained with an anti-idiotypic antibody to angiotensin II antibodies. Our results provide strong evidence that angiotensin-containing hypothalamic neurons are innervated by angiotensinergic neurons.

Introduction

The brain renin-angiotensin system, which, among other effects, plays an important role in cardiovascular function and maintenance of body fluid homeostasis, has been the subject of many reviews (Saavedra, 1992; Bunnemann et al., 1993; Wright and Harding, 1994). By the use of immunocytochemical methods (Lind et al., 1985; Imboden et al., 1987) positive angiotensin II-like immunoreactive cells have been demonstrated in the magnocellular part of the paraventricular nucleus (PVN), the accessory nucleus and the supraoptic nucleus (SON). Autoradiographic binding studies (Tsutsumi and Saavedra, 1991) and electrophysiological investigations (Ambühl et al., 1992) have shown that in the paraventricular nucleus angiotensin II is mediated predominantly by the AT_1 receptor subtype. Furthermore, the existence of angiotensin receptors in paraventricular nucleus and supraoptic nucleus have been confirmed by immunocytochemical studies (Pfister et al., 1993; Phillips et al., 1993).

Despite the knowledge of the existence of angiotensin itself as well its receptors in these hypothalamic nuclei, the coexistence of the components in individual cells have not been reported. Therefore, the aim of this study was to investigate such a possible coexistence by the use of immunocytochemistry. In the present study three different antibodies were used: 1) An affinity-purified anti-angiotensin II antibody, 2) an antibody to a partial sequence of the angiotensin AT_1 receptor subtype, and 3) an anti-idiotypic antibody against angiotensin II-antibodies.

Material and methods

The specific anti-angiotensin II serum "BODE" has been raised in our laboratory and has been described earlier in detail (Imboden et al., 1989). The anti-idiotypic serum was prepared and kindly provided by Dr. P.O. Couraud (1987). For anti-AT_1 serum a peptide responding to amino acids 225-237 in the third intracellular loop of the angiotensin AT_1 receptor was crosslinked to thyroglobulin, emulsified in complete Freund`s adjuvant and injected in rabbits. Specificity tests showed that these antibodies bind to the AT_1 receptor subtype for angiotensin II (Zelezna et al., 1992).

The immunocytochemical method has been published earlier (Imboden et al., 1987; Imboden and Felix, 1995). Male Wistar Kyoto (WKY) rats between 6 and 12 weeks of age were used. The paraformaldehyde fixed, 5 μm thick cryosections were picked up on gelatine-coated slides and were inserted into disposable coverplates (SHANDON). Inside these coverplates the immunocytochemical incubations was performed by using the peroxidase anti-peroxidase method together with diaminobenzidine. The slides were separated from the coverplates and the sections were counterstained with toluidine blue.

Results

The lack of pretreatment of the animals with colchicine and the purification of the specific antibodies against angiotensin were important steps in our investigation. With this procedure it was possible to stain not only neurons in the paraventricular nucleus, the accessory nucleus and in the supraoptic nucleus, but also to visualize angiotensinergic fibers in the paraventricular-hypophysial pathway. The approximate diameter of the neurons in these hypothalamic nuclei varied from 15 to 30 μm. For our investigation we have used 5 μm thick sections allowing a chance to get at least 3 adjacent sections out of one single cell. With serial sections we were able to carry out immunoreaction studies using the three antibodies in one single neuron. As an example, Fig. 1 illustrates the coexistence of angiotensin and its receptor subtype AT_1 in individual neurons of the supraoptic nucleus.

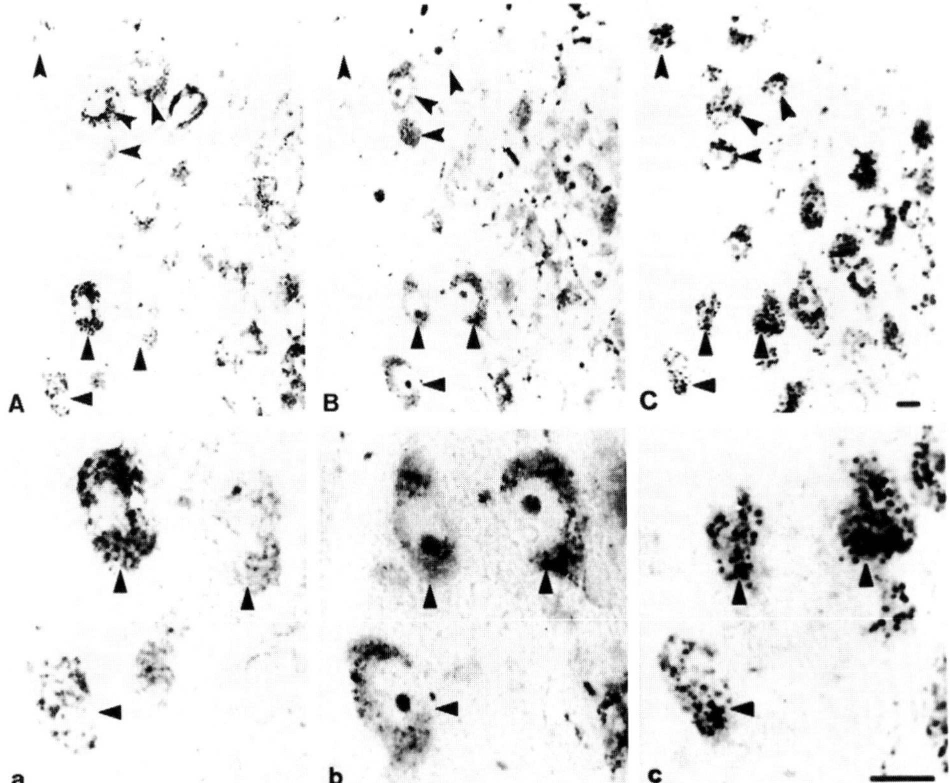

Fig. 1. Coexistence of angiotensin and angiotensin receptors in neurons of the supraoptic nucleus.
A: Immunoreactive staining with the anti-idiotypic antibody for angiotensin II receptor.
B: Immunoreactive staining with the affinity purified antibody to angiotensin II.
C: Immunoreactive staining with the antibody to the angiotensin AT_1 receptor oligo-sequence.
The lower row (a - c) are high-magnification micrographs of A, B and C. Scale bars: 10 μm.

Similarly, immunoreactive staining was also obtained with the anti-idiotypic antibody. For both, AT_1 receptor and anti-idiotypic antibodies, dot-like staining was confined to cell bodies and dendrites. In nerve fibres, however, only labelling for angiotensin II was observed.

It has to be pointed out, that in all nuclei investigated, there were more neurons marked for the angiotensin receptors than for the angiotensin peptide.

Discussion

Our immunocytochemical results demonstrate the coexistence of angiotensin receptors with angiotensin containing neurons in different hypothalamic nuclei. Although several individual studies using immunocytochemical (Pfister et al., 1993; Phillips et al., 1993), autoradiographic (Tsutsumi and Saavedra, 1991) or electrophysiological (Ambühl et al., 1992) techniques reported the existence of angiotensin or angiotensin receptors, this is, to our knowledge, the first report of coexistence of the peptide and its receptors within the same neuron. The fact that angiotensin containing neurons possess receptor sites as well, raises the possibility of an angiotensinergic input onto the same neuron. The physiological role of these receptors, however, has to be further elucidated. An interaction with other peptidergic systems cannot be ruled out, since angiotensin and vasopressin coexist in hypothalamic neurons (Imboden and Felix, 1991).

The advantage of using thin sections allowed to test the different antibodies in individual slices of the same neuron, thereby preventing any possible interaction with each other. To summarize, the chosen approach offers a good tool for any future studies on colocalization of the components of hypothalamic peptidergic systems.

Conclusion

The results of our study indicate for the first time that angiotensin containing neurons in the hypothalamus might be innervated by angiotensinergic neurons.

Acknowledgements
We are especially grateful to P.O. Couraud, J. Marie and S. Jard of the Centre CNRS-INSERM de Pharmacologie-Endocrinologie in Montpellier, France, who supplied us with the anti-idiotypic antibody. This work was supported by Grant No. 31-40469.94 from the Swiss National Science Foundations, the "Stiftung zur Förderung der wissenschaftlichen Forschung an der Universität Bern" and the NIH grant (HL 33610). We wish to thank Susanne Gygax for technical assistance and Ruth Schweizer for typing the manuscript.

References

Ambühl, P., Felix, D., Imboden, H., Khosla, M.C. and Ferrario, C.M. (1992) Effects of angiotensin analogues and angiotensin receptor antagonists on paraventricular neurons. *Regul. Peptides* 38: 111-120.

Bunnemann, B., Fuxe, K. and Ganten, D. (1993) The renin-angiotensin system in the brain: an update 1993. *Regul. Peptides* 46: 487-509.

Couraud, P.O. (1987) Anti-angiotensin II anti-idiotypic antibodies bind to angiotensin II receptor. *J. Immunol.* 138: 1164-1168.

Imboden, H., Harding, J.W., Abhold, R.H., Ganten, D. and Felix, D. (1987) Improved immunohistochemical staining of angiotensin II in rat brain using affinity purified antibodies. *Brain Res.* 426: 225-234.

Imboden, H., Harding, J.W. and Felix, D. (1989) Hypothalamic angiotensinergic fibre systems terminate in the neurohypophysis. *Neurosci. Lett.* 96: 42-46.

Imboden, H. and Felix, D. (1991) An immunocytochemical comparison of the angiotensin and vasopressin hypothalamo-neurohypophysial systems in normotensive rats. *Regul. Peptides* 36: 197-218.

Imboden, H. and Felix, D. (1995) Immunocytochemistry in Brain Tissue. *In:* M.I. Phillips and D. Evans (eds.): *Methods in Neurosciences* Volume 24, *Neuroimmunology*, Academic Press , London, pp. 236-260.

Pfister, J., Felix, D. and Imboden, H. (1993) Immunohistochemical demonstration of angiotensin II receptors in rat brain by use of an anti-idiotypic antibody. *Regul. Peptides* 44: 109-117.

Lind, R.W., Swanson, L.W. and Ganten, D. (1985) Organization of angiotensin II immunoreactive cells and fibers in the rat central nervous system. *Neuroendocrinology* 40: 2-24.

Phillips, M.I., Shen, L., Richards, E.M. and Raizada, M.K. (1993) Immunhistochemical mapping of angiotensin AT_1 receptors in the brain. *Regul. Peptides* 44: 95-107.

Saavedra, J.M. (1992) Brain and pituitary angiotensin. *Endocrine Reviews* 13(2): 329-380.

Tsutsumi, K. and Saavedra, J.M. (1991) Differential development of angiotensin II receptor subtypes in the rat brain. *Endocrinology* 128: 630-632.

Wright, J.W. and Harding, J.W. (1994). Brain angiotensin receptor subtypes in the control of physiological and behavioural responses. *Neuroscience and Biobehavioural Reviews* 18: 21-53.

Zelezna, B., Richards, E.M., Tang, W., Lu, D., Sumners, C. and Raizada, M.K. (1992) Characterization of a polyclonal antipeptide antibody to the angiotensin II type-1 (AT_1) receptor. *Biochem. Biophys. Res. Commun.* 183: 781-788.

The Peptidergic Neuron
B. Krisch and R. Mentlein (eds)
© 1996 Birkhäuser Verlag Basel/Switzerland

Bradykinin binding sites on isolated cultured dorsal root ganglion cells demonstrated with gold-labelled bradykinin

G. Segond von Banchet, M. Petersen, A. Eckert and B. Heppelmann

Institute of Physiology, University of Würzburg, Röntgenring 9, D-97070 Würzburg, Germany

Summary. Bradykinin is a powerful mediator to activate primary afferents that contain a great variety of neuropeptides. To study the distribution of bradykinin binding sites on cultured dorsal root ganglion (DRG) cells we used a recently developed non-radioactive method. Bradykinin that had been covalently linked to a small gold particle bound in a proportion of dorsal root ganglion cells. The binding was evenly distributed across the soma membrane. The proportion of cells that bound bradykinin markedly depended on the length of time in culture. After 18 hours, bradykinin bound to about 43% of the neurons. This proportion increased to about 85% after 1.75 days and slowly decreased to about 7% after 6.75 days. Bradykinin bound to cells of all sizes. In some cells a binding was also seen along the total length of their processes. The addition of specific B_1 and B_2 receptor ligands, separately or in combination with bradykinin, markedly reduced the binding indicating that most of the binding sites belong to the B_2 and B_1 subtype. The data allow to hypothesize that the transient increase in bradykinin receptor expression might be caused by cell injury due to disruption of the axon. Injury induced upregulation of the receptor in vivo could cause dramatic physiological reactions.

Introduction

A large proportion of group III and IV primary afferent neurons contain a great variety of neuropeptides that may be involved in different trophic and regulatory processes. The release of neuropeptides from peptidergic afferents seems to be an important component in inflammations. Electrophysiological studies revealed that the majority of primary afferents can be activated by the nonapeptide bradykinin indicating that those neurons express bradykinin receptors (see e.g. Kanaka et al., 1985). The somata of dorsal root ganglion (DRG) cells serve as a model to investigate the electrophysiological and biochemical events underlying the bradykinin response in sensory neurons (see e.g. McGuirk and Dolphin, 1992). Therefore, in the present study we investigated the distribution of bradykinin receptors in cultured dorsal root ganglion cells of the rat.

To demonstrate the binding of bradykinin in cultured dorsal root ganglion cells, we used a recently developed non-radioactive method (Segond von Banchet and Heppelmann, 1995). Using a gold labeling reagent (NHS-Nanogold), Bradykinin was covalently linked with a single 1.4-nm gold particle at the primary amino group. This technique combines the advantages of a receptor ligand affinity-labelling system with a non-radioactive detection system.

Material and methods

Dorsal root ganglions were dissociated from 12 adult rats of either sex after they had been sacrificed with a lethal dose of sodium pentobarbital. After dissociation cells were plated on 12-mm glas coverslips coated with poly-L-lysine (200 µg/ml) and maintained for 18 hours to 6.75 days at 37°C in a humified incubator gassed with 3.5% CO_2 and air.

The bradykinin-gold conjugate was prepared as described by Segond von Banchet and Heppelmann (1995). Cultured cells were shortly fixed. Thereafter, the coverslips were incubated with glycine and with bovine serum albumin and gelatin to block free aldehyde groups and nonspecific binding sites. After washing, cells were incubated with 3 nmol bradykinin-gold at 4°C in a moist chamber. Finally, cells were postfixed, and the gold particles were intensified with silver enhancer.

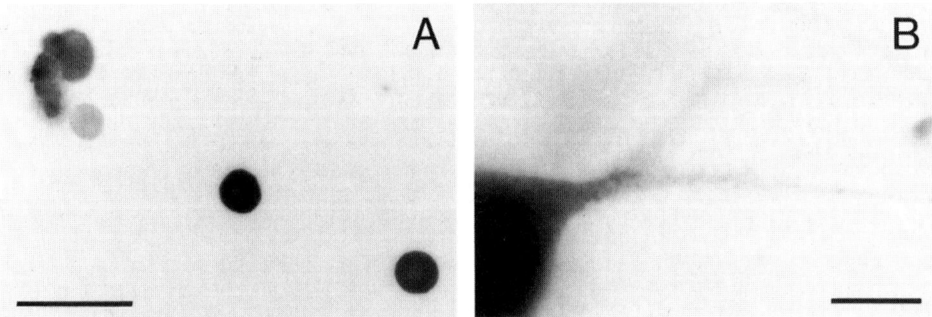

Fig. 1. Photomicrographs of rat dorsal root ganglion cells in culture after an incubation with gold labelled bradykinin. **A:** After 1.75 days in culture dorsal root ganglion cells show a wide range in binding intensities that are evenly distributed across the soma membrane. **B:** In 2.75 days old cultures most cells have developed processess. Some of these cells show an intense binding both on the soma and along the total length of their processes. (Bar = 50 µm; x 250)

In control incubations cells were incubated with 3 nmol bradykinin in the presence of 1 μmol native bradykinin or with 3 nmol NHS-Nanogold (NHS, N-hydroxy-succinimide) alone. To examine whether the binding was related to B_1 or B_2 receptors, bradykinin-gold was incubated in the presence of 1 μmol [des-Arg^{10}]-Lys-Bradykinin, a B_1 receptor ligand, or 1 μmol D-Arg[Hyp^3-β-(2-thienyl)^{5,8}-D-Phe^7]-bradykinin, a B_2 receptor ligand.

The area and the mean gray value of all cell somata were determined with an image analyzing system (Optimas). To exclude possible differences in the silver enhancement reaction a relative gray value of each soma was calculated by dividing the mean gray value of the soma by the mean gray value of the corresponding background.

Results

Numerous dorsal root ganglion cells showed a gray staining reaction of different intensities indicating different quantities of bradykinin binding sites (Figs. 1A, 2). Some cells showed an intense binding both on the somata and along the total length of their processes (Fig. 1B) that appear after about 1 day in culture.

To determine the proportion of cells with or without a bradykinin-binding, cells with a relative gray value within the range of those from the control incubations were considered as cells with no bradykinin-binding (Fig. 2, white bars).

The length of time under culture conditions markedly affected the bradykinin-binding at the cells. After 18 hours 43 ± 13 % (± standard derivation) of the cells showed a positive binding (Fig. 3A). This proportion increased to 85 ± 14 % at day 1.75. In cultures of 2.75 - 6.75 days incubation time, the proportion of positive cells decreased to 7 ± 5 %. The soma size of neurons classified as positive cells did not show any preferential size at all points in time (not shown).

The addition of the B_1 receptor ligand caused a partial blockade of the bradykinin-gold binding in 1.75 old cultures (Fig. 3B). At this point in time 52 ± 4 % of the neurons showed a bradykinin-binding. Blocking the B_2 receptor only 27 ± 7 % remained positive. A combined

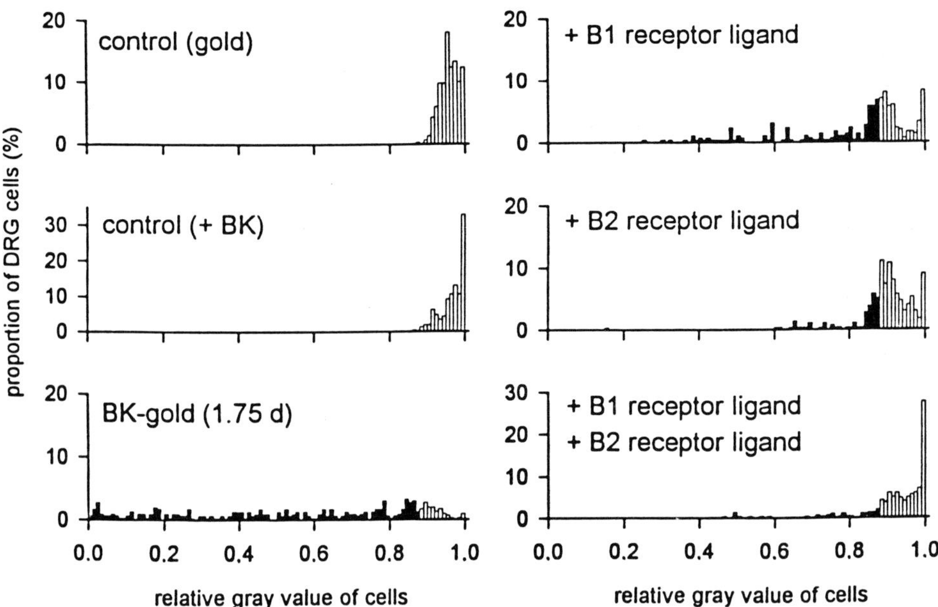

Fig. 2. Distribution of the relative gray values (mean gray value of cells / mean gray value of the background) of dorsal root ganglion neurons after incubation with bradykinin-gold. Each diagram shows the pooled data of 3 cultures and 300 neurons. The discrimination between cells with no binding (white bars) and cells with a bradykinin-binding (black bars) is based on the distribution of the relative gray values of cells examined in control incubations. The presence of B_1 and B_2 receptors is determined by blocking the binding sites by a B_1 and/or B_2 receptor ligand.

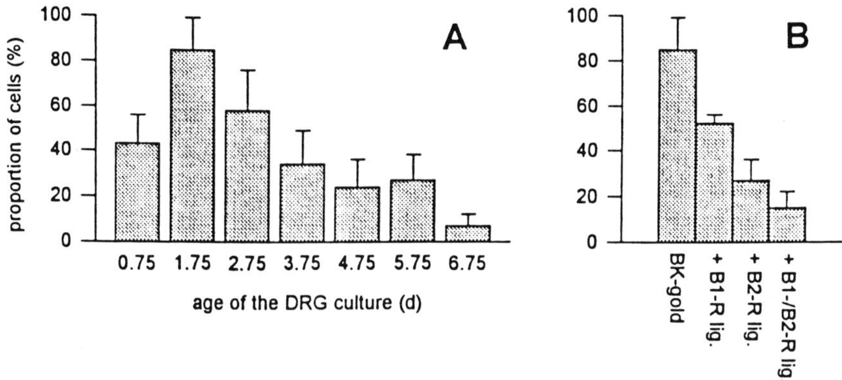

Figure 3. A: Proportion of neurons with a positive bradykinin-binding on dorsal root ganglion cells of different times under culture conditions. **B:** Proportion of positive cells in 1.75 days old cultures after the addition of a B_1 and/or B_2 receptor ligand. All values are expressed as the mean ± standard derivation (n=3 cultures and 300 neurons).

incubation of bradykinin-gold together with both bradykinin analogues further reduced the proportion of positive cells to 15 ± 7 %.

Discussion

The present study reveals that the proportion of rat dorsal root ganglion cells in culture that bind bradykinin markedly depends on their time in culture. It is tempting to assume that the increase at shorter incubation times is due to the injury sustained by the disruption of the axon during isolation of the somata. Interestingly, this observation seems to be independent of outgrowth of processes. The factors that induced the decline in bradykinin-binding in older cultures are unknown.

In 1.75-day-old cultures bradykinin predominantly bound via the B_2 receptor. However, a substantial proportion of B_1 receptors was also present. Whether the small proportion of bradykinin-binding cells that are still visible after the addition of both receptor ligands may indicate the presence of other bradykinin binding sites is unknown. Using the cobalt uptake technique an *in vitro* study revealed that in sections of rat dorsal root ganglion a bradykinin-induced uptake was evoked by B_2 receptors only. *In vivo* studies showed that under pathological conditions like inflammation, B_1 receptors contribute to the developed hyperalgesia (see e.g. Davis and Perkins, 1994).

Conclusion

This study shows that Nanogold-labelled peptides can be used for quantitative studies of the distribution of peptide receptors in cultured dorsal root ganglion cells. Using this method, we showed on the cellular level that sensory neurons express the B_1 receptor subtype in addition to the B_2 one. As the distribution of bradykinin receptors markedly varies with the time in culture, it seems to be necessary to take into account that the age of the culture may influence the results of *in vitro* studies.

Acknowledgements
This work has been supported by the Deutsche Forschungsgemeinschaft (PE 299/3-1 and HE 1919/2-2).

References

Davis, A.J. and Perkins, M.N. (1994) Induction of B_1 receptors in vivo in a model of persistent inflammatory mechanical hyperalgesia in the rat. *Neuropharmacology* 33: 127-133.

Kanaka, R., Schaible, H.-G. and Schmidt, R.F. (1985) Activation of fine articular afferent units by bradykinin. *Brain Res.* 327: 81-90.

McGuirk, S.M. and Dolphin, A.C. (1992) G-protein mediation in nociceptive signal transduction: an investigation into the exitatory action of bradykinin in a subpopulation of cultured rat sensory neurons. *Neuroscience* 49: 117-128.

Segond von Banchet, G. and Heppelmann, B. (1995) Non-radioactive localization of substance P binding sites in rat brain and spinal cord using peptides labelled with 1.4-nm gold particles. *J. Histochem. Cytochem.* 43: 821-827.

IV. Comparative aspects

The invertebrate neurosecretory cell: state of the art

J. W. Truman, J. Ewer, S. Gammie and S. McNabb

Department of Zoology, University of Washington, Box 351800, Seattle, WA 98195, USA

Summary. The study of neurosecretion in invertebrates is not only particularly stimulating and effective because neuropeptides are among the phylogenetically oldest messenger substances yielding results are of general validity. Moreover, the often rather large neurosecretory cells can be unequivocally tagged by modern techniques and their developmental fate observed during the animal's life cycle. Genetic manipulation and single cell recording in invertebrate neurosecretory cells offer insights in interaction and regulation of these cells. The present review is focused on insect neurosecretory processes, particularly on their genetic regulation and hormonal influences of excitability.

Introduction

Invertebrates have played a seminal role in the development of the field of neurosecretion. The extirpation experiments by Kopec (1917, 1922) on pupae of the gypsy moth provided the first indication that the brain could regulate physiology through a pathway that involve the blood. Later experiments by Wigglesworth (1940) on the bug *Rhodnius* and by Berta Scharrer (e.g. Scharrer and Scharrer, 1944) on the cockroach *Leucophaea maderae* firmly established the secretory function of the brain and the transport and release of material from specialized neurohemal sites. Such studies laid the foundation that underlie current concepts of the neurosecretory neuron.

As in other areas of neurobiology, the study of neurosecretion in invertebrates has benefitted from the fact that many invertebrate neurons are large cells that are identifiable as unique individuals. Consequently it is possible to study the same cell in successive animals and how it "behaves" under different physiological and developmental conditions. This identified neuron approach continues to be a key to the study of neurosecretion in invertebrates. It has become especially powerful when combined with two areas that have shown explosive growth in recent years. One is the molecular revolution that is revealing the details of the genetic organization of organisms. Also, molecular biology is providing techniques to experimentally manipulate these genetic programs. The second area involves new methods in cell biology such as patch clamp recording and various types of imaging techniques which provide detailed pictures of how excitable cells interact with their environment.

This brief essay will focus on the invertebrate neuron in the context of these cellular and molecular issues. Because of space limitations and the inclinations of the authors our comments will focus primarily on insect neurosecretory systems.

Molecular advances in invertebrate neurosecretion

The advent of molecular studies of invertebrate neuropeptides lagged behind that in the vertebrates. A major reason for this lag was the paucity of information, until about 10 years ago, on the primary structure of invertebrate neuropeptides. The primary sequence of the egg-laying hormone in the mollusc *Aplysia*, provided the avenue for the isolation of the first neuropeptide gene from an invertebrate (Scheller et al., 1983). This work, along with that for a number of vertebrate neuropeptide genes (e.g. Herbert and Uhler, 1982) showed that neuropeptides were cleaved from larger precursors that often encoded a number of other biologically active peptides. The molecular approach has resulted in an explosion in the known number of potentially bioactive sequences in both molluscs and arthropods. Physiologists are now faced with the challenge of unraveling the physiological properties of this myriad of peptides products and how each peptide contributes to complex physiological and developmental responses. In some of the earliest examples, such as the pro-opiomelanocortin gene in mammals (Herbert et al., 1981) or the egg-laying hormone precursor in *Aplysia* (Scheller et al., 1983) the prohormone is cut into a number of dissimilar peptides. Different actions could be ascribed to the different peptide products, and components of the peptide blend then worked together to orchestrate a coordinated physiological response (e.g. Mayeri and Rothman, 1985). More perplexing are the precursors that encode families of very similar peptides such as the extended FRMFamides in *Drosophila* (Nambu et al., 1988; Schneider and Taghert, 1988) or the allatostatins in the cockroach *Diploptera* (Donly et al., 1993). Does each variant have its own unique action or are we simply looking at the evolutionary drift of a less constrained portion of the molecule? Probably the answer lies between the two extremes.

Molecular and immunological techniques have also shown that secreted peptides are evident in more that just the classic neurosecretory neurons. A given peptide may be found in a bewildering array of cells. How then does this peptide distribution relate to the overall functioning of the nervous system? An elegant example of using molecular approaches to unravel the expression pattern of a particular neuropeptide gene is provided by the studies by Paul Taghert and colleagues on the expression of the FMRFamide gene in *Drosophila* (O'Brien et al., 1991; Schneider et al., 1993). This gene is expressed by about 120 neurons comprising about 15 discrete cell types in the central nervous system of *Drosophila* in a complex spatial and temporal pattern. Using *lacZ* reporter gene constructs and germline transformation, Schneider et al. (1993) analysed the promoter region of the *FMRFamide* gene. An 8 kb DNA fragment including the first intron and 5' upstream sequences of the FMRFamide gene was sufficient to drive *lacZ* expression in a pattern similar to that shown by the native FMRFamide gene. Subsequent analysis of this 8 kb piece by deleting regions and by using them as enhancers for heterologous promoters then identified specific portions that were needed to direct FMRFamide expression in particular subsets of neurons. Consequently, the complex pattern of FMRFamide expression appears to result "from the activity of discrete, cell type-specific enhancers that are independently regulated" (Schneider et al., 1993).

This type of promoter/enhancer analysis, though, has greater research value than simply describing the DNA "address" that targets neuropeptide expression to a given cell. For example, although the FMRFamide system in *Drosophila* lacks a mutant that is deficient in FMRFamide expression, once such a mutant is isolated, the techniques developed by Schneider et al. (1993) will provide the means to restore FMRFamide expression either to the whole group of neurons or to selected neuronal sets within the group. By determining which neurons rescue which aspects of a mutant phenotype, it may then be possible to relate the roles of FMRFamide to the diverse sets of cells that express it. Moreover, the FMRFamide gene encodes for a number of different members of an N-terminally extended family as well as an unrelated peptide (Schneider and Taghert, 1988). Rather than using a wild-type copy of the gene, the flies devoid of FMRFamide could be transformed with an altered structural gene that lacks one or more of the peptides that normally occur on the precursor. Therefore one can determine if any member of the extended family can restore a particular function or whether a blend of peptides is actually required for normal function. This molecular genetic approach in *Drosophila* provides an important avenue to examine neuropeptide systems that are complex in both their spatial distribution and in the types of products that they produce.

A recombinant DNA approach has benefits for simple neurosecretory systems as well as complex ones. The eclosion hormone (EH) system in insects is the antithesis of FMRFamide. In *Drosophila*, eclosion hormone transcripts and immunostaining are found in a single pair of ventromedial neurons (other insects such as the moths have 2 pairs of identical ventromedial neurons). In moths, the eclosion hormone precursor contains only a single peptide product, eclosion hormone (Horodyski et al., 1989), although in *Drosophila* an apparent insertion event has resulted in a longer precursor that includes a potential 9 amino acid peptide in addition to eclosion hormone (Horodyski et al., 1993). Analysis of the promoter region of eclosion hormone in *Drosophila* is in its early stages. A piece of DNA extending 830 bases upstream of the transcription start site restricts the expression of a reporter gene to only the single pair of ventromedial neurons (McNabb, Riddiford and Truman, unpublished). Thus, this region has all the information needed to direct the spatial expression of the eclosion hormone gene.

Genetic constructs that direct gene expression to specific neurosecretory cells such as the eclosion hormone cells or subsets of the FMRFamide cells provide the means for the selective modification of specific neurosecretory cells. The reporter gene of choice for such as approach is the one that encodes the yeast transcription factor GAL4 (Brand and Perrimon, 1993). The GAL4 protein binds to a unique DNA sequence, the upstream activating sequence (UAS) that is present in yeast but not found in flies. Therefore, although GAL4 might be expressed in specific cells in transformed flies, there is no target for it to activate transcription. The power of the approach is that these flies carrying GAL4 driven by a selective promotor can then be crossed to flies which have been transformed with a desired gene that is downstream of a UAS control sequence (Fig. 1). The cross brings together in the F1 progeny the GAL4 protein and the UAS regulated target in the same cell. For example, by crossing the eclosion hormone-promoter/GAL4 line to flies that have a UAS sequence upstream of *reaper*, a gene that encodes a protein that induces programmed cell death (White et al., 1994), one would direct expression of *reaper*, and, hence, cell death, selectively in the ventromedial cells of the F1 progeny. Besides this precise spatial resolution, the binding by GAL4 is apparently temperature sensitive so the timing of gene activation can be regulated by shifting flies at the desired time from 18 to 25ºC (Staeling-Hampton et al., 1994).

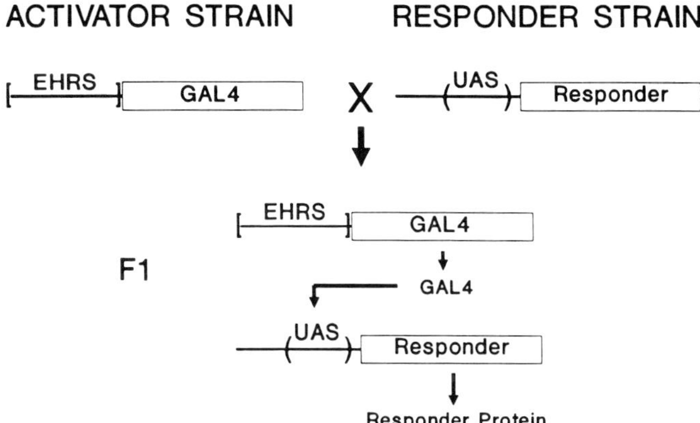

Fig. 1. The scheme for using the GAL4 system in *Drosophila* to express genes selectively in the ventromedial neurons, that produce eclosion hormone. The "activator" strain is transformed with the gene for the yeast GAL4 transcription factor under the control of eclosion hormone regulatory sequences (eclosion hormoneRS). This strain is crossed to a "responder" strain that carries a responder gene (e.g. one that encodes a protein that causes cell death) that is downstream of a UAS regulatory sequence (the binding site for the GAL4 transcription factor). In the F1 generation, the responder gene is then expressed only in the neurons that normally express the eclosion hormone gene.

Invertebrates, with their relatively simple nervous systems provide useful model systems for studying factors that regulate the activity of peptidergic cells. Peptidergic neurons often require sustained excitation in order to release significant amounts of their peptide. This problem of maintained excitability is especially intriguing in the case of neurons that are only episodically active; cells that may go through long periods of inactivity punctuated by a bout of intense secretory activity. What are the mechanisms by which these cells are roused out of inactivity to release massive quantities of their product? How do these cells maintain this activated state? Is it only from synaptic drive or is it due to physiological adjustments within the neurosecretory cells themselves?

The neuroendocrine cascade that regulates ecdysis in insects is one of the most complex cascades found in invertebrates and it involves the abrupt activation of networks of neurons that are typically quiescent. The activation of this neuroendocrine cascade is coordinated with the molt cycle, specifically through the action of the steroid hormones, the ecdysteroids (Truman, 1992). The ventromedial cells, with their release of eclosion hormone, stand at the beginning of this neuroendocrine cascade. These cells accumulate peptide throughout the intermolt and molt period and then release 90 to 95% of their stored product over a span of about 15-20 min (Riddiford et al., 1994). Prior to these release episodes, the ventromedial cells are relatively inexcitable, with a threshold 40-45 mV above the resting potential. With the approach of the normal time of secretion, however, the threshold of the ventromedial cells rapidly drops and the cells show repetitive firing in response to moderate depolarization (Hewes and Truman, 1994). The changes in threshold occur without alteration in resting potential or input resistance.

Studies on larvae of the moth, *Manduca sexta*, show that the excitability of the ventromedial cells is regulated by the titer of ecdysteroids that bring about molting. The increased excitability of the cells requires the presence of ecdysteroids followed by their withdrawal. If the steroid titers are maintained at a high level, the cells remain refractory but the withdrawal of the steroid is then followed by the appearance of excitability within about 6-8 h (Hewes and Truman, 1994). This delayed onset of excitability is blocked by inhibiting RNA synthesis with actinomycin D (Hewes and Truman, 1994). Hence, the changes in excitability appears to be mediated through a classical genomic action of the steroid. Since these analyses were done on cells that were treated *in vivo* it is not known whether this represents the action of steroid directly on the ventromedial cells themselves or on other cells that then act on the ventromedial cells to regulate their excitability.

Whole-cell patch clamp analysis of the ventromedial cells in *Manduca* revealed only 3 currents in the soma: a voltage-sensitive Ca^{2+} current (I_{Ca}), a calcium sensitive K^+ current ($I_{K(Ca)}$), and an A-type K^+ current (I_A). The change in excitability of the ventromedial cells appears to be due to the enhancement of I_{Ca} (Hewes, 1993). How the genomic actions of ecdysteroids result in the change in I_{Ca} is not known.

Indications of how eclosion hormone relates to the second tier of neuropeptide release has come from studies of the neurons that release crustacean cardioactive peptide (CCAP). This neuropeptide was originally described from crustaceans but subsequent chemical and immunological work showed that it is also widely distributed throughout the insects (Dircksen, 1994). The neurons that contain CCAP are distributed primarily through the ventral nervous system and include paired type 1 and type 2 cells found in each segmental ganglion. CCAP is one of the cardioacceleratory peptides that are released around the time of ecdysis in *Manduca* (Cheung et al., 1992).

A striking feature of the CCAP cells is their pronounced accumulation of intracellular cyclic 3',5' guanosine monophosphate (cGMP) at the time of ecdysis (Fig. 2). This was first observed for the ecdyses of the moth, *Manduca sexta* (Ewer et al., 1994), but it has since been documented for the CCAP cells of ecdysing insects ranging from the primitive silverfish through the advanced groups such as beetles, Lepidoptera and lower flies (Ewer and Truman, 1996). Presumably, CCAP acts at ecdysis in these insects to cause the circulatory changes that move hemolymph anteriorly to aid in rupturing the old cuticle and expansion of the new one.

Early treatment of molting *Manduca* with eclosion hormone brings about the precocious cGMP accumulation in the CCAP cells in association with an early ecdysis (Ewer et al., 1994). The long latency (25-30 min) between the time of eclosion hormone injection and the cGMP increase made it seem unlikely that the response in these cells is directly to eclosion hormone. However, recent experiments show that the injection of the hormone eventually results in the release of endogenous stores of eclosion hormone within the central nervous system (Truman and Ewer, unpublished). The timing and location of this central release is consistent with it being the trigger for the cGMP increase.

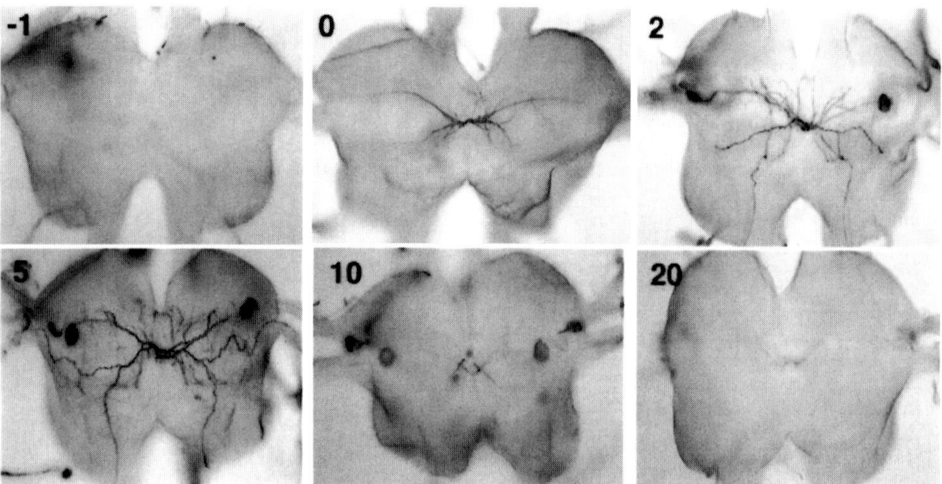

Fig. 2. Photomicrographs showing the appearance of cGMP immunoreactivity in the CCAP neurons in the 1st thoracic ganglion of the grasshopper, *Locusta migratoria*. The times are in minutes relative to the onset of ecdysis behaviour. From Truman et al., 1996.

Other studies on *Manduca* (Ewer et al., 1994) as well as recent ones on the ecdysis of newly hatched locusts, *Locusta migratoria* (Truman et al., 1996), indicate that once it is given, the response of the CCAP cells to the descending command is relatively rapid. Hatching locusts are covered by an embryonic cuticle as they dig through the soil but then shed it as soon as they arrive at the surface. Digging larvae show no detectable cGMP in the CCAP cells, but within 60 sec of arriving at the surface, they initiate ecdysis behaviour and their CCAP cells have acquired moderate levels of cGMP. Unfortunately, in the locust the timing of eclosion hormone release is not known. Likely, it occurs earlier to initiate the hatching/digging behaviour (the equivalent to pre-ecdysis behaviour in moths [Truman, 1992]), but then stimuli associated with being free from the soil switches the animal into the next phase that involves both the activation of the CCAP network and the onset of the ecdysis motor program.

The cGMP increase in the CCAP cells appears to be due to the activation of guanylate cyclase, rather than the suppression of cGMP metabolism by the inhibition of phosphodiesterase. This conclusion is based on the failure of inhibitors of phosphodiesterase to induce detectable cGMP in these cells. The guanylate cyclase that is being activated, though, is likely not a soluble guanylate cyclase. Treatment with activators of soluble guanylate cyclase, such as the nitric oxide donor, sodium nitroprusside, causes dramatic increases in cGMP in some neurons in the ventral part of the central nervous system but not in the CCAP cells (Ewer et al., 1994).

In most instances, the activation of second messengers is relatively brief and confined to the site of stimulation. Prolonged cellular responses typically result from downstream events such as protein phosphorylation. The cGMP changes that occur in the CCAP cells are unusual in both their spatial and temporal characteristics. In some of the cells in *Manduca*, for example, the cGMP increase persists for over 6 h. While a descending signal is necessary to initiate this cGMP increase, it is not necessary to maintain it. Indeed, when the nerve cord is transected after the cGMP increase begins, it is then maintained in the cells on either side of the cut for at least an hour (Ewer et al., 1994). The exact time-course, though, has yet to be determined for the posterior section.

Many of the CCAP neurons that are involved in this response are projection neurons that have long axons that extend the posterior length of the ventral nervous system or out into the periphery. Despite the fact that the cells are apparently activated by synaptic events, cGMP accumulation occurs rapidly throughout the cell including a few centimeters of axon. The mechanism by which a local signal results in a cGMP response that involves the entire extent of the cell is not known.

What are the effects of the cGMP on the physiology of the CCAP cells? In *Manduca* an increase in heart rate and the release of cardioactive peptides is associated with ecdysis (Tublitz and Truman, 1985). Likewise in ecdysing locusts, changes in blood pressure (Bernays, 1971) mirror the appearance and loss of cGMP in the CCAP neurons. Consequently, the cGMP increase is likely to be associated with some aspect of CCAP release. Intracellular recording from these cells prior to larval ecdysis in *Manduca* show that like the eclosion hormone cells, the CCAP cells also have a very high spike threshold during the intermolt and most of the molt period. Simultaneous cGMP immunocytochemistry and electrophysiological recording through the time of eclosion hormone action and ecdysis showed an abrupt drop in the threshold of these cells correlated with the increase in intracellular cGMP (Gammie et al., 1994). This increase in excitability is rapidly and prematurely induced by bathing the neurons with a membrane permeant form of cGMP, 8-bromo-cGMP. The corresponding analogue of cAMP, by contrast, does not lower the threshold but rather raises it slightly (Gammie and Truman, unpublished). Whether the increase in excitability caused by the increase in cGMP is in itself sufficient to result in the tonic firing of the cells or renders the cell responsive to synaptic inputs has not yet been determined.

An intriguing problem is the mechanism by which cGMP affects the excitability of the CCAP cells. Results from ion substitution experiments suggest that the effects of the cGMP may be on a Ca^{2+} current (Gammie and Truman, unpublished). This would be consistent with the effects of cGMP on the excitability of molluscan neurons (Paupardin-Tritsch et al., 1986). Is the effect occuring, though, through activation of cGMP-dependent protein kinase as in molluscs (Paupardin-Tritsch et al., 1986) or by direct interaction of the cGMP with channels in the cell membrane (e.g. Yau and Baylor, 1989)? The latter option is especially attractive because it would explain the high and sustained levels of cGMP that are maintained throughout the cell.

Maintaining sustained high levels of cGMP is a novel mechanism by which neurons can alter their excitability. Besides the CCAP cells, we see other neurons that also show this phenomenon. In most cases, these cells are also peptidergic neurons, but they display their cGMP increases in physiological and behavioural contexts other than ecdysis. Thus, this mechanism may be widely used amongst the peptidergic cells and it may a new way of identifying novel neurons that become activated in different physiological or behavioural contexts.

Acknowledgements
Unpublished results were obtained on grants from the National Science Foundation to J.W.T. and L.M. Riddiford.

References

Brand, A.H. and Perrimon, N. (1993) Targeted gene expression as a means of altering cell fates and generating dominant phenotypes. *Development* 118: 401-415.

Bernays, E.A. (1971) The intermediate moult (first ecdysis) of *Schistocerca gregaria* (Forskal) (Insecta, Orthoptera). *Z. Morph. Tiere* 71:160-179.

Cheung, C.C., Loi, P.K., Sylwester, A.W., Lee, T.D. and Tublitz, N.J. (1992) Primary structure of a cardioactive neuropeptide from the tobacco hornworm, *Manduca sexta. FEBS Lett.* 313:165-168.

Dircksen, H. (1994) Distribution and physiology of crustacean cardioactive peptide in arthropods. *In*: K.G. Davey, R.E. Peter, and S.S. Tobe (eds.): *Perspectives in Comparative Endocrinology*, National Research Council of Canada, Ottawa, pp. 139-148.

Donly, B.C., Ding, C., Tobe, S.S. and Bendena, W.G. (1993) Molecular cloning of the gene for the allatostatin family of neuropeptides from the cockroach *Diploptera punctata. Proc. Natl. Acad. Sci. USA* 90: 8807-8811.

Ewer, J., de Vente, J., and Truman J.W. (1994). Neuropeptide induction of cyclic GMP increases in the insect CNS: resolution at the level of single identifiable neurons. *J. Neurosci.* 14: 7704-7712.

Ewer, J. and Truman, J.W. (1996) Increases in cyclic GMP occur at ecdysis in an evolutionarily conserved insect neuronal network. *J. Comp. Neurol.; submitted.*

Gammie, S.C., Ewer, J. and Truman, J.W. (1994) An endogenous increase in cGMP is associated with increased excitability in identified neurosecretory cells in *Manduca sexta. Soc. Neurosci. Abstr.* 24:315.

Kopec, S. (1917) Experiments on metamorphosis of insects. *Bull. Int. Acad. Cracov* B, pp. 57-60.

Kopec, S. (1922) Studies on the necessity of the brain for the inception of insect metamorphosis. *Biol. Bull.* 142: 323-342.

Herbert, E. and Uhler, M. (1982) Biosynthesis of polyprotein precursors to regulatory peptides. *Cell* 30:1-2.

Herbert, E., Birnberg, N., Lissitsky, J.-C., Civelli, O. and Uhler, M. (1981) Pro-opiomelanocortin: a model for the regulation of expression of neuropeptides in pituitary and brain. *Neurosci. Comm.* 1: 16-27.

Hewes, R.S. (1993) *The Regulation and Distribution of Eclosion Hormone Release in the Tobacco Hornworm, Manduca sexta.* Doctoral Dissertation, University of Washington, Seattle, pp. 155.

Hewes, R.S. and Truman, J.W. (1994) Steroid regulation of excitability in identified insect neurosecretory cells. *J. Neurosci.* 14: 1812-1819.

Horodyski, F.M., Riddiford, L.M. and Truman, J.W. (1989) Isolation and expression of the eclosion hormone gene from the tobacco hornworm, Manduca sexta. *Proc. Natl. Acad. Sci. USA* 86: 8123-8127.

Horodyski, F.M., Ewer, J., Riddiford, L.M. and Truman, J.W. (1993) Isolation, characterization, and expression of the eclosion hormone gene of *Drosophila melanogaster. Eur. J. Biochem.* 215: 221-228.

Mayeri, E. and Rothman, B.S. (1985) Neuropeptides and the control of egg-laying behaviour in *Aplysia. In*: A.I. Selverston (ed.): *Model Neural Networks and Behaviour,* Plenum, New York, pp. 285-301.

O'Brien, M.A., Schneider, L.E. and Taghert, P.H. (1991) In situ hybridization analysis of the *FMRFamide* neuropeptide gene in *Drosophila*. II. Constancy in cellular pattern of expression during metamorphosis. *J. Comp. Neurol.* 304: 623-638.

Paupardin-Tritsch, D., Hammond, C., Gerschenfeld, H.M., Nairn, A.C. and Greengard, P. (1986) cGMP-dependent protein kinase enhances Ca^{2+} current increase in snail neurons. *Nature* 323:812-814.

Riddiford, L.M., Hewes, R.S. and Truman, J.W. (1994) Dynamics and metamorphosis of an identifiable peptidergic neuron in an insect. *J. Neurobiol.* 25: 819-830.

Scharrer, B. and Scharrer, E. (1944) Neurosecretion. VI. A comparison between the intercerebralis-cardiacum-allatum system of the insects and the hypothalamo-hypophyseal system of the vertebrates. *Biol. Bull.* 87: 242-251.

Scheller, R.H., Jackson, J.F., McAllister, L.B., Rothman, B.S., Mayeri, E. and Axel, R. (1983) A single gene encodes multiple neuropeptides mediating a stereotyped behaviour. *Cell* 32: 7-22.

Schneider, L.E. and Taghert, P.H. (1988) Isolation and characterization of a *Drosophila* gene that encodes multiple neuropeptides related to Phe-Met-Arg-Phe-NH$_2$ (FMRFamide). *Proc. Natl. Acad. Sci. USA* 85: 1993-1997.

Schneider, L.E., Roberts, M.S. and Taghert, P.H. (1993) Cell type-specific transcriptional regulation of the Drosophila *FMRFamide* neuropeptide gene. *Neuron* 10: 279-291.

Staehling-Hampton, K., Hoffmann, F.M., Baylles, M.K., Rushton, E. and Bate, M. (1994) dpp induces mesodermal gene expression in *Drosophila. Nature* 372: 783-786.

Truman, J.W. (1992) The eclosion hormone system of insects. *Progress Brain Res.* 92: 361-374.

Truman. J.W., Ewer, J. and Ball, E.E. (1996) Dynamics of cyclic GMP changes in identified neurones during ecdysis behaviour in the locust, *Locusta migratoria. J. Exp. Biol.; in press.*

Tublitz, N.J. and Truman, J.W. (1985) Insect cardioactive peptides: II. Neurohormonal control of heart activity by two cardioacceleratory peptides (CAPs) in the tobacco hawkmoth, *Manduca sexta. J. Exp. Biol.* 114: 381-395.

White, K., Grether, M.E., Abrams, J.M., Young, L., Farrell, K. and Steller, H. (1994) Genetic control of programmed cell death in *Drosophila. Science* 264: 677-683.

Wigglesworth, V.B. (1940) The determination of characters at metamorphosis in *Rhodnius prolixus* (Hemiptera). *J. Exp. Biol.* 17: 201-222.

Yau, K.-W. and Baylor, D.A. (1989) Cyclic GMP-actiavted conductance of retinal photoreceptor cells. *Ann. Rev. Neurosci.* 12:289-327.

The Peptidergic Neuron
B. Krisch and R. Mentlein (eds)
© 1996 Birkhäuser Verlag Basel/Switzerland

Neurotrophin-like immunoreactivity in the nervous system of the earthworm *Eisenia foetida* (Annelida, Oligochaeta)

C. Davoli, A. Serafino, A. Marcheggiano[1], C. Iannoni[1], A. Marconi and G. Ravagnan

Institute of Experimental Medicine, CNR, Viale Marx 15-43, I-00137, Rome, Italy
[1]*Department of Gastroenterology I, "La Sapienza" University, Rome, Italy*

Summary. Nerve growth factor (NGF) is the best characterized neurotrophic factor in a family of structurally and immunologically related proteins (neurotrophins) identified in vertebrates. In the present study, using rabbit antisera and affinity purified antibodies to mouse 2.5S NGF, neurotrophin-like immunoreactive neurons were detected in *Eisenia foetida* (Annelida, Oligochaeta) earthworms by light and electron microscopy. By light immunocytochemistry NGF-like immunoreactivity was detected in many neuronal cell bodies and some nerve fibres in both the cerebral ganglion and the ventral nerve cord of the earthworms. In particular, numerous positive neurons were located in the posterodorsal zone of the brain and in the cephalic portions of the circumoesophageal connectives. Moreover, some positive cells and immunoreactive fibres were observed in segmental ganglia of the ventral nerve cord. The presence of NGF-related peptides in *Eisenia foetida* nervous system was confirmed by ultrastructural studies showing NGF-like immunoreactivity localized in numerous secretory granules within the perikarya of positive neurons. In the cerebral ganglion, release via omega figures of immunoreactive granules was detected in the extracellular space adjacent to perikaryal profiles, suggesting a possible autocrine and/or paracrine role of the molecules immunologically related to NGF in earthworms. A similar pattern of distribution of NGF-like immunoreactivity in earthworm tissues was observed using two anti-NGF sera displaying different specificities but recognising all neurotrophins on Western blots. The occurrence of molecules immunologically related to at least one member of the neurotrophin family in neurosecretory cells of earthworms supports the hypothesis that NGF-like proteins first evolved in the invertebrates.

Introduction

The neurotrophins are a family of structurally related proteins, recently identified in vertebrates, displaying distinct as well as overlapping neuronal specificities. Nerve growth factor (NGF), the first discovered and the best characterised neurotrophic factor, represents the prototypical member of this family that now includes brain-derived neurotrophic factor (BDNF), neurotrophin-3 (NT-3) and NT-4 (Levi-Montalcini, 1987; Thoenen, 1991; Lindsay et al., 1994). All family members share 50-60% amino acid identity, retaining six cysteine residues involved in the intrachain disulfide bonds in NGF molecule, interact with the same low-affinity NGF receptors, and probably have similar tertiary structure (Bradshaw et al., 1993; Narhi et al., 1993). In addition to being structurally similar, these proteins are also immunologically related since antibodies to mouse NGF have been found to react with other members of the neurotrophin family (Acheson et al., 1991; Murphy et al., 1993).

Recent studies in *Drosophila* and *Lymnaea* indicate that NGF-like neurotrophic factors may be also present in invertebrates (Ridgway et al., 1991; Hayashi et al., 1992). Moreover, the evolutionary conservation of the NGF-like molecules in primitive vertebrates suggests that these factors first evolved in invertebrates (Hallböök et al., 1991; Ebendal, 1992). In the present study, using rabbit antisera and affinity purified antibodies to mouse 2.5S NGF, neurotrophin-like immunoreactive neurons have been detected in *Eisenia foetida* earthworms by light and electron microscopy.

Materials and methods

Sexually mature earthworms (*Eisenia foetida*) kept under controlled laboratory conditons were used in these studies. Two antisera raised in rabbits against mouse submandibular gland 2.5S NGF (mNGF) (Angeletti et al., 1973) and antibodies purified by affinity chromatography were provided by Dr. D. Mercanti (Institute of Neurobiology, CNR, Rome); a commercial rabbit anti-mNGF 2.5S serum was purchased from Sigma Chem. Co., USA. The specificity of all antisera has been previously determined by neutralization assays of NGF biological activity and by Western blot analysis.

Immunohistochemical studies were performed by the indirect immunofluorescence technique using anti-mNGF sera on serial sections from Bouin-fixed, paraffin-embedded, cephalic and body segments of *Eisenia foetida* earthworms. Mouse submandibular glands were used as a positive control. Antigen-preabsorbed antiserum (for the first anti-mNGF serum), pre-immune serum (for the second) and normal rabbit serum (for the commercial), at the same working dilution as the specific antisera (1:500 for 24 h at 4°C incubation), were used as a negative control. Tissue sections were examined on a Leitz Orthoplan microscope equipped for epifluorescence.

For electron microscopy, dissected cerebral ganglia and ventral nerve cords were fixed in 4% paraformaldehyde and 0.05% glutaraldehyde in Millong's phosphate buffer, and embedded in LR White acryl resin (The London Resin Co., England). Ultrastructural immunocytochemical studies were performed on ultra-thin sections by the indirect immunogold technique using anti-NGF sera and purified antibodies. The working dilutions of primary antisera, as well as of pre-immune and normal rabbit sera, were 1:2000 (24 h at 4°C). Gold (10 nm) conjugates F(ab')$_2$ goat anti-rabbit IgG (BioCell Res. Lab., England) were used at 1:20 dilutions (1 h at room temperature). The sections counterstained with uranyl acetate and lead citrate were observed on a Philips CM12 transmission electron microscope at 100 kV.

The reactivity of anti-mNGF sera to human recombinant (hr) β-NGF, hrBDNF, hrNT-3, hrNT-4 (PeproTech. Inc., USA) and, as a control, to mNGF 2.5S (Sigma) was tested on Western blots (15% SDS-PAGE).

Results

In immunohistochemical studies, by the use of different rabbit anti-mNGF sera, intensely fluorescent cells and fibres were detected in the central nervous system of *Eisenia foetida* earthworms (Fig. 1). In particular, numerous positive perikarya and some immunoreactive nerve fibres were observed in the cerebral ganglion. These cells were mainly located in the cortico-dorsal zones of the lateral portions of the brain and displayed positive axons which projected toward the neuropile (Fig. 2). Positive plexiform nerve fibres and several immunostained nerve cells were also observed in the cephalic portion of the circumoesophageal connectives. In the ventral nerve cord a few positive perikarya were located at the roots of the segmental nerves and scattered along ganglia where filiform immunoreactive fibres were also detected. This pattern of staining was obtained with all antisera whereas no staining was obtained using either antigen-preabsorbed anti-mNGF serum or pre-immune and normal rabbit sera.

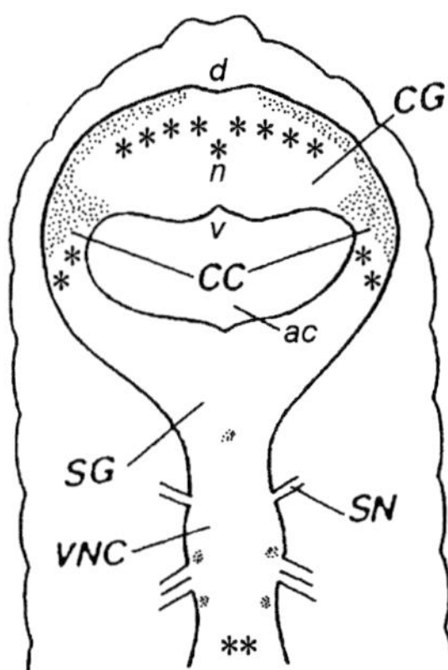

Fig. 1. Schematic drawing showing the localization of the NGF-like immunoreactive cells (dotted) and nerve fibres (asterisks) in the central nervous system of the earthworm. CC, circumoesophageal connectives; CG, cerebral ganglion; SG, suboesophageal ganglion; SN, segmental nerve; VNC, ventral nerve cord; ac, alimentary canal; d, dorsal zone; n, neuropile; v, ventral zone.

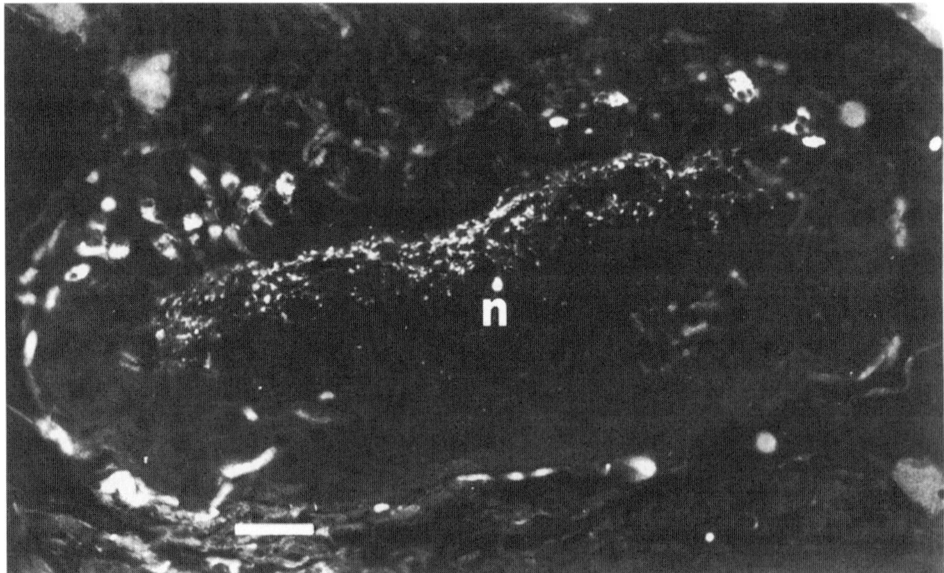

Fig. 2. Immunofluorescent cells and cross-sectioned nerve fibres in midportion of cerebral ganglion, transversal section. n, neuropile. x200; bar = 64 µm.

Ultrastructural studies confirmed the presence of NGF-like immunoreactivity in *Eisenia foetida* nervous system. Using both anti-mNGF sera and purified antibodies, immunogold labelling was detected in numerous secretory granules within the perikarya of positive neurons (Fig.3). In the cerebral ganglion, gold-labelled electron-dense material associated to the omega figure indicating exocytosis was occasionally observed in the extracellular space adjacent to perikaryal profiles.

In immunoblot analyses, essentially the same distribution of electrophoretic bands was detected by anti-mNGF sera from mNGF and hrNGF samples. Moreover, the two polyclonal antisera – although showing a similar pattern of distribution of NGF-like immunoreactivity in earthworm tissues –, displayed different reactive properties, but recognized all neurotrophin molecules on Western blots.

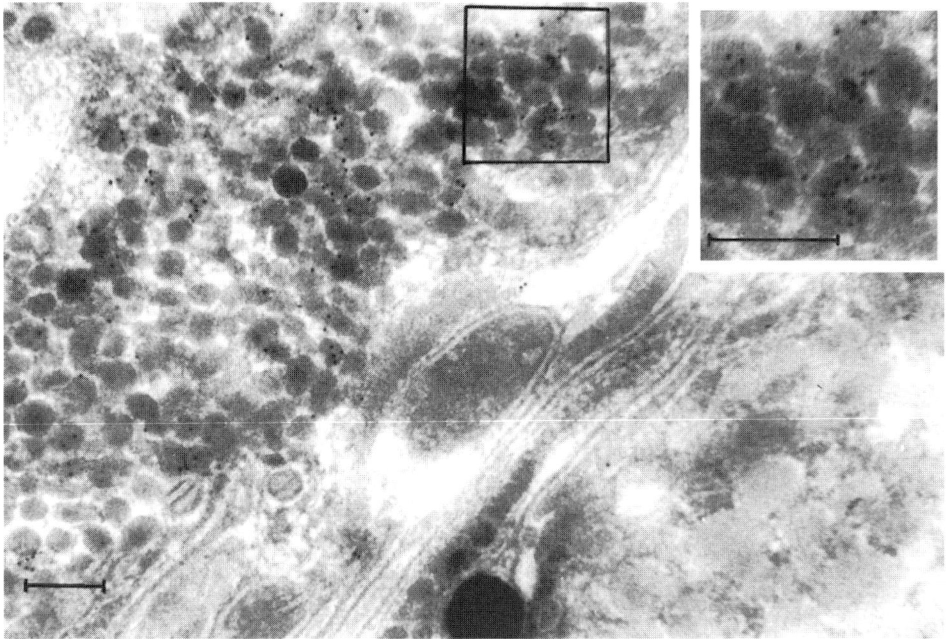

Fig. 3. Electron micrograph of two paired neurosecretory cells in the cerebral ganglion: the positive cell shows immunostained secretory granules. x66,000; bar = 200 nm. Inset: a detail at higher magnification x110,000; bar = 200 nm.

Discussion

Neurosecretory cells have long been known in annelids and ultrastructural aspects of neurosecretion have been carefully investigated in earthworms (Golding and Pow, 1988). In the central nervous system of these animals, exocytotic release of neurosecretory granules is a widespread phenomenon. In particular, in the cerebral ganglion, neurosecretory granules have been found to be released by exocytosis from cell bodies as well as from axon terminals (Al-Yousuf et al., 1992).

In the present study NGF-like immunoreactivity has been demonstrated in the neurosecretory cells of the nervous system of *Eisenia foetida* earthworms. Ultrastructural evidences for exocytotic release of immunoreactive granules in the extracellular space adjacent to perikaryal profiles suggest a possible autocrine and/or paracrine role of NGF-like immunoreactive molecules in earthworms. The NGF-like immunoreactivity has been detected by the use of rabbit anti-mNGF sera containing antibodies with different specificities to the different parts of NGF molecule, but displaying the ability to recognize all neurotrophins. It has to be pointed out that many antisera generated against purified NGF molecules have been found to react with other neurotrophins (Acheson et al., 1991; Murphy et al., 1993; Zhou and Rush, 1993).

The reactive properties of our antisera raise the possibility that the molecules immunologically related to NGF in the earthworms may be more similar to other members of the vertebrate neurotrophin family than to NGF itself or be quite unique with only some sequences in common with all vertebrate neurotrophins. Further studies are required to achieve definitive conclusions about their possible relationship with the vertebrate NGF-like neurotrophic factors and also to evaluate the extent of their structural homology to NGF molecule. However, our results demonstrate the presence of molecules immunologically related to at least one member of the neurotrophin family in *Eisenia foetida* earthworms. This suggests that some conservative sequences of NGF structure arose very early in evolution and supports the hypothesis that NGF-like proteins first evolved in invertebrates.

Acknowledgements
The authors thank Dr. Delio Mercanti for the generous supply of specific antisera and purified antibodies to mouse 2.5S NGF and also for his helpful criticism and advice.

References

Acheson, A., Barker, P.A., Alderson, R.F., Miller, F.D. and Murphy, R.A. (1991) Detection of brain-derived neurotrophic factor-like activity in fibroblasts and Schwann cells: inhibition by antibodies to NGF. *Neuron* 7: 265-275.

Al-Yousuf, S., Konishi, A., Nomura, S. and Mizuno, N. (1992) An electron microscopic study of neurosecretion in the cerebral ganglion of the earthworm. *Neurosci. Lett.* 140: 189-191.

Angeletti, R.H., Mercanti, D. and Bradshaw, R.A. (1973) Amino acid sequences of mouse 2.5S nerve growth factor. I. Isolation and characterization of the soluble tryptic and chymotryptic peptides. *Biochemistry* 12: 90-100.

Bradshaw, R.A., Blundell, T.L., Lapatto, R., McDonald, N.Q. and Murray-Rust, J. (1993) Nerve growth factor revisited. *Trends Biochem. Sci.* 18: 48-52.

Ebendal, T. (1992) Fuction and evolution in the NGF family and its receptors. *J. Neurosci. Res.* 32: 461-470.

Golding, D.W. and Pow, D.V. (1988) The new neurobiology - ultrastructural aspects of peptide release as revealed by studies of invertebrate nervous systems. *In:* M.C. Thorndyke and G.J. Goldsworthy (eds.): *Neurohormones in Invertebrates,* Cambridge University Press, Cambridge, pp. 7-18.

Hallböök, F., Ibáñez, C.F. and Persson, H. (1991) Evolutionary studies of nerve growth factor family reveal a novel member abundantly expressed in Xenopus ovary. *Neuron* 6: 845-858.

Hayashi, I., Perez-Magallanes, M. and Rossi, J.M. (1992) Neurotrophic factor-like activity in Drosophila. *Biochem. Biophys. Res. Commun.* 184: 73-79.

Levi-Montalcini, R. (1987) The nerve growth factor 35 years later. *Science* 237: 1154-1162.

Linsday, R.M., Wiegand, S.J., Altar, C.A. and DiStefano, P.S. (1994) Neurotrophic factors: from molecule to man. *Trends Neurosci.* 17: 182-190.

Murphy, R.A., Acheson, A., Hodges, R., Haskins, J., Richards, C., Reklow, E., Chlumecky, V., Barker, P.A., Alderson, R.F. and Linsday, R.M. (1993) Immunological relationships of NGF, BDNF, and NT-3: recognition and functional inhibition by antibodies to NGF. *J. Neurosci.* 13: 2853-2862.

Narhi, L.O., Rosenfeld, R., Talvenheimo, J., Prestrelski, S.J., Arakawa, T., Lary, J.W., Kolvenbach, C.G., Hecht, R., Boone, T., Miller, J.A. and Yphansis, D.A. (1993) Comparison of the biophysical characteristics of human brain-derived neurotrophic factor, neurotrophin-3 and nerve growth factor. *J. Biol. Chem.* 268: 13309-13317.

Ridgway, R.L., Syed, N.I., Lukowiak, K. and Bulloch, A.G.M. (1991) Nerve growth factor (NGF) induces sprouting of specific neurons of the snail, Lymnaea stagnalis. *J. Neurobiol.* 22: 377-390.

Thoenen, H. (1991) The changing scene of neurotrophic factors. *Trends Neurosci.* 14: 165-170.

Zhou, X. and Rush, R.A. (1993) Localization of neurotrophin-3-like immunoreactivity in peripheral tissues of the rat. *Brain Res.* 621: 189-199.

The Met-callatostatins of the blowfly *Calliphora vomitoria* : post-translational modifications, neuronal mapping and functional significance

H. Duve, A.H. Johnsen[1] and A. Thorpe

School of Biological Sciences, Queen Mary and Westfield College, University of London, Mile End Road, London E1 4NS, UK
[1]*Department of Clinical Biochemistry, State University Hospital, Copenhagen, DK-2100, Denmark*

Summary. The extraction, isolation and characterization of four members of the Met-callatostatin family of neuropeptides from the blowfly *Calliphora vomitoria* is described. The parent compound has the amino acid sequence Gly-Pro-Pro-Tyr-Asp-Phe-Gly-Met-NH$_2$ and the other three are post-translational modifications derived by prolyl-hydroxylation in either the Pro2 or Pro3 positions and by N-terminal truncation to give the hexapeptide. Immunocytochemical studies have been carried out to determine the localization of the peptides which are found in various neurons within the central nervous system as well as in mid-gut endocrine cells. Physiological studies reveal a potent inhibitory effect of the peptides on the spontaneous peristaltic contractions of the ileum.

Introduction

The callatostatins are neuropeptides produced by a variety of neurons in the central and peripheral nervous system of the blowfly *Calliphora vomitoria* (Duve et al., 1993). Since they are also produced in endocrine cells of the midgut (Duve and Thorpe, 1994; East et al., 1995), they may be categorised as brain/gut peptides, analagous to the well-known regulatory peptides of vertebrates such as cholecystokinin. The callatostatins are homologues of the allatostatins initially found in cockroaches (Woodhead et al., 1989) and more recently in other insects (Lorenz et al., 1995). The general term, allatostatin, was used to indicate a specific function in cockroaches, e.g. the inhibition of juvenile hormone,

                                H. Duve et al.

the secretory product of the corpus allatum, important in larvae for the retention of juvenile characteristics during metamorphosis, and also in adults in the process of reproduction.

However, the initial terminology, based on a function in cockroaches, may appear inappropriate for homologous 'allatostatins' in other insects. Thus, in blowflies, the callatostatins appear not to be allatostatic (Duve et al., 1993) although they do have other potent physiological effects (Duve and Thorpe, 1994). The generic term 'allatostatin' is retained for the blowfly peptides in the modified form of 'callatostatins' to indicate the species of origin and the sharing of common amino acid sequences. Thus, all members of this family of peptides, whatever the function, are characterised structurally by the common C-terminal pentapeptide sequence Tyr-Xaa-Phe-Gly-Yaa-NH$_2$, with an N-terminal extension that is variable both in sequence and length. Amino acid residues in the Xaa position appear to be restricted to Asn, Asp, Ser, Gly or Ala. The Yaa residue is almost always Leu-NH$_2$, with the blowfly showing an important difference in possessing Met-NH$_2$ as well as Leu-NH$_2$ variants. In the same way in which the terminology for the Leu- and Met-enkephalins was coined, we have referred to the two groups of peptides in *C. vomitoria* as the Leu- and Met-callatostatins (Duve and Thorpe, 1994).

The apparently unique nature of the Met-callatostatins of *C. vomitoria* (compared with all other allatostatins) prompts an examination of the specific problems associated with their isolation and characterization, cellular localization and physiological roles.

Peptide isolation and characterization

The parent molecular form of Met-callatostatin has the structure Gly-Pro-Pro-Tyr-Asp-Phe-Gly-Met-NH$_2$ and was originally discovered in extracts of whole flies (Duve et al., 1992; 1993). Antibodies raised against the synthetic peptide have subsequently been used in radioimmunoassay to confirm the abundant presence of Met-callatostatin in extracts of blowfly heads. It is significant that there is negligible cross-reactivity between Met-callatostatin and our wide range of antibodies against several Leu-callatostatin variants. Thus, in chromatographic profiles at all stages in the purification, peaks of Met-callatostatin remain undetected in Leu-callatostatin radioimmunoassays and *vice versa*. In the final chromatographic step we have regularly made use of a Vydac C_{18} 5 , 300 Å column with a gradient of acetonitrile/0.1% trifluor-acetic acid at a flow rate of 0.2 ml/min and a rate of change of 0.5%/min. Under these conditions Met-callatostatin, in the reduced state (molecular mass $M^* = 883$), elutes at approximately 17% acetonitrile. However, the methionine residue tends to become oxidised to methionine sulfoxide with the result that a variable amount is always found in the oxidised state. This oxidised variant ($M^* = 899$) elutes almost 3% earlier than the reduced, 'parent' compound (around 14% acetonitrile). There appears to be no differentiation in radioimmunoassay between the reduced and oxidised peptides. In material eluting approximately 1% in front of the reduced form of Met-callatostatin we identified a variant in which the second of the two prolyl residues was hydroxylated. This peptide, designated [Hyp^3]Met-callatostatin was in the reduced state since it had an M^* of 882. However, during the repurification of this material, a small amount of oxidised [Hyp^3]Met-callatostatin appeared in the profile eluting at 12.2% acetonitrile compared with 16% for the reduced form. The situation became even more complex when we analysed Met-callatostatin material appearing on the back edge of, and immediately adjacent to, that which contained [Hyp^3]-Metcallatostatin. This material, eluting less than 1% later, contained the oxidised form of another Met-callatostatin post-translational modification in which the first, rather than the second, of the two prolyl residues was hydroxylated. We refer to this peptide as [Hyp^2]Met-callatostatin. The M^* of 914 showed that it was the oxidised form of the peptide. Furthermore, in the neighbouring two fractions (also less than 1% later-eluting than that containing oxidised [Hyp^2]Met-callatostatin, we identified the truncated Met-callatostatin, desGly-Pro-Met-callatostatin (recorded M^* 744.5 and, therefore, oxidised) and another small peak of [Hyp^2]Met-callatostatin, which most probably represented the reduced form of this peptide.

In summary, analysis of Met-callatostatin-immunoreactive material extracted from some 12,000 heads of *C. vomitoria* has revealed no fewer than four variants of the parent peptide Met-callatostatin, Gly-Pro-Pro-Tyr-Asp-Phe-Gly-Met-NH$_2$, each of which is capable of existing in the reduced or oxidised state (Table I). The proportional abundance of the four main types of peptides is Met-callatostatin 100: [Hyp^3]Met-callatostatin 20: and both

[Hyp^2]Met-callatostatin and des Gly-Pro Met-callatostatin < 5.

Table I. Variants of Met-callatostatins isolated from extracts of heads of the blowfly *Calliphora vomitoria*.

			M*	
Peptide designation	Amino acid sequence	CH$_3$CN %	Found	Calculated
Met-callatostatin (red)	Gly-Pro-Pro-Tyr-Asp-Phe-Gly-Met-NH$_2$*	17	882.7	882.0[a,b]
Met-callatostatin (ox)	Gly-Pro-Pro-Tyr-Asp-Phe-Gly-Met-NH$_2$	14	898.6	898.0[b]
[Hyp^2]Met-callatostatin (red)	Gly-Hyp-Pro-Tyr-Asp-Phe-Gly-Met-NH$_2$**	10.8	-	898.0[c]
[Hyp^2]Met-callatostatin (ox)	Gly-Hyp-Pro-Tyr-Asp-Phe-Gly-Met-NH$_2$	10.4	915.1	914.0[c]
[Hyp^3]Met-callatostatin (red)	Gly-Pro-Hyp-Tyr-Asp-Phe-Gly-Met-NH$_2$	16	899.1	898.0[b]
[Hyp^3]Met-callatostatin (ox)	Gly-Pro-Hyp-Tyr-Asp-Phe-Gly-Met-NH$_2$	10.5***	915.0	914.0[b,c]
des Gly-Pro Met-callatostatin (ox)	Pro-Tyr-Asp-Phe-Gly-Met-NH$_2$	10.2	744.5	744.0[c]

*The carboxyamidation of the peptides was demonstrated by determining the total number of free carboxyl groups by methylation of an aliquot and remeasurement of the molecular mass (Talbo et al., 1991). In each case the mass increased by 14, thereby demonstrating the presence of only the single free carboxyl group, that of the aspartyl residue.

**Identification of reduced form of [Hyp^2]Met-callatostatin inferred from amino acid sequence and later retention time compared with that of oxidised form.

***The value for the % acetonitrile at which the oxidised form of [Hyp^3]Met-callatostatin elutes, is taken from the purification procedure which led to the discovery of [Hyp^2]Met-callatostatin (reduced and oxidised forms) and des Gly-Pro Met-callatostatin (Duve et al., 1995). A comparison of these four values above shows the close similarity of their elution characteristics (all within 0.6% acetonitrile). In other purifications [Hyp^3]Met-callatostatin (oxidised) has eluted slightly later e.g. 12.2 % acetonitrile (Duve et al., 1994).

[a] Duve et al., 1993; [b] Duve et al., 1994; [c] Duve et al., 1995.

The presence of hydroxyproline in regulatory peptides is extremely rare judged by the sparse number of reports that present evidence of such a post-translational modification (for references see Duve et al., 1994). The example of the prolyl-hydroxylated Met-callatostatin variants of *C. vomitoria* is particularly interesting in that it represents the first report of a peptide possessing successive prolyl residues, either of which has the potential to be hydroxylated. This suggests a versatility of the prolyl hydroxylase enzymes that may not have been recognised before.

The provision in blowflies of such a complex array of peptides suggests that they may have equally complex and varied functions, and in order to discover clues concerning the nature of these functions we have carried out extensive immunocytochemical studies with Met-callatostatin antisera.

Immunocytochemical studies with Met-callatostatin antisera

Antisera against Met-callatostatin were raised in Dutch rabbits as previously described (Duve and Thorpe, 1994). Tissues, including brain, corpora cardiaca/corpora allata complex, thoracico-abdominal ganglion and the entire gut, were fixed in Bouin's fluid, embedded in paraffin wax and sectioned at 6 μm for use in the peroxidase-antiperoxidase (PAP) technique (Sternberger, 1974). For whole-mount immunocytochemistry tissues were fixed in 4% paraformaldehyde using the indirect immunofluorescence technique. For tissue sections antisera were applied at 1:2000 and for whole-mounts 1:500, both overnight at 4°C. Full details of these protocols have been given previously (Duve and Thorpe, 1994). The specificity of the antisera relative to the two main types of callatostatins (Leu- and Met-) present in the tissues has been tested by liquid-phase absorption. In these experiments, Met-callatostatin added at a concentration of 50 μM to the diluted antisera (1:2000) resulted in complete abolition of staining of Met-callatostatin antisera but had no effect on Leu-callatostatin immunostaining. The reverse was also shown to be true i.e. Leu-callatostatin at 50 μM was able to absorb out the Leu-callatostatin immunostaining, but not that induced by Met-callatostatin antisera. Although these results suggest that Met-callatostatin antisera will only reveal the presence of Met-callatostatin peptides, it cannot be ruled out that some cross-reactivity may exist with Leu-callatostatin type of peptides slightly different (in the Xaa) position from the specific peptide (which had Gly in the Xaa position) against which the Leu-callatostatin antisera were raised.

Results from extensive immunocytochemical studies in *C. vomitoria* have shown that whereas some tissues and cells respond positively to both Met- and Leu-callatostatin antisera, others are uniquely stained with Met-callatostatin antisera. In view of the cross-absorption experiments referred to above, the results where the same tissues are immunostained with both types of antisera could indicate either co-expression of the two separate Met- and Leu-callatostatin genes, or cross-reactivity by peptide(s), the nature of which we have not yet characterized. Co-immunoreactivity is shown strikingly by inter-neurons of the visual system (Fig. 1B), endocrine cells of the midgut, and also by up to five neurosecretory neurons in the abdominal ganglion, the axons of which innervate the hindgut (Fig. 1C). Unique Met-callatostatin immunostaining, on the other hand, is shown by some specific glomeruli in the antennal lobes (Fig. 1A) and by certain lateral neurosecretory cells (Fig. 1D). The observation of Met-callatostatin immunoreactivity in such a variety of different cells and tissues is interesting from a functional point of view, the only possible conclusion being that these peptides subserve a wide range of different physiological roles within the body.

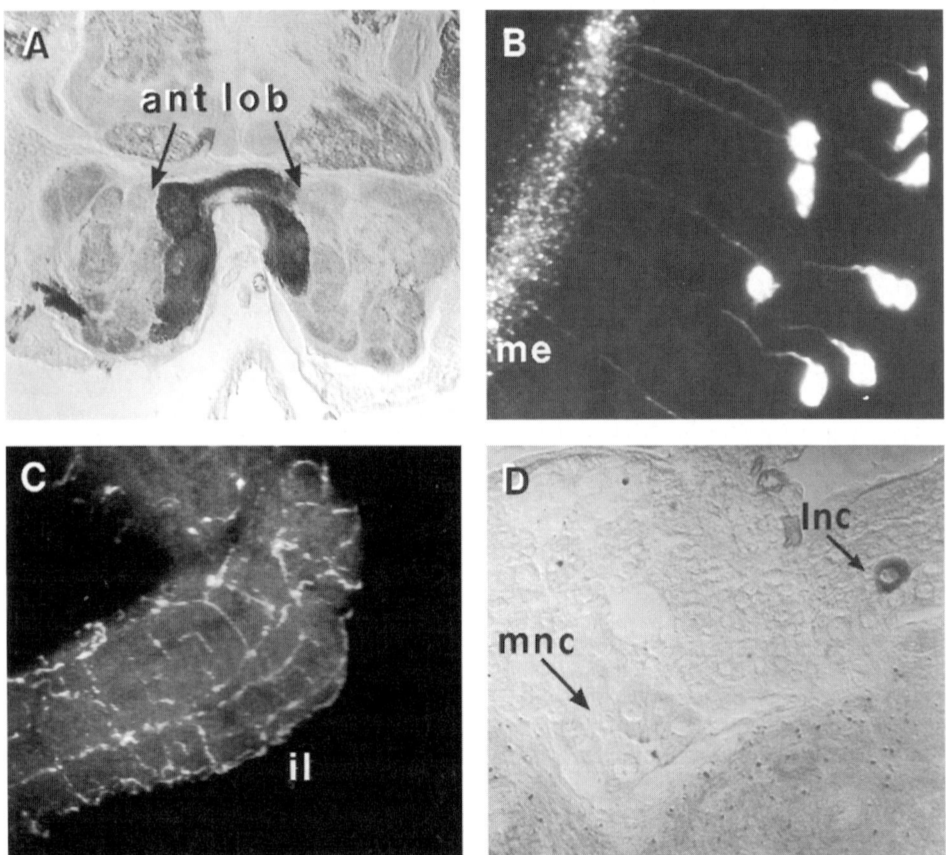

Fig. 1. Met-callatostatin immunoreactivity in tissues of brain and gut of *Calliphora vomitoria*.
A: Frontal anterior view of brain showing immunoreactivity in certain glomeruli of the antennal lobes.
B: Detail of one of the optic lobes showing immunoreactivity in intrinsic neurones of the medulla.
C: Innervation of the hindgut by axons from neurones in the abdominal ganglia via the median abdominal nerve.
D: Frontal view of lateral neurosecretory cells in dorsal medial view of the brain.
ant lobe, antennal lobe; *il*, ileum; *lnc*, lateral neurosecretory cell; *me*, medulla; *mnc*, median neurosecretory cells.
Magn.: A: x 200; B: x 500; C: x 210; D: x 300.

The action of Met-callatostatin and its post-translationally modified forms on the control over gut motility

The major immunocytochemical clue leading to physiological experimentation was the identification of Met-callatostatin immunoreactivity in neurosecretory neurons in the abdominal ganglion (Duve et al., 1994). The axons of these neurons have been shown in whole-mount studies to project to the hindgut where they provide an intense network of innervation especially over the ileum (Fig. 1C), and also over the rectal valve and pouch. The ileum is particularly responsive in *in vitro* experiments, the regular spontaneous peristaltic contractions being inhibited over a range of concentrations from 10^{-16} to 10^{-6} M (Fig. 2). It is significant that the post-translationally modified forms of Met-callatostatin in which either of the two prolyl residues are hydroxylated are much more active than the 'parent' compound itself (50% inhibition of peristalsis at 10^{-12} M with [Hyp^3]Met-callatostatin and even at 10^{-14} M with [Hyp^2]Met-callatostatin, compared with the inactivity of Met-callatostatin at the same concentration). This suggests that prolyl hydroxylation provides for a higher receptor-binding affinity, maybe even indicating that there are specialised receptor subtypes. In contrast, the naturally occurring hexapeptide, des Gly-Pro Met-callatostatin has approximately the same potency as Met-callatostatin itself and the pentapeptide desGly-Pro-Pro-Met-callatostatin (not, as far as we know, a naturally-occurring peptide) diminished potency (Fig. 2).

In addition to variations in receptor-binding characteristics provided by prolyl-hydroxylation, it is interesting to consider that this post-translational modification might also lead to a delay in enzymatic degradation, also a factor of possible adaptive value. In order to test this hypothesis, it is necessary to determine the types of degradative enzymes used by *C. vomitoria* in the breakdown of the Met-callatostatin family of peptides. In view of the apparent rarity of hydroxyprolyl residues in regulatory peptides and their relative 'abundance' in the blowfly, it would seem that this species provides a useful model in which to study in particular those enzymes capable of cleaving pre-, post- and inter-prolyl bonds.

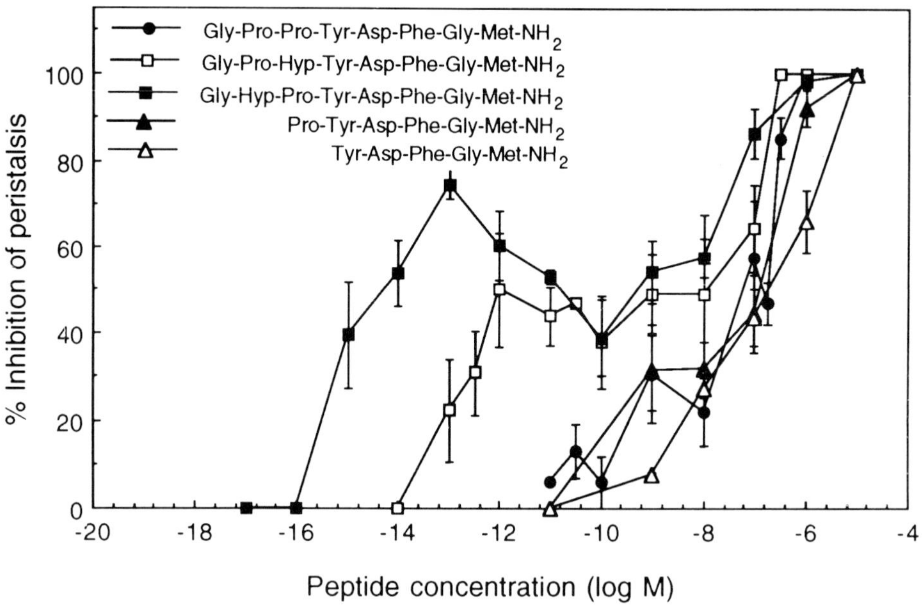

Fig. 2. Dose responses for inhibition of peristalsis of the hindgut of *Calliphora vomitoria* by members of the callatostatin neuropeptide family. Percentage inhibition relative to basal spontaneous peristaltic movements. Each point represents the mean of 5-16 measurements from a number of vitellogenic female flies.

Concluding comments

Peptidergic neurons and endocrine cells confront biologists with many challenges and although our understanding of the nature of the products of these cells and the processes which they control is increasing, there are still many intriguing problems. Firstly in order to have a complete understanding of the form and function of neuropeptide families it is important to have more amino acid sequence data. Generalizations made from a few sequences from only a few animals can be misleading. This is particularly true for entomologists confronted with so many different species. Furthermore, it is essential not to rely solely on putative amino acid sequences that may be deduced from the DNA. The need to carry out simultaneous isolation and characterization studies of the peptides from tissue extracts is apparent from the range of post-translationally-modified forms of the 'parent'

Met-callatostatin molecule reported here from *C. vomitoria*. However, although structural information on insect neuropeptides is an important issue, perhaps the major challenge in the future lies in ascertaining their physiological roles and exploiting this knowledge in the development of environmentally-safe insecticides.

Acknowledgements
The original research was supported by a research grant from the Agricultural and Food Research Council (AG 68022). We are pleased to acknowledge the support of the Royal Society who provided Conference Grants to both HD and AT.

References

Duve, H. and Thorpe, A. (1994) Distribution and functional significance of Leu-callatostatins in the blowfly *Calliphora vomitoria*. *Cell Tissue Res.* 276:367-379.

Duve, H., Johnsen, A.H., Sewell, J.C., Scott, A.G., Orchard, I., Rehfeld, J.F., and Thorpe, A. (1992) Isolation, structure, and activity of -Phe-Met-Arg-Phe-NH$_2$ neuropeptides (designated calliFMRFamides) from the blowfly *Calliphora vomitoria*. *Proc. Natl. Acad. Sci. USA,* 89: 2326-2330.

Duve, H., Johnsen, A.H., Scott, A.G., Yu, C.G., Yagi, K.J., Tobe, S.S., and Thorpe, A. (1993) Callatostatins: neuropeptides from the blowfly *Calliphora vomitoria* with sequence homology to cockroach allatostatins. *Proc. Natl. Acad. Sci. USA,* 90:2456-2460.

Duve, H., Johnsen, A.H., Scott, A.G., East, P., and Thorpe, A. (1994) [Hyp^3]Met-callatostatin. Identification and biological properties of a novel neuropeptide from the blowfly *Calliphora vomitoria*. *J. Biol. Chem.* 269: 21059-21066.

Duve, H , Johnsen A. H., Scott, A.G. and Thorpe, A. (1995) Isolation, identification and functional significance of [Hyp^2]Met-callatostatin and des Gly-Pro Met-callatostatin, two further post-translational modifications of the blowfly neuropeptide Met-callatostatin. *Regul. Pept.* 57: 237-246.

East, P.D., Thorpe, A. and Duve, H. (1995) Leu-callatostatin gene expression in the blowflies *Calliphora vomitoria* and *Lucilia cuprina* studied by *in situ* hybridisation: comparison with Leu-callatostatin confocal laser scanning immunocytochemistry. *Cell Tissue Res.* 280: 355-364.

Lorenz, M.W., Kellner, R. and Hoffmann, K.H. (1995) Identification of two allatostatins from the cricket, *Gryllus bimaculatus* de Geer (Ensifera, Gryllidae): additional members of a family of neuropeptides inhibiting juvenile hormone biosynthesis. *Regul. Pept.* 57: 227-236.

Sternberger, L.A. (1974) *Immunocytochemistry*. Prentice-Hall, Englewood Cliffs, NJ.

Talbo, G., Hrjrup, P., Rahbek-Neilsen, H., Andersen, S.O. and Roepstorff, P. (1991) Determination of the covalent structure of an N- and C-terminally blocked glycoprotein from endocuticle of *Locusta migratoria*. Combined use of plasma desorption mass spectrometry and Edman degradation to study post-translationally modified proteins. *Eur. J. Biochem.* 195: 495-504.

Woodhead, A.P., Stay, B., Seidel, S.L., Khan, M.A. and Tobe, S.S. (1989) Primary structure of four allatostatins: Neuropeptide inhibitors of juvenile hormone synthesis. *Proc. Natl. Acad. Sci. USA* 86: 5997-6001.

Secretory stimulation induces the preferential release of newly synthesized peptide hormones by the neuroendocrine adipokinetic cells of *Locusta migratoria*

J.H.B. Diederen, H.E. Sharp-Baker, R.C.H.M. Oudejans, H.G.B. Vullings and D.J. Van der Horst

Department of Experimental Zoology, Utrecht University, Padualaan 8, NL-3584 CH Utrecht, The Netherlands

Summary. The influence of secretory stimulation on the release of adipokinetic hormones from the adipokinetic cells in the corpus cardiacum of *Locusta migratoria* was investigated both at the level of the secretory granules and at the level of the hormone molecules. The results indicate that the adipokinetic cells preferentially release newly formed secretory granules containing newly synthesized hormones over older granules with older hormones.

Introduction

The adipokinetic hormone-producing cells (AKH cells) in the migratory locust are neuron-like cells, with their cell bodies and cell processes intermingled within the glandular part of the corpus cardiacum. As a consequence, the hormones which are produced by the AKH cells are released within the corpus cardiacum; there is no separate neurohemal area for these hormones elsewhere. The AKH cells produce three peptide hormones, the adipokinetic hormones AKH I, II and III, which control the release of carbohydrates and lipids from the fat body during flight (Beenakkers et al., 1985; Oudejans et al., 1991). Three prohormones, one for each of the three AKHs (Bogerd et al., 1995), are synthesized in the cell bodies, packaged into secretory granules at the *trans*-Golgi network (Jansen et al., 1989) and processed to the bioactive AKHs. The secretory granules are transported to the cell processes, where they are stored or released into the hemolymph by exocytosis (Diederen et al., 1992, 1993).

All three AKHs are continuously synthesized by the AKH cells, and as the locust ages, the amounts of all three AKHs within the AKH cells increase (Oudejans et al., 1993), as does the number of secretory granules (Diederen et al., 1992). Flight is the only known natural stimulus for the release of AKHs, and then only a fraction - approximately 2% per hour of flight - of the total available store is released (Cheeseman et al., 1976). As a result, the store of hormones in the AKH cells consists of a mixture of young and old granules. This raises the question: is there a relationship between the age of a secretory granule and its chance of release during flight? To answer this question, the influence of secretory stimulation on the release of newly synthesized AKHs from the AKH cells was investigated, both at the level of the secretory granules (*in vivo*) and at the level of the hormone molecules (*in vitro*).

Material and methods

Two labeling methods were applied to distinguish young, newly formed secretory granules from older, pre-existing ones. One method was based on the endocytotic incorporation of the enzyme horseradish peroxidase conjugated with the lectin wheat-germ agglutinin (WGA-HRP). This lectin binds to the plasma membrane due to its high affinity for certain sugar residues there. Endocytosed fragments of plasma membrane with bound label are transported to the *trans*-Golgi network and used for the production of new secretory granules, which thus become labeled with peroxidase (Fig. 1A) (Diederen and Vullings, 1995). WGA-HRP was administered by injection of 10 µl physiological insect saline containing 1% WGA-HRP into the abdomen of adult male locusts, 12 days after the imaginal ecdysis. During a 31.5 h period, locusts (n=5) were injected 3 times with WGA-HRP, flight-stimulated 4 times and then decapitated, according to the time schedule depicted in Fig. 2A. Flight stimulation was achieved by suspending locusts from a motor-driven carousel. Control locusts (n=5) only received the WGA-HRP injections. The corpora cardiaca were dissected and fixed, and the HRP was visualized enzyme-cytochemically at the electron-microscopic level as previously reported (Diederen et al., 1992).

Per locust, 15 electron micrographs at a final magnification of 11, 700x were analyzed quantitatively. On each photograph, cell bodies and cell processes were analyzed separately. The cytoplasmic area of the analyzed cell bodies (without the nuclei) and cell processes were determined using a planimeter. The number of WGA-HRP-labeled (sectional profiles of) secretory granules as well as the total number of labeled and unlabeled (sectional profiles of) secretory granules were manually counted in cell bodies and processes. The data from all 15 photographs were then pooled and expressed per animal as number of granules per 100 μm^2 cytoplasmic area. The percentages of the (sectional profiles of) secretory granules which were labeled with WGA-HRP in the cell bodies and cell processes were also calculated per animal. The arithmetic means and standard deviations of the data for the groups of flown and unflown locusts were then calculated from the values per animal.

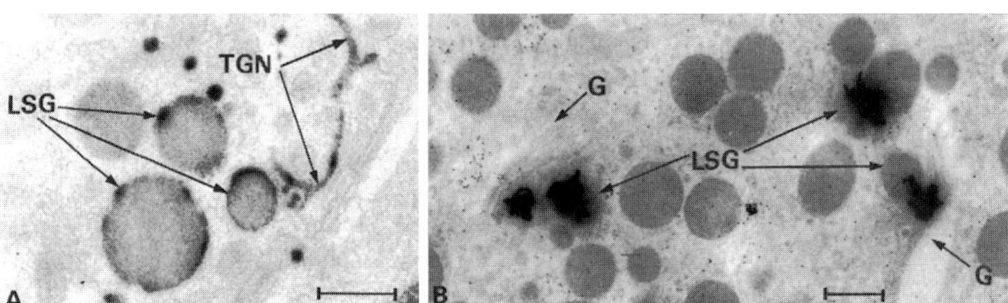

Fig. 1. The ultrastructural appearance of labeled, newly formed secretory granules.
A: WGA-HRP labeling: HRP reaction product situated at the inner surface of the secretory granule membrane.
B: Radioactive labeling: silver grain situated on a secretory granule or at a distance ≤ HR (=318nm) from it; a silver grain was allotted to one secretory granule only. LSG=labeled secretory granules; TGN=*trans*-Golgi network; G=Golgi complex. Bars = 250 nm.

The other method for labeling young, newly formed secretory granules was based on the biosynthetic incorporation of ^{3}H-amino acids into newly synthesized substances at the rough endoplasmic reticulum. These substances are then packaged into new secretory granules at the *trans*-Golgi network, which results in the formation of radioactively labeled new secretory granules. 2.96 MBq of a ^{3}H-amino acid mixture (TRK 440, Amersham Internat.) in 20 µl insect saline were injected into the abdomen of male adult locusts, 12 days after the imaginal ecdysis. During a 5 h period, locusts (n=7) were injected once with the ^{3}H-amino acids, flight-stimulated 2 times and then decapitated, according to the time schedule depicted in Fig. 2B. Controls (n=5) only received the ^{3}H-amino acid mixture. The corpora cardiaca were dissected and fixed, and the incorporated radioactive amino acids were visualized autoradiographically at the electron-microscopic level (Fig. 1B) according to methods described by Sharp-Baker et al. (1995) with slight modifications.

Per locust, 15 electron micrographs at a final magnification of 12, 480x were analyzed quantitatively. On each photograph, cell bodies and cell processes were analyzed separately. The circle method according to Williams (1969) was used as an aid to determine the radioactive labeling of the secretory granule compartment. The half-distance (HD) value was determined from literature (187 nm), from which the half-radius (HR) value (= 1.7 times HD = 318 nm) was calculated (Kopriwa et al., 1984). The HR value corresponded to 3.7 mm on the electron micrographs at the above-mentioned magnification. A circle with radius HR placed around a silver grain gives a probability of 50% that the radioactive source which induced the silver grain is located within this circle. Since the HR value and the mean diameter of the (sectional profiles of the) secretory granules (273 nm in cell bodies and 319 nm in cell processes; Diederen et al., 1987) are approximately the same, it is not possible to be sure whether the radioactive source of a particular silver grain is situated within a certain secretory granule. It was decided, therefore, to count the silver-associated secretory granules, defined by us as those (sectional profiles of) secretory granules with silver on them or at a distance ≤ HR from them. In the latter case, the silver grain was allotted to one secretory granule only. The number and percentage of silver-associated secretory granules as well as the total number of secretory granules, both in cell bodies and cell processes, were determined as described above for the WGA-HRP labeling experiment.

The statistical significance of the observed differences between flown and unflown locusts was established by Student's t-test and the Mann-Whitney U-test.

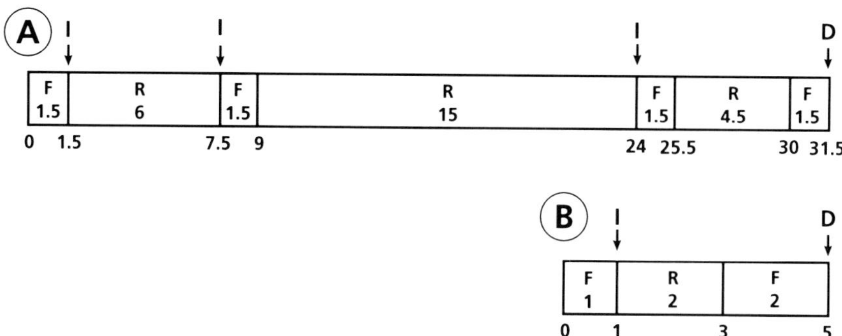

Fig. 2. Time schedules (in hours) of the experiments at the level of the secretory granules.
A: WGA-HRP labeling.
B: Radioactive labeling. *F*=flight stimulation; *I*=injection; *R*=rest period; *D*=decapitation.

The influence of secretory stimulation on the release of newly synthesized AKHs at the level of the molecules was studied in an *in vitro* system. Newly synthesized AKH molecules were labeled by means of biosynthetic incorporation of ^{3}H-phenylalanine, since all three AKHs contain this amino acid. This made the newly synthesized (radioactive) AKH molecules distinguishable from the older (non-radioactive) AKH molecules. A combination of pulse-chase and time-course experiments was used to follow the labeled AKH molecules over time. Pools of 15 isolated corpora cardiaca of adult male locusts, 12 days after the imaginal ecdysis, were exposed to 370 kBq ^{3}H-phenylalanine in locust incubation medium, according to methods reported previously (Oudejans et al., 1993), with slight modifications. The AKH cells were stimulated with 100 µM 3-isobutyl-1-methylxanthine (IBMX, a phosphodiesterase inhibitor), which raises the intracellular cAMP level, which in turn stimulates the release of AKHs (Passier et al., 1995). Four separate experiments were performed according to the schedule in Table I, with the total incubation time varying from 2.25 to 16.25 h. The released and non-released AKHs, the latter extracted from the tissue, were separated and quantified by reversed-phase high-performance liquid chromatography (HPLC), and their radioactivity was measured by liquid scintillation counting of the column eluate as previously reported (Oudejans et al., 1993). Synthetic AKHs were used for calibration of the HPLC in order to identify and quantify the released and non-released AKHs.

Table I. Time schedules (in hours) of the 4 radioactive labeling experiments at the level of the AKH molecules.

Experiment	Pulse	Rinse	Chase	Secretory stimulation	Total incubation time
I	1.5	0.25	0	0.5	**2.25**
II	3.5	0.25	0	0.5	**4.25**
III	3.5	0.25	4	0.5	**8.25**
IV	3.5	0.25	12	0.5	**16.25**

The specific radioactivities of both released and non-released AKHs were calculated, after which their ratio was determined. This ratio was used to reduce the effects of the variation between single pools of corpora cardiaca. The total amount of radioactivity in released and non-released AKHs was taken as a measure for the total amount of newly synthesized AKHs. The percentage of newly synthesized AKHs which were actually released could then be calculated from the amount of radioactivity in the released AKHs and the total amount of radioactivity in both released and non-released AKHs. The arithmetic mean and standard deviation of each variable was calculated. The experiments were analyzed by one way analysis of variance (ANOVA t-test) and Scheffe's procedure.

Results

The results from the experiment with WGA-HRP-labeled secretory granules are shown in Fig. 3A for cell bodies and cell processes separately. The total number of secretory granules, labeled as well as unlabeled ones, was almost equal in flight-stimulated and non-stimulated cells. The number of labeled granules, however, was significantly lower in flight-stimulated than in non-stimulated cells, both in cell bodies and in cell processes. Consequently, the ratio between the numbers of labeled and unlabeled secretory granules was also lower in flight-stimulated cells.

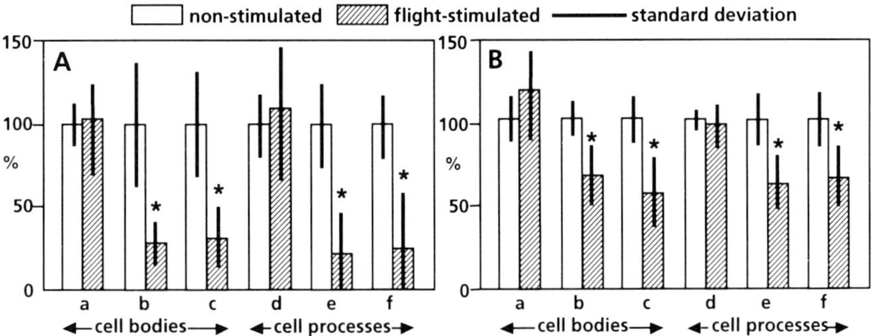

Fig. 3. Results of the experiments at the level of the secretory granules.
A: WGA-HRP labeling.
B: Radioactive labeling. The values of the flight-stimulated cells were normalised with respect to the values of the non-stimulated cells, which were set to 100%. * = result is statistical significant (P≤0.01). *a,d*=total number of secretory granules; *b,e*=number of labeled secretory granules; *c,f*=percentage of labeled secretory granules.

The results from the experiment with radioactive-labeled secretory granules are shown in Fig. 3B. These results agree well with those obtained from the WGA-HRP labeling experiment. The total number of secretory granules again was nearly equal, while the number as well as the percentage of silver-associated secretory granules were significantly lower in flight-stimulated than in non-stimulated cells, both in cell bodies and in cell processes.

The results from the experiments with radioactive-labeled AKH molecules are shown in Tables II and III. Since the amounts of AKH I, II and III are present in a molar ratio of approximately 14:2:1 (Oudejans et al., 1993), the results obtained for AKH I are less subject to variation than those for AKH II and III. The ratio between the specific radioactivities of the AKHs released into the incubation medium and of the non-released AKHs was significantly higher than 1.0, at all four incubation times, for all three AKHs (ANOVA *t*-test, P≤0.05).

The percentages of newly synthesized (radioactive) AKH I which was released differed significantly between different incubation times (ANOVA t-test, P<0.01; Scheffe's procedure, P<0.01). This resulted in a pattern of increase and decrease in the percentage of AKH I which was released. Up to 8.25 h, this percentage increased, and between 8.25 and 16.25 h, it decreased. The percentages of newly synthesized AKH II which was released, tended to follow a similar pattern. The percentages of newly synthesized AKH III which was released, tended to peak at 8.25 h.

J.H.B. Diederen et al.

Table II. Ratios between the specific radioactivities of released and non-released AKHs.

Incubation time(hours)	Number of incubations	AKH I	AKH II	AKH III
2.25	3	2.4±0.8	2.7±0.8	6.8±4.8
4.25	5	1.8±0.6	2.9±1.9	3.1±2.9
8.25	6	3.2±0.4	2.6±0.4	3.6±0.4
16.25	3	2.2±0.7	3.0±1.5	4.9±2.3

Table III. Percentages of newly synthesized AKHs which had been released.

Incubation time(hours)	Number of incubations	AKH I	AKH II	AKH III
2.25	3	7.0±1.9	6.1±1.6	7.5±3.7
4.25	5	12.7±6.5	13.5±12.2	7.1±8.1
8.25	6	17.1±3.1	15.2±2.6	12.2±2.6
16.25	3	4.1±2.3	11.4±11.3	7.1±1.9

Discussion

Both methods used to label young secretory granules demonstrated that flight-stimulated AKH cells had significantly fewer WGA-HRP-labeled or silver-associated secretory granules than unstimulated cells. At the same time, the total number of secretory granules was almost unaffected. Consequently, the percentage of labeled secretory granules was significantly lower in flight-stimulated than in unstimulated cells. This means that during secretory stimulation, young secretory granules are preferentially released over older granules.

The observation that the total number of secretory granules did not diminish in the flight-stimulated cells in comparison to the unstimulated cells points to an increase in the formation of new secretory granules in flight-stimulated cells, to compensate for the flight-induced exocytosis of secretory granules. This supposition is strengthened by findings described elsewhere (Diederen and Vullings, 1995) which point to a flight-induced enhancement of secretory granule production by the *trans*-Golgi network. Secretory stimulation of AKH cells thus results in an enhancement of the production of new secretory granules, which are then released preferentially.

At all four *in vitro* incubation times, the ratio of specific radioactivities of the released and non-released AKHs was greater than 1.0. This indicates that newly synthesized (radioactive) hormones are preferentially released over older (non-radioactive) ones, which is in line with the observed preferential release of newly formed secretory granules.

The percentage of newly synthesized AKH I which was released, increased between 2.25 and 8.25 h and decreased between 8.25 and 16.25 h. The percentage of newly synthesized AKH II which was released, tended to follow the same pattern. This is quite conceivable, since AKH I and II are co-localized in the same secretory granules (Diederen et al., 1987). The localization of AKH III within the AKH cells is not yet known. The synthesis, packaging into secretory granules and processing of the AKH prohormones to bioactive AKHs takes less than 1.25 h (Oudejans et al., 1990,1991), while the transport of secretory granules along the 20 to 200 µm long axonal processes (Orchard and Shivers, 1986) by means of fast anterograde transport takes but a few minutes (Friedel and Loughton, 1980). Thus, the observed pattern of increase and decrease in the percentage of newly synthesized AKH I (and II) which is released, suggests that apart from the time needed for the synthesis, packaging and processing of the hormones and for the transportation of the secretory granules to the cellular processes for release, the secretory granules require considerable time for some further maturation before they can actually release their content. If only the very youngest secretory granules with the most recently synthesized (radioactive) hormones were releasable, then the ratio of specific radioactivities would be greatest at the shortest incubation time and would thereafter decrease as the secretory granules age. The results, however, do not support this. Between 8.25 and 16.25 h secretory granules containing radioactive AKH I (and II) are apparently entering a non-releasable pool consisting of older secretory granules, This may explain why the percentage of radioactive AKH I (and II) which is released at 16.25 h is less than at 8.25h.

References

Beenakkers, A.M.T., Bloemen, R.E.B., De Vlieger, T.A., Van der Horst, D.J. and Van Marrewijk, W.J.A. (1985) Insect adipokinetic hormones. *Peptides* 6 (Suppl.3): 437-444.
Bogerd, J., Kooiman, F.P., Pijnenburg, M.A.P., Hekking, L.H.P., Oudejans, R.C.H.M. and Van der Horst, D.J. (1995) Molecular cloning of three distinct cDNAs, each encoding a different adipokinetic hormone precursor, of the migratory locust, *Locusta migratoria*. *J. Biol. Chem.* 270: 23038-23043.

Material and methods

Immunohistochemistry. The immunohistochemical staining of tachykinin- and FMRFamide-like immunopositive neurons was performed on Vibratome sections. Brains were fixed in 4% paraformaldehyde for 4 h. Vibratome sections were preincubated in 0.1 M phosphate-buffered saline (pH 7.4), containing 2% normal goat serum, 0.03% NaN_3 and 1% Triton-X-100 and incubated in the primary antiserum for 24 h at room temperature. The following antisera (diluted in the same buffer) were used: anti-locusta-tachykinin II, 1:20,000 (polyclonal; rabbit, H. Agricola, Jena); anti-substance P, 1: 500 (monoclonal, rat, Serva, Heidelberg, Germany); anti-FMRFamide, 1: 7,000 (polyclonal, rabbit; Incstar, Stillwater, USA). Tertramethylrhodamine isothiocyanate (TRITC) or fluoresceine isothiocyanate (FITC) coupled secondary antibodies were used, and in some cases the peroxidase-antiperoxidase method was applied. As controls, the antibodies were preabsorbed with the respective synthetic peptide, at 20 nmol for substance P and locusta-tachykinin II (Bachem, Heidelberg, Germany) and 50 nmol for FMRFamide (Sigma) per ml diluted antiserum, overnight at 4°C.

Postinjection-Method. Single cell analysis of locusta-tachykinin immunopositive neurons was performed as previously described by Wegerhoff and Breidbach (1994).

Results

Specificity of the antisera. In all preabsorption experiments with the respective peptides, the antibodies did not produce any immunolabelling. A preabsorption of the substance P antibody with the locusta-tachykinin II peptide does not effect immunostaining. For the locusta-tachykinin II antiserum, no crossreactivity to peptides other than locusta-tachykinins was found (H. Agricola, per. com.).

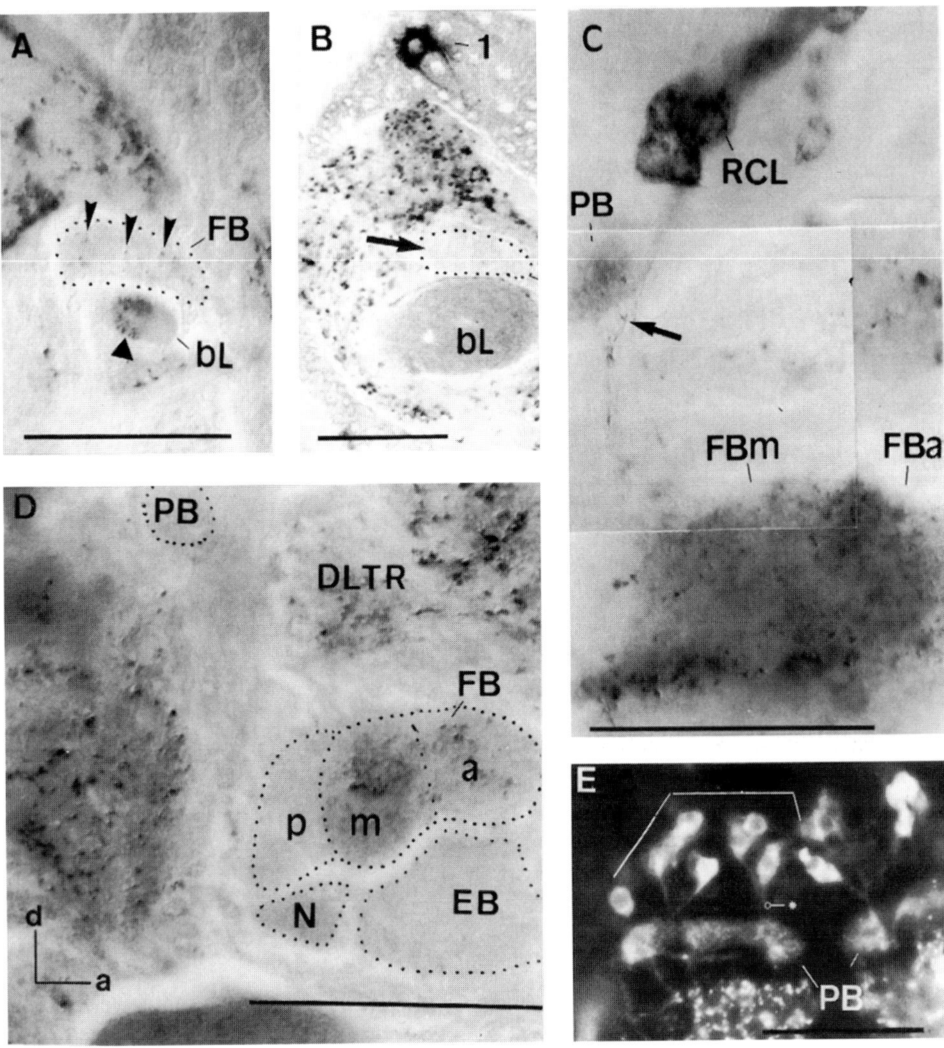

Fig. 1. A,B,D: FMRFamide-like immunoreactivity in the central body. **C,E:** Locusta-tachykinin-like immunoreactivity in the central complex. **A:** 90% embryo, parasagittal section, showing weak immunoreactive arborizations in the fan-shaped body (FB, arrow heads), the triangle points to the beta lobe (bL). **B:** 5th larval instar, sagittal section, note the absence of immunoreactivity in the FB (arrow). **C:** Imago, sagittal section, showing the projections of locusta-tachykinin immunoreactive neurons from the repetitive clusters (RCL) to the protocerebral bridge (PB) and the anterior and median FB. Arrow points to the origin of PB arborizations. **D:** 24h old pupa, sagittal section, recurrence of FMRFamide immunoreactivity in the anterior and median FB (aFB, mFB), the PB remains unstained. **E:** Prepupa, glomerular-like arborizations of RCL neurons in the PB, asterisk points to the median RCL 1 projection, white bar indicates RCL 1,2 neurons from one hemisphere a: anterior, d: dorsal, N: Nodulli, p: posterior. Scale bars = 50 mm.

Tachykinin-like immunoreactivity. Using the monoclonal rat antibody against substance P and the polyclonal rabbit antiserum against locusta-tachykinin II (Lom-TK II), all immunopositive neurons and processes with connections to the central complex are recognized with both antibodies. Even in the brain of an embryo (90% development), the fan-shaped body (FB) neuropil is innervated by immunopositive fibres, which fill up the eight subcompartments of the fan-shaped body. The projections of these neurons terminate in the dorsal lateral part of the ventral bodies. From the 5th larval instar onwards, the respective somata are detectable. They form four groups, the repititive clusters (RCL 1-4), which are located in the posterior dorsal protocerebrum of each hemisphere. The pattern of immunopositive fibres created by these neurons in the fan-shaped body and the ventral bodies does not change during larval development. These neurons of repetitive clusters persist from the larval to the adult stage and, with the onset of the metamorphosis, they enlarge their arborization fields and increase in number from 32 to 64 during pupation. Additional invasions of immunoreactive fibres are found in the newly formed median part of the fan-shaped body and in the protocerebral bridge (Fig. 1C). The fibre plexus in the protocerebral bridge are extensive up to the late prepupa, forming eight glomerular-like subdivisions in each hemisphere (Fig. 1E). The midline of the protocerebral bridge remains unstained. These dense arborizations become less conspicious during pupation and dissplay a more homogeneous immunostaining. The staining intensity of the fibre plexus in the ventral bodies is not altered. Using the technique of intracellular injection of Lucifer yellow in previously immunolabeled neurons, it can be shown that the neurons of the repetitive clusters in prepupa display dense arborizations within the protocerebral bridge. In pupae and adult beetles, these neurons outline projections with short side branches into the protocerebral bridge..

FMRFamide immunohistochemistry. In the late embryo (80-100% development), weak FMRFamide-like immunoreactivity is found in the fan-shaped body (Fig. 1A). This distribution pattern is maintained up to the fourth larval instar and disappears in the late larval instars (Fig. 1B). From the prepupa onwards, faintly labeled immunopositive fibres can be identified in the anterior and the newly formed median fan-shaped body. This pattern is slightly expanded during pupation (Fig. 1D) and remains constant up to the stage of imago.

In the protocerebral bridge weak immunopositive fibres can be detected only in the adult animals. The ellipsoid body, the posterior fan-shaped body and the nodulli are free of immunopositive fibres. The respective somata of fan-shaped body invading neurons could not be verified.

Discussion

As known from *Drosophila*, the expression of the FMRFamide gene in the nervous system shows stage-dependent variations and provides evidence for stage-specific functions (O´Brien et al., 1991). In *Tenebrio,* the distribution of immunopositive structures in the central body also changes, whereas most of the identified FMRFamide- immunopositive neurons of the larval brain persist throughout metamorphosis without changing their pattern of immunostaining (Breidbach and Wegerhoff, 1994). The distribution of immunopositive material in embryonic and early larval stages is in accord with developmental features in the fan-shaped body (Breidbach et al., 1992; Wegerhoff and Breidbach, 1992). The prepupal reappearance of immunopositive sites is in agreement with the extension, new-formation and remodelling within the central body and the nervous system in general. This could depend on a change in peptide content of persisting neurons or on a new invasion of immunopositive neurons. In contrast to FMRFamide-like immunoreactivity, the tachykinin-immunopositive neurons do not show a decrease of immunoreactivity in the fan-shaped body. They progressively extend in their target areas during development. The tachykinin-like immunopositive neurons outline a persisting distribution in the anterior fan-shaped body and the ventral bodies and possesses intrinsic side branches in the median fan-shaped body and the protocerebral bridge. The arborizations of the protocerebral bridge show an intense sprouting in the prepupa and an evident loss of side branches during the pupal stage. This formation of precise neural adult connections emerges from an immature pattern during pupation and seems to reflect a refinement, as it is characterized by activity-dependent processes (Schatz, 1995).

Conclusion

Tachykinin- and FMRFamide immunoreactive neurons of *Tenebrio molitor* are already present from the embryo onwards. They are involved in the formation of the central complex. FMRFamide-immunoreactive neurons appear to have stage-specific functions, whereas tachykinin-immunopositive neurons display a more constant distribution. The combination of both immunohistochemical reactions allows a detailed analysis of developmental features during brain development.

Acknowledgements
We thank Dr. Hans Agricola (Jena) for generous gifts of the Lom-TK II antisera, Dr. Lee Shaw for providing valuable comments on the manuscript and Miss Marlies Rusch for exellent technical assistance.

References

Breidbach, O., Dennis, R., Marx, J., Görlach, C., Wiegand, H. and Wegerhoff, R. (1992) Insect glial cells show differential expression of a glycolipid-derived, glucuronic acid-containing epitope throughout neurogenesis:Detection during postembryogenesis and regeneration in the central nervous system of *Tenebrio molitor* L. *Neurosci. Lett.* 147: 5-8.

Breidbach, O. and Wegerhoff, R. (1994) FMRFamide-like immunoreactive neurons in the brain of the beetle, *Tenebrio molitor* L. (Coleoptera: Tenebrionidae): Constancies and variations in development from the embryo to the adult. *Int. J. Insect. Morph. Embryol.* 23: 383-404.

Nässel, D.R. (1993) Neuropeptides in the insect brain: A review. *Cell Tissue Res.* 273: 1-29.

O'Brien, M.A., Schneider, L.E. and Taghert, P.H. (1991) In situ hybridisation analysis of the FMRFamide neuropeptide gene in *Drosophila*. II. Constancy in the cellular pattern of expression during metamorphosis. *J. Comp. Neurol.* 304: 623-638.

Panov, A.A. (1959) Bau des Insektengehirns während der postembryonalen Entwicklung. II. Zentralkörper. *Entomol. Obozr.* 38: 301-311.

Schatz, C.J. (1995) Brain waves and brain wiring. *In:* N. Elsner and R. Menzel (eds.): *Learning and Memory, Göttingen Neurobiology Report,* Thieme, Stuttgart, New York, pp. 159-187.

Wegerhoff, R. and Breidbach, O. (1992) Structure and development of the larval central complex in a holometabolan insect, the beetle *Tenebrio molitor*. *Cell Tissue Res.* 268: 341-359.

Wegerhoff, R. and Breidbach, O. (1994) Intracellular dye injection of previously immunolabeled insect neurons in fixed brain slices. *J. Neurosci. Meth.* 53: 87-93.

Comparative aspects of neurohypophyseal hormone genes

A. Urano, K. Kubokawa[1], M. Suzuki[1] and H. Ando

Hokkaido University, Graduate School of Science, Division of Biological Science, Sapporo 060, Japan
[1] *Ocean Research Institute, University of Tokyo, Laboratory of Molecular Biology, Nakano, Tokyo 164, Japan*

Summary. Based on the nucleotide and deduced amino acid sequences, we analyzed comparative aspects of neurohypophyseal hormone precursors and genes encoding them in vertebrates from cyclostomes to mammals. The evolutionary pathway of neurohypophyseal hormone precursors was estimated by use of the neighbour-joining method. The pathway is obviously composed of two branches: one of teleosts, and the other of tetrapods, implying that teleost isotocin and amphibian mesotocin precursors were independently derived from ancestral vasotocin precursor by gene duplication in each lineage. Interestingly, salmonid fishes have two genes for each, vasotocin and isotocin. Salmon isotocin genes have 4 exon/3 intron structure like hagfish vasotocin gene, whereas salmon vasotocin genes have 3 exon / 2 intron structure as mammalian genes. Despite a probable deletion of the first intron from the ancestral vasotocin gene, the structure of neurohypophyseal hormone precursors has been well conserved in gnathostomes. In contrast to the conserved structure of the coding regions, the sequences and also the structure of 5' upstream regions responsible for regulation of gene expression are highly variable not only among different vertebrate classes, but also between the genes for vasopressin and oxytocin families even in the same species. Biological or adaptational meanings of such diversity in the promoter and enhancer regions remain to be clarified.

Introduction

Neurohypophyseal hormones are nonapeptides synthesized mainly by hypothalamic neurosecretory neurons. More than ten distinct neurohypophyseal principles have been characterized in a wide variety of vertebrates since the first isolation and characterization of mammalian hormones in 1953 (Acher, 1993). They can be classified into two groups: the vasopressin and the oxytocin families, which are believed to emerge from a common ancestral molecule by gene duplication. Numerous evolutionary pathways were proposed depending on the amino acid sequences of these nonapeptides. Recent insight into cDNAs encoding neurohypophyseal hormone precursors in both invertebrates and vertebrates enabled us to estimate the most probable evolutionary pathway of precursor molecules in vertebrates (Urano et al., 1992, 1994; Suzuki et al., 1995). Here, we have estimated the probable evolutionary pathway of neurohypophyseal hormone precursors by the neighbour-joining method. Further, we have compared the structures of genes encoding neurohypophyseal hormones, since evolution of a certain family of peptide hormones is a result of accumulation of mutations in the hormonal genes.

Structure of cDNAs and precursors

Hitherto known cDNAs and/or genes encoding neurohypophyseal hormone precursors have been obtained from American hagfish (Heierhorst et al., 1992), Japanese hagfish and lamprey (Suzuki et al., 1995), several salmon species (see Urano et al., 1994), white sucker (Heierhorst et al., 1989), Japanese toad (Nojiri et al., 1987), chicken (Hamann et al., 1992) and several mammalian species (see Gainer and Wray, 1994). All precursors deduced from them are composed of three moieties: signal peptide, hormone, and neurophysin that is connected to the hormone by the Gly-Lys-Arg residues as a cleavage and amidation signals.

The deduced amino acid sequences of hormone precursors are strikingly similar to each other, particularly in the hormone domain and the central part of neurophysin. The positions of cysteine residues in the hormone and neurophysin domains are highly conserved in all species examined. Nonetheless, the sizes of salmon vasotocin and isotocin neurophysins, which have extended C-terminus regions of similar length, are larger than those of toad mesotocin and mammalian oxytocin, suggesting that the oxytocin gene derived from the mesotocin gene. Similarity of the hydropathy profile of rat oxytocin precursor to that of toad mesotocin precursor further indicates that amphibian mesotocin neurophysin is ancestral to both oxytocin and vasopressin neurophysins (data not shown).

The idea mentioned above coincides well with the topology showing a probable evolutionary pathway of neurohypophyseal hormone precursors in gnathostomes estimated by the neighbour-joining method using the lamprey sequence as the outgroup (Fig. 1).

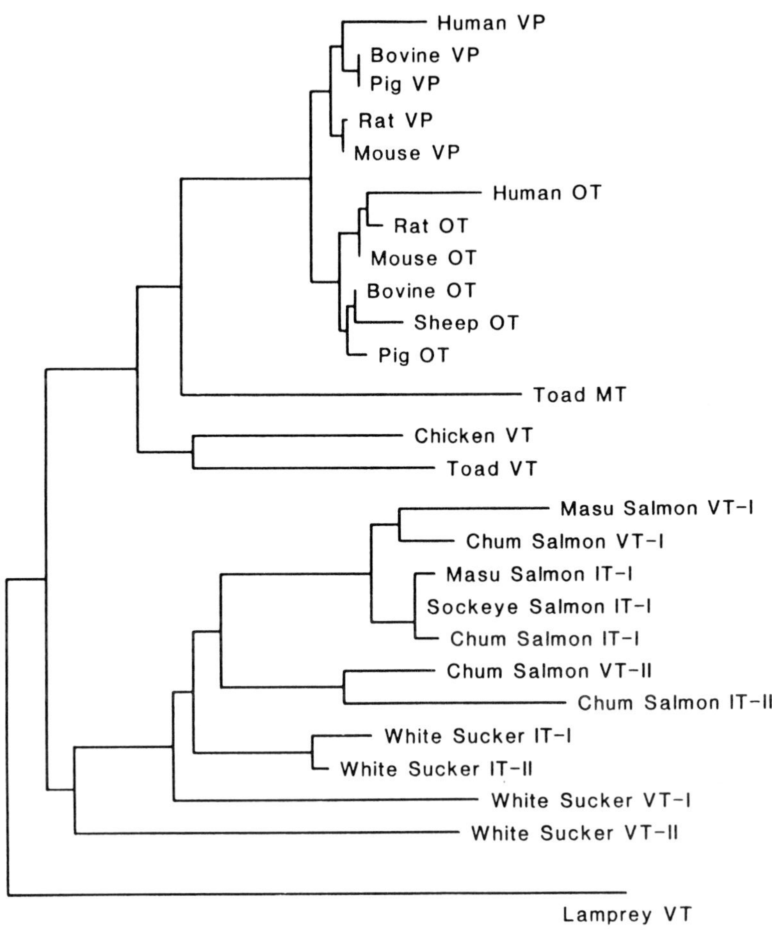

Fig. 1. An evolutionary tree of highly conserved parts of gnathostome neurohypophyseal hormone precursors which were predicted by analyses of cDNAs and/or genes. The horizontal lengths of branches are proportional to the numbers of amino acid substitutions. VP, vasopressin; OT, oxytocin; MT, mesotocin; IT, isotocin. Reproduced from Suzuki (1994).

The most remarkable feature in the evolutionary pathway is that the teleost and tetrapod lineages were divided at their origin immediately after the diversion from the lamprey lineage, indicating that teleost neurohypophyseal hormone precursors have their own evolutionary history other than that of tetrapodes. It is evident from the topology in Fig. 1

that, in tetrapods, mammalian vasopressin and oxytocin precursors were derived from mesotocin precursor which had diverged from ancestral amphibian vasotocin precursor. We sustain this hypothesis, by similar results from our previous estimation of molecular evolution of neurohypophyseal hormones obtained by different methods (Urano et al., 1992, 1994).

Structure of genes, coding region

Genes coding for the neurohypophyseal hormone precursors were studied mostly in several mammalian species, such as the cow, sheep, rat, mouse and human (see Gainer and Wray, 1994), and in a few species of lower vertebrates (see for fish, Urano et al., 1994; chicken, Hamann et al., 1992). Hence, we have cloned and analyzed the nucleotide sequences of genes encoding chum salmon vasotocin and isotocin precursors (Satomi et al., 1994; Kuno et al., 1995), and have compared the structure of both coding and non-coding regions with the previous studies.

The characteristic features of mammalian vasopressin and oxytocin genes are that they are located on the same chromosome locus in opposite transcriptional orientations. The intergenic distances between vasopressin and oxytocin genes vary from 3.5 to 11 kb. However, we could not detect a presence of vasotocin and isotocin genes on the same locus within the range of 20 kb by Southern blot analysis of the genomic DNA of chum salmon, despite the presence of two genes for each of the vasotocin and isotocin precursors in salmonids (Hiraoka et al., 1993). Further, restriction analysis of the genomic DNA from a masu salmon showed the presence of at least three different vasotocin genes and two isotocin genes, in contrast to mammals which have only one copy each of vasopressin and oxytocin genes.

As is shown in Fig. 2, most of the genes coding for neurohypophyseal hormone precursors, including teleost genes for two vasotocin precursors (provasotocin-I and -II) in both chum salmon and white sucker, consist of three exons and two intervening introns (3E/2I structure), whereas genes for hagfish vasotocin precursor and chum salmon isotocin precursors (proisotocin-I and -II) are composed of 4 exons and 3 introns (4E/3I structure). White sucker isotocin genes are intronless.

In the neurohypophyseal hormone genes of 3E/2I structure, the first exon (usually referred to as exon A) encodes a signal peptide, the hormone and the N-terminal portion of neuro-physin; the second exon (exon B)codes for the conserved central part of neurophysin; and the third exon (exon C), the C-terminal part of neurophysin. In spite of the presence of three introns in the hagfish vasotocin and salmon isotocin genes, and the lack of introns in the white sucker isotocin genes, similar structures are well conserved in the coding regions of neurohypophyseal hormone precursor genes. Corresponding sequences in the coding regions between genes for neurohypophyseal hormone precursors from different species are highly homologous.

Interestingly, the arrangement of exons and introns, except for the lengths and sequences of introns, is almost the same in the hagfish vasotocin gene and the salmon isotocin genes (Fig. 2), suggesting that a framework of genes encoding the ancestral neurohypophyseal hormone precursor is conserved in the salmon isotocin genes. Similarity of hydropathy profiles of deduced precursor molecules between hagfish vasotocin and salmon isotocin seems to support this hypothesis. Deletion of three introns during teleost evolution, probably after the diversion of Clupeiformes and Cypriniformes, but before the tetraploidization which may occur in many cyprinid fishes, may account for the lack of introns in the isotocin genes of white sucker. Since the neurohypophyseal hormone gene contains the sequence encoding vasotocin in the hagfish, and vasotocin was the only identified neurohypophyseal hormone in cyclostomes, isotocin may not be an ancestral molecule of fish neurohypophyseal hormones.

These analyses of the structure of genes for neurohypophyseal hormone precursors strongly support our hypothesis that teleost isotocin and amphibian mesotocin genes were independently derived from the ancestral vasotocin gene in each lineage by gene duplication event. In salmonid, the first intron in the ancestral teleostean vasotocin gene may have been conserved in the isotocin genes, but may have been deleted from vasotocin genes probably after the divergence of ancestral vasotocin gene to vasotocin and isotocin

                                   A. Urano et al.

genes. Such divergence should happen before tetraploidization of salmonid species which occurred about 100 million years ago.

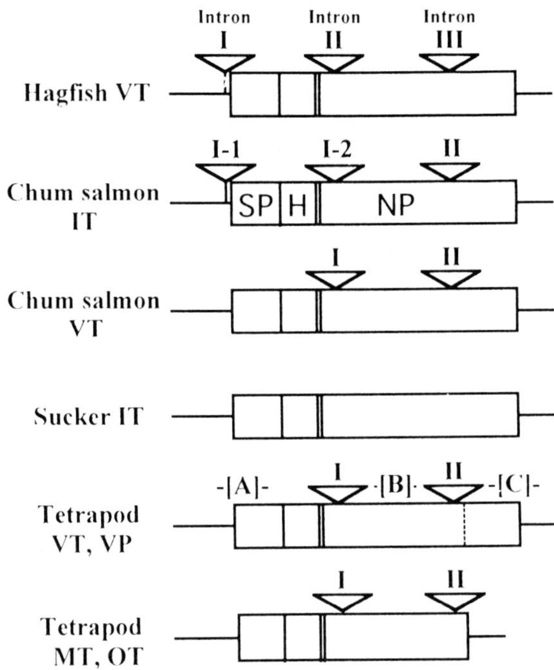

Fig. 2. Structure of neurohypophysial hormone precursors and their relations to encoding genes. The positions where introns intervene exons are shown by open triangles. The exon A locates upstream to the intron I in chum salmon vasotocin (VT) and tetrapod neurohypophyseal hormone genes; exon B, between introns I and II; and exon C, downstream to the intron II. The genes for hagfish vasotocin and chum salmon isotocin (IT) have three introns, whereas that for white sucker does not have any introns. VP, vasopressin; MT, mesotocin; and OT, oxytocin. SP, signal peptide; H, hormone; NP, neurophysin.

Structure of genes, 5' upstream region

The pattern of increased expression of vasopressin and oxytocin genes by hyperosmotic stimulation was apparently coincident in the rat brain (see Hyodo and Urano, 1991; Gainer and Wray, 1994), whereas expression of salmon vasotocin genes, but not isotocin genes, was increased by hypo-osmotic stimulation (see Urano et al., 1994). We have presumed that such difference in the expression of the genes in the same superfamily of neurohypophyseal hormones reflects evolution of the structural organization of the 5' upstream region, which is responsible for regulation of transcription to mRNAs. If so, structural differences in the regulatory regions of hormonal genes in the same family may yield explanation for molecular mechanisms of adaptation. Therefore, we compared the structure of 5' upstream regions of hitherto known chicken and mammalian sequences with that of salmon provasotocin-I gene (Satomi et al., 1994). For this analysis, we used computer software for harplot analysis (Genetyx, Software Development, Tokyo) and signal search to localize *cis*-acting elements (Sscan) referring to transcription factor database (TFD v.7.4, by Dr. David Ghosh, National Institute of Health, USA).

Harplot analysis for similarity of nucleotide sequences clearly showed that the sequences in the exons A and B are considerably homologous among the salmon provasotocin-I gene, chicken vasotocin gene, rat and bovine vasopressin and oxytocin genes, whereas the exon C is not so highly conserved. Despite rather conservative features in the coding region, i.e., in the exons, the upstream region and introns hardly share any noticeable homologies between genes for vasopressinthe family of and oxytocin peptides even in the same species. Actually, vasopressin and oxytocin genes do not share any homologous sequences in their promoter regions. The rat vasopressin and oxytocin genes only share a similar sequence of about 30 bases in the enhancer region which is located 1.5 kb upstream to the exon A in the oxytocin gene, although any particular *cis*-acting elements registered in the transcription factor database were not found in this site.

It is exceptional that the 5' upstream regions of rat and bovine vasopressin genes have considerable homology with one another in their promoter regions. It is also true for the oxytocin genes between the rat and bovine. Generally, 5' upstream regions seem to be highly variable, since no sequence homologies were shown among genes for neurohypophyseal peptides of the same family in different vertebrate classes. Even the precursor genes encoding the same hormone do not share similarity. The salmon and chicken vasotocin genes really do not show any sequence similarity with each other nor with mammalian genes.

Here, the question arises whether genes encoding neurohypophyseal hormone precursors include similar *cis*-elements in their promoter and enhancer regions. Although the location and the numbers are variable, they possess similar *cis*-acting elements, such as an estrogen responsive element, glucocorticoid responsive element, AP-1 site where phosphorylated products of immediate early genes bind, and cAMP responsive element in their upstream and coding regions.

At present, we do not know, or cannot conclude any general rules for the arrangement of *cis*-acting elements in terms of regulation of gene expression, probably because positions of regulatory elements are not critical for gene expression. Although we do not have any experimental data in lower vertebrates, the present analysis suggests that neurohypophyseal hormone genes in vertebrates are under the influences of similar regulatory factors, whose roles in physiological or adaptational functions remain to be examined.

Acknowledgments
We thank our collaborators, Mr. Y. Satomi, Ms. S. Kuno, Dr. S. Hyodo, Dr. H. Kishino, Dr. J. Adachi and Professor M. Hasegawa for their invaluable help and advice. Part of our study referred to in this article was supported by the Ministry of Education, Science and Culture, and the Fisheries Agency, Japan.

References

Acher, R. (1993) Neurohypophysial peptide systems: processing machinery, hydroosmotic regulation, adaption and evolution *Regul. Pept.* 45: 1-13.

Gainer, H. and Wray, S. (1994) Cellular and molecular biology of oxytocin and vasopressin. In: K. Knobil and J.D. Neill (eds.): *The Physiology of Reproduction. Second Edition,* Raven Press, New York, pp. 1099-1129.

Hamann, D., Hunt, N. and Ivell, R. (1992) The chicken vasotocin gene. *J. Neuroendocrinol.* 4: 505-513.

Heierhorst, C., Lederis, K. and Richter D. (1992) Presence of a member of the Tel-like transposon family from nematodes and Drosophila within the vasotocin gene of a primitive vertebrate, the Pacific hagfish, Eptatretus stouti. *Proc. Natl. Acad. Sci. USA* 89: 6798-6802.

Heierhorst, C., Morley, S.D., Figueroa, J., Krentler, C., Lederis, K. and Richter, D. (1989) Vasotocin and isotocin precursors from the white sucker, *Catostomus commersoni*: Cloning and sequence analysis of the cDNAs. *Proc. Natl. Acad. Sci. USA* 86: 5242-5246.

Hiraoka, S., Suzuki, M., Yanagisawa, T., Iwata, M. and Urano, A. (1993) Divergence of gene expression in neurohypophyseal hormone precursors among salmonids. *Gen. Comp. Endocrinol.* 92: 292-301.

Hyodo, S. and Urano, A. (1991) Expression of neurohypophyseal hormone precursor genes in the mammalian hypothalamus. *Zool. Sci.* 8: 1005-1022.

Kuno, S., Kubokawa, K., Urano, A., Ishii, S. and Nagasawa, H. (1995) Cloning and sequence analysis of isotocin-I gene of chum salmon. *Zool. Sci.* 12 (Suppl.): 7.

Nojiri, H., Ishida, I., Miyashita, E., Sato, M., Urano, A. and Deguchi, T. (1987) Cloning and sequence analysis of cDNAs for neurohypophyseal hormones vasotocin and mesotocin for the hypothalamus of toad, Bufo japonicus. *Proc. Natl. Acad. Sci. USA* 84: 3043-3046.

Satomi, Y., Kubokawa, K., Ando, H., Ono, M. and Urano, A. (1994) Cloning and sequence analysis of vasotocin-I gene of chum salmon. *Zool. Sci.* 11 (Suppl.): 9.

Suzuki, M. (1994) *Molecular evolution of neurohypophyseal hormone genes.* Ph.D. Thesis, University of Tokyo.

Suzuki, M., Kubokawa, K., Nagasawa, H. and Urano, A. (1995) Sequence analysis of vasotocin cDNAs of the lamprey, *Lampetra japonica*, and the hagfish, *Eptatretus burgeri*: evolution of cyclostome vasotocin precursors. *J. Mol. Endocrinol.* 14: 67-77.

Urano, A., Hyodo, S. and Suzuki, M. (1992) Molecular evolution of neurohypophyseal hormone precursors. *Prog. Brain Res.* 92: 39-46.

Urano, A., Kubokawa, K. and Hiraoka, S. (1994) Expression of the vasotocin and isotocin gene family in fish. *Fish Physiology.* 13: 101-132.

V. Signal transduction and integrative systems

Signal transduction and second messengers in neurosecretory cells

J. Meldolesi, E. Clementi, F. Codazzi, R. Pezzati, G. Racchetti and F. Grohovaz[1]

DIBIT, Scientific Institute San Raffaele, and [1]Department of Pharmacology, CNR Cytopharmacology and B. Ceccarelli Centers, University of Milan, Italy

Summary. The article summarizes general information about signal transduction and second messenger generation and functions in neurosecretory cells. Special attention is given to Ca^{2+} homeostasis events: release from intracellular stores and influx through voltage-insensitive channels opened either following activation of receptors coupled to polyphosphoinositide hydrolysis or as a consequence of internal store discharge. The spatio-temporal aspects of dynamic Ca^{2+} events, i.e. oscillations and waves, are discussed. Studies in progress on total calcium distribution, revealed at the ultrastructural level by electron spectroscopic imaging (ESI), are presented.

Introduction

The importance of transmembrane signalling studies in biological research has increased progressively during the last decades. Initially the definition of the process was interpreted *sensu strictu*, i.e. it was believed to include only the events generated at the internal face of the plasmalemma following activation of surface receptors. Thus, for many years work in the transmembrane signalling field consisted of the investigation of a few types of events, such as the receptor-induced changes of cytosolic cAMP and Ca^{2+} concentration and the mechanisms of G-protein interaction and of tyrosine phosphorylation, the latter induced by growth factor receptors. With time, however, it became more and more clear that the events generated inside a cell do not only operate *per se*, but participate in complex networks which account for the so called cross talking of signals. The understanding of synapse function first, and the development of knowledge on Ca^{2+} homeostasis more recently, have called attention to the heterogeneity of these processes within the cell, due to the variable distribution not only of surface receptors (and thus of second messenger generation), but also of second messenger targets. At present signalling needs therefore to be conceived as a vast area, governed by dynamic equilibria continuously adjusted by variously distributed factors and events. These events can travel between (regions of) the plasmalemma and the underlying cytoplasm, or consist of microevents that are extinguished locally. The signalling area is therefore not limited at the plasmalemma but extends profoundly into the cell, including critical events such as decision making, uphill of the operational activities of cell life.

In view of the complexity mentioned so far, signalling studies require a multidisciplinary approach in which classical biochemistry, pharmacology and electrophysiology can be integrated with each other and accompanied by other techniques including microscopy of living and fixed cells and molecular biology. Such studies should by characterized by high resolution both in terms of time (down towards the milliseconds level of classical electrophysiology) and in terms of space (μm and below). Up to now studies of this complexity have been carried out primarily in model cell systems. Expansion of the results to other physiologically more interesting cell types remains a challenge for future work.

Two cell models: rat chromaffin cells and their tumoural counterparts, PC12 cells

So far, studies on chromaffin cells have been carried out most often with bovine cells, rat cells being employed only rarely primarily for recovery problems. This is of course not the case of PC12 cells that can be obtained in unlimited amounts and are therefore amply used as a neurosecretory cell model. Cell lines, however, have their drawbacks, the first being the variability of the populations investigated in the various laboratories. Description of this last property of PC12 cells has rarely gone beyond personal communications. For many years, therefore, the real heterogeneity levels of these populations have remained difficult to evaluate without specific experimental examination.

During the last 5 years, the transmembrane signalling and secretion properties of a group of permanent PC12 clones, isolated by the neomycin selection procedure, have been investigated in our laboratory (Fasolato et al., 1991; Zacchetti et al., 1991; Clementi et al., 1992a). The observed variability was considerable and concerned almost all properties investigated. In terms of transmembrane signalling, the clones exhibited different levels of expression of various receptors, in particular of M_3 muscarinic and P_{2x} purinergic (Zacchetti et al., 1991) receptors. In no case did our clones exhibit nicotinic responses (a classical property of other PC12 cells), whereas the B_2 bradykinin and the P_{2u} purinergic receptors (Zacchetti et al., 1991) were always expressed. All clones responded to inositol triphosphate (IP_3) generation by releasing Ca^{2+} from intracellular stores, thus confirming the ubiquity of the corresponding intracellular receptor. Responses to caffeine and ryanodine were in contrast restricted to about half of them. In the positive clones this response was due to the expression of a type II ryanodine receptor which is located in the same store responsive to IP_3, i.e. in areas of the endoplasmic reticulum replenished of Ca^{2+} by sarcoplasmic-endoplasmic reticulum Ca^{2+} ATPases (SERCAs), (Zacchetti et al., 1991). Accurate studies of PC12 cells excluded localization of the above intracellular Ca^{2+} channels in other (putative) stores such as secretion granules, as it has been reported recently in other cell types (Blondel et al., 1995). Even in the latter cells, however, such a localization has been questioned (Meldolesi and Pozzan, 1995 and L. Orci, personal communication). The exclusive localization of the rapidly exchanging Ca^{2+} stores in the endoplasmatic reticulum appears therefore to be generally valid for all cells, including those competent for neurosecretion.

Receptor endowment of PC12 cells

In the preceding section we have already mentioned a few receptor types expressed by PC12 cells. Here we would like to rediscuss the problem systematically, paying attention not only to receptors *per se* but also to the intracellular responses triggered by their activation. Much of the initial popularity of PC12 cells depended on their ability to undergo a neuronal-like differentiation when exposed to nerve growth factor NGF (Greene and Tischler, 1976). Although not general, this property is shared by most PC12 clones inasmuch as they express at their surface both NGF receptors: the high affinity tyrosine kinase trkA and the low affinity gp75 receptors. trkA, however, is not the only growth factor tyrosine kinase receptor expressed by PC12. Responses induced by epidermal growth factor (EGF) and fibroblast growth factor (FGF) have in fact also been reported, with variable effects on cell

                                    J. Meldolesi et al.

differentiation (Togari et al., 1983; Mark et al., 1995).

Among adrenergic receptors, β-receptors have been shown to induce the classical response of increased [cAMP], which was partially inhibited by α_2 stimulation (Gatti et al., 1988). Polyphosphoinositide (PPI) hydrolysis was induced by the B_2 bradykinin, the M_3 muscarinic and the P_{2u} purinergic (ATP-UTP) receptors (Zacchetti et al., 1991). Among the ionophoric receptors, only P_{2x} (Fasolato et al., 1990) has been found to be frequently expressed by our PC12 clones and by the parental mixtures. This receptor channel appears to be permeable not only to Na^+ but also to Ca^{2+}, capable therefore to induce $[Ca^{2+}]_i$ responses also independently from depolarization and activation of voltage-gated channels, such as those of the L type, that are abundant in PC12 cells (Fasolato et al., 1990).

Generalities on $[Ca^{2+}]_i$ responses: from single spikes to oscillations and waves

To our knowledge PC12 were the first neuroendocrine cells to be investigated for $[Ca^{2+}]_i$ responses in the early eighties. At that time, cells were loaded in suspension with the "pioneer" Ca^{2+} dye, quin-2, and then exposed to either depolarization (by high $[K+]_o$) or activation of PPI hydrolysis-coupled receptors. In both cases, the $[Ca^{2+}]_i$ responses (those originally recorded from quin-2 loaded cells and more recently confirmed in cells loaded with fura-2) appeared composed by an initial spike followed by a slowly declining plateau. However, the significance of these two phases is profoundly different. In depolarized cells, both the spike and the plateau are in fact sustained primarily by the activation of the same voltage-gated Ca^{2+} channels, occurring synchronously at the beginning of the stimulation (spike) and becoming less frequent and asynchronous later on (plateau). Consistent with this interpretation the responses can be prevented, or immediately blocked, by either the removal of extracellular Ca^{2+} (by application to the medium of excess EGTA) or the blockade of Ca^{2+} channels (by verapamil and dihydropyridines). With PPI hydrolysis-coupled receptors, on the other hand, the two phases of the responses are profoundly different in nature. The spike is in fact sustained primarily by IP_3-induced intracellular release of Ca^{2+}, while the plateau depends largely on influx, occurring however through channels different from voltage-gated ones and therefore insensitive to the above blockers (Fig. 1; Pozzan et al., 1994). A discussion about these channels will be presented in the following section of this article. Here we wish to emphasize that, in the case of PPI hydrolysis-coupled receptors, the spike-plateau response revealed by the analysis of cell suspensions represents nothing but an overall summary of complex events occurring within individual cells.

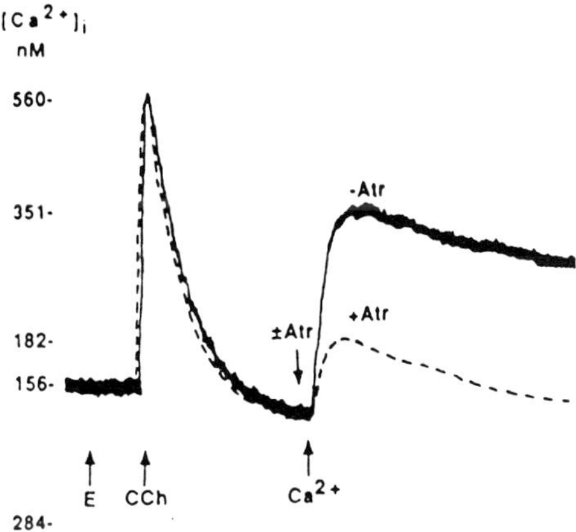

Fig. 1. $[Ca^{2+}]_i$ response induced in a suspension of PC12 cells by the administration of carbachol (CCh, 500 µM) in a typical Ca^{2+}-free (E=EGTA), Ca^{2+} reintroduction experiment. The initial peak increase of $[Ca^{2+}]_i$ is due to release from intracellular stores whereas the increases after Ca^{2+} reintroduction are due to Ca^{2+} influx. Solid and dashed lines show the effect of the muscarinic antagonist atropine (Atr, 10 µM) on the latter process. The atropine-resistant influx is due to activation of the store-operated Ca^{2+} channels. From Clementi et al., 1992b.

Progress of studies at the microscopical (single cell) level have demonstrated in fact, in many cell types (excluding however PC12), that these events consist in $[Ca^{2+}]_i$ oscillations, i.e. rhythmic spikes separated by periods (many seconds) during which $[Ca^{2+}]_i$ remains near the resting level (Fig. 2). So far, two different mechanisms have been described to sustain $[Ca^{2+}]_i$ oscillations. The first is based on the rhythmic activation of channels in the plasmalemma. Studies of this type of process are still limited, and therefore our present information about the channels involved is still incomplete. In bovine chromaffin cells, however, involvement of voltage-gated Na^+ channels appears likely inasmuch as exposure to

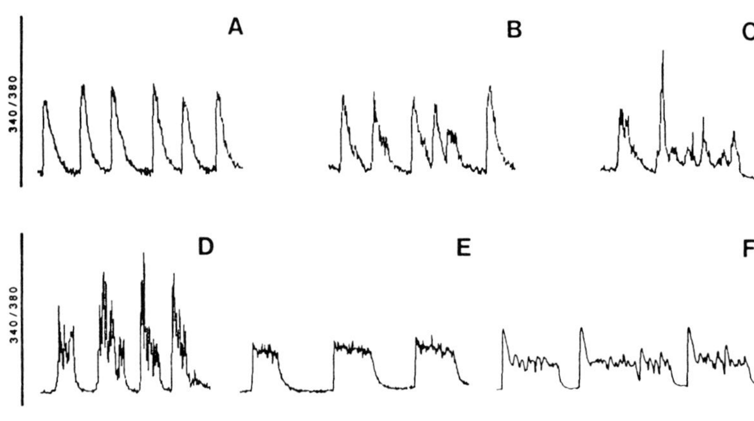

Fig. 2. Spontaneous $[Ca^{2+}]_i$ oscillations of different morphology in rat chromaffin cells. The regular patterns shown in A and B are the most frequently observed. From D'Andrea et al., 1993.

veratridine (a drug that decreases the threshold and delays the inactivation of these channels) has been shown to induce the appearance of surface-generated $[Ca^{2+}]_i$ oscillations (López et al., 1995). Much more is known about the oscillations largely sustained by Ca^{2+} discharge from intracellular stores. Mechanistically, one may wonder how can oscillations, which are multiple and may be asynchronous in the various cells, account for the spike-and-plateau response of cell suspensions. However, when induced by strong receptor stimulations which force IP_3 levels to increase steeply in all cells, the responses are first sustained by Ca^{2+} discharge occurring more or less synchronously from all IP_3-sensitive stores (spike). The subsequent plateau is due to the ongoing influx and to the continuous, fibrillating discharge of the various stores replenished as a consequence of the influx itself, i.e. a condition in which subtle equilibria among the components of the system are profoundly perturbed. Under resting or weak stimulation conditions, on the other hand, equilibria are preserved and the oscillation process can thus develop based on two fundamental properties of the IP_3 receptor: its heterogeneous distribution and its modulability by $[Ca^{2+}]_i$ (see Berridge, 1993 and Pozzan et al., 1994).

Heterogeneous distribution means accumulation of higher IP_3 receptor concentrations in the stores of a single cytoplasmic area, now identified as the pacemaker or trigger zone of oscillations (Fig. 3; D'Andrea et al., 1993; Hirose and Ino, 1994). These stores can be stimulated to release their Ca^{2+} at a concentration of IP_3 that remains subthreshold for the rest of the system. An alternative possibility to explain the existence of the pacemaker, i.e. that the differential sensitivity among the stores is due to differential distribution of various types, I-III, of IP_3 receptors, characterized by a different affinity for the second messenger, appears unlikely. Recent results have revealed in fact that within single cells the molecular heterogeneity of these receptors concerns subunits rather than functional tetramers. Tetramer assembly appears in fact to take place at random by the participation of all subunits expressed in that specific type of cell, probably yielding single, Gaussian-type assortments of complete receptors (Monkawa et al., 1995).

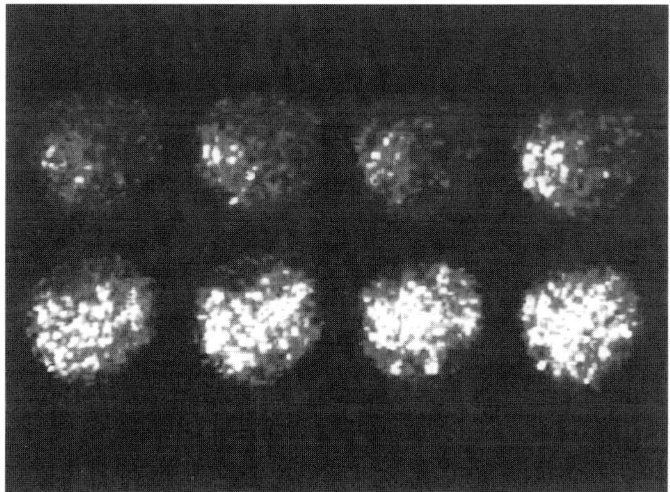

Fig. 3. Initial phase of a $[Ca^{2+}]_i$ oscillation in a rat chromaffin cell. Notice the beginning at a discrete trigger zone with subsequent expansion of the signal to the rest of the cell. From D'Andrea et al., 1993.

Once pacemaker stores are activated, discharge takes place and $[Ca^{2+}]_i$ therefore increases in the surrounding cytosol, then diffuses, although slowly, to the adjacent areas. At this point it is the $[Ca^{2+}]_i$ modulability of IP_3 receptors, positive from resting, 0.1 μM, to 0.3 μM, and negative at higher values, that plays the fundamental role in the spiking process. By this biphasic process, and with no need of any additional IP_3 changes, the stores located first close and then more and more away from the pacemaker are recruited in sequence, first to discharge (completion of the spike) and then to be inhibited. This latter event creates the conditions by which Ca^{2+} ions flowing into the cell by activation of voltage-insensitive channels can be quickly reaccumulated within the stores. This reaccumulation subtracts Ca^{2+} from the cytosol and thus maintains $[Ca^{2+}]_i$ near the resting levels throughout most of the interspike period. Only when the stores approach their full state does the accumulation slow down, and the ensuing slow increase of $[Ca^{2+}]_i$ leads to the stimulation of the pacemaker and thus to the initiation of the subsequent spike (Berridge, 1993; Pozzan et al., 1994).

Based on this description it is clear that, in order to be competent for oscillations, the cell needs to possess a pacemaker and an appropriate distribution of the rapidly exchanging Ca^{2+} stores. In the absence of the first such property oscillations are impossible, yet other dynamic events can still occur. When the generation of IP_3 is concentrated primarily at a discrete site of the plasmalemma, the initial Ca^{2+} release and the ensuing $[Ca^{2+}]_i$-modulated recruitment of stores in the surrounding areas can occur in a precise direction, resulting in the establishment of non-decremental waves within the cell. So far, the most spectacular waves have been described in *Xenopus* eggs (Leichleiter et al., 1991). Others, depending on the differently heterogeneous distribution of PPI hydrolysis-coupled receptors, have been observed in PC12 cells (Lorenzon et al., 1995). They will be discussed in a following section.

The plasmalemma Ca^{2+} channels

In the previous sections we have already mentioned that the plasmalemmal channels participating in the $[Ca^{2+}]_i$ responses are numerous and markedly heterogeneous. Voltage-gated channels have already attracted considerable attention in a variety of excitable systems. They therefore will not be discussed in detail here. In contrast other types of channels, insensitive to voltage, have been neglected until recently. At the moment, however, they represent a hot spot in Ca^{2+} research because their study has revealed a number of unexpected and highly interesting events. At least two main channel families appear to be involved (see

Fig. 1). The first, which participates directly in the transmembrane coupling of the PPI-hydrolysis receptors, exhibits differential expression in the various cell types: abundant in excitable – rare or absent in non excitable cells. Although knowledge about these channels is still limited, growing experimental evidence suggests their regulation to be mediated by G-proteins different from those responsible for receptor coupling to phospholipase C, i.e. from the G-proteins that govern PPI hydrolysis. Activation of the first class of voltage-insensitive channels appears therefore to be a process parallel to PPI hydrolysis, occurring only in the cells (such as PC12) that express both the channels and the regulating G-proteins (Fig. 1; Clementi et al., 1992b).

The second class, the so called store-operated Ca^{2+} channels, appear to be ubiquitous (or almost) among cells (see Fig. 1). The activity of these channels (some of which, of very low conductance, indicated as I_{CRAC} channels) is controlled by the state of filling of the rapidly-exchanging IP_3-sensitive Ca^{2+} stores. Thus, activation of Ca^{2+} influx through these channels occurs not directly, but rather indirectly following receptor activation. Moreover, any treatments causing intracellular discharge, even independently of receptor function, appear competent for channel activation. The rationale for the existence of these channels is clear: cells should never be depleted of their rapidly-exchanging Ca^{2+} pool. When discharge of this pool from the stores takes place, reaccumulation is therefore favoured by increased influx at the plasmalemma and, concomitantly, by the high $[Ca^{2+}]_i$ -induced inhibition of IP_3 sensitive receptors (see the preceding section and Clementi et al., 1992b). What in contrast is unclear is the mechanism by which an activation signal can flow contrariwise to the usual direction, i.e. backwards, from the stores to the plasmalemma, rather than inwards, from the plasmalemma to the stores. Recent developments have identified a channel (named trp) initially cloned from the eye of drosophila, that appears a good candidate for playing the role of the store-operated Ca^{2+} channel. Mammalian trp homologues have been identified and are now being intensely investigated both molecularly and physiologically. In the meantime drugs have been shown to block store-operated channels with some specificity over voltage-gated or PPI hydrolysis-coupled channels. Further developments, including therapeutical approaches developed by the use of these drugs, are expected to take place in the near future.

The cytoplasm is an excitable milieu

Until recently the definition of 'excitable' was attributed almost exclusively to membranes, specifically to the plasma membrane of the cells that express voltage-gated channels. When these cells are adequately stimulated, they can in fact give rise to autoregenerative ion flux events sustained by the successive activation of plasmalemmal channels. Recent results, parts of which have been already alluded to in a previous section, demonstrate however that excitability is not specific to membranes but is also shared by the cytoplasm (Leichleiter et al., 1991; Berridge, 1993). This property is based on the complex equilibria established between cytosol and Ca^{2+} stores, sustained by the activation of the voltage-insensitive channels in the plasmalemma. Notice that such a cytoplasmic excitability is not restricted to specific types of cells, as it occurs with membrane excitability, but is, at least potentially, ubiquitous, as is the IP_3 receptor. The degree of expression, however, is certainly quite variable. Some cells in fact exhibit clear waves whereas many others don't, even if micro-autoregenerative events most probably also occur in these cells.

Results recently obtained in a large fraction of differentiated, neuron-like PC12 cells (Lorenzon et al., 1995) delineate a type of cytoplasmic excitability that exhibits impressive aspects. In these cells both the phosphatidyl inositoldiphosphate hydrolysis-coupled receptors and the L type Ca^{2+} channels appear to be often heterogeneously distributed. In particular, the activation of both receptors, such as M_3 and B_2, and of channels was found to induce initial $[Ca^{2+}]_i$ increases localized at the tip of neurites. The subsequent evolution of the process was however different. In depolarized cells the heterogeneous distribution of $[Ca^{2+}]_i$ was maintained, and the rise in the cell body was very limited. In contrast, when receptors were activated, autoregenerative $[Ca^{2+}]_i$ waves did initiate, moving then centripetally to finally invade the whole cell body. This second type of response appeared entirely sustained by intracellular Ca^{2+} release occurring at the IP_3 receptor. In fact, responses of this type were observed also when cells devoid of ryanodine receptors were incubated in Ca^{2+}-free medium (Lorenzon et al., 1995) to exclude the contribution of influx across the plasmalemma. In order to explain the difference between depolarized and receptor-stimulated cells, one may argue that in resting PC12 cells the concentration of IP_3 is low. A simple increase of Ca^{2+}, as it occurs after depolarization, would thus be insufficient to get the IP_3 receptors activated, whereas this could happen when it is the generation of the second messenger to be first stimulated, as it occurs at the receptor level.

Two properties of the receptor-triggered waves occurring in PC12 cells are worth emphasising (see Lorenzon et al., 1995 for details). First, their direction was found to be receptor-specific. As already mentioned, a centripetal direction was observed after cell stimulation with either bradykinin or acetylcholine. However with ATP (or UTP), working at the P_{2u} receptor, the direction was the other way around, the beginning being in the cell body, followed by a subsequent invasion of the fibres. These results demonstrate that the cytoplasmic excitability does not occur always in one direction, but depends on the localization of its starting area. Once again, this is a property similar to those of membrane excitability. Second, the process was found to be temperature dependent, with considerable (10-20x) deceleration going from 37°C to 18°C. This property was important because experiments carried out in the cold revealed properties that could have not been distinguished at the physiological temperature.

Feedback control of transmembrane signalling responses: the example of the NO-G kinase system

As it is the case for most cellular events, also those participating in transmembrane signalling occur under the control established by products of the reaction chains initiated at the receptor level. As a whole, this idea is not new. In a preceding section we have already mentioned the positive and negative modulation of the IP_3 receptor dependent on moderate and high increases of $[Ca^{2+}]_i$ in the surrounding environment. Even more sensitive to Ca^{2+} is the ryanodine receptor, whose activation is induced by $[Ca^{2+}]_i$ rises (CICR = Ca^{2+}- induced Ca^{2+} release). At a very high $[Ca^{2+}]$, however, the ryanodine receptor is also inhibited (Berridge, 1993; Pozzan et al., 1994). Other important negative modulations of plasmalemma channels (voltage-gated) and receptor activities are mediated by protein kinases C, a family of enzymes that operate under the control of a second messenger, diacylglycerol, generated by the hydrolysis of polyphosphoinositides (together with IP$_3$) and also of other phospholipides.

A new exciting area that is now being developed in this field concerns nitric oxide, NO, and the G kinase. Many cell types, especially neurosecretory cells and neurons, are known to express a constitutive form of NO synthase, turned on by the increase of [Ca^{2+}]$_i$. The classical, first recognized function of the gas thus generated is the activation of the two G-kinases I, α and β. Recent studies have revealed that these kinases modulate the activity of two of the key components that sustain the [Ca^{2+}]$_i$ responses triggered by G protein-coupled as well as by tyrosine kinase possessing receptors: the phospholipases C (and families) that hydrolyze polyphosphoinositides – and the voltage-independent channels. In the first case, however, the modulation is negative while in the second it is positive (Clementi et al., 1995 a and b). The final effects of these modulations consist in the rapid turning off of the IP$_3$ generation, and thus of the Ca^{2+} release response induced by the second messenger, accompanied by the rapid refilling of the stores. These NO-induced regulatory effects can only occur co-ordinately with those mediated by [Ca^{2+}]$_i$, and are therefore expected to participate in the control of oscillations and waves, and of cytoplasmic excitability in general.

Free versus total calcium in the various cell compartments

The discussion about calcium carried out so far has concerned primarily the free, ionised fraction present in the cytosol. In the latter, as well as in all other cell compartments, free Ca^{2+} is present in equilibrium with another fraction bound to specific Ca^{2+} binding molecules. The properties of these molecules depend on the conditions that predominate in their compartment. For example in the cytosol, in which [Ca^{2+}] is low, the binding molecules are of high affinity. Among these are the so called E-F hand proteins, whose K$_D$ is around resting [Ca^{2+}]$_i$. Due to these molecules the cytosol is buffered, i.e. the amounts requested for its changing are much greater (50-200 fold) than the change finally achieved. Another consequence of the buffer is the slow rate of Ca^{2+} diffusion, which in the cytosol is 1-2 orders of magnitude slower of that in pure water.

Within a membrane-bound compartment, the endoplasmic reticulum, $[Ca^{2+}]$ is expected to be much higher than in the cytosol, due to the presence of low affinity lumenal Ca^{2+} binding proteins. Recently, these values have been evaluated to be 1 mM (Bastianutto et al., 1995; Montero et al., 1995), with approximately the same amount bound to proteins (Bastianutto et al., 1995). These values have been deduced from experimental data based on indirect criteria rather than obtained from direct measurements. In fact, of the methods available for assaying total calcium, some are of low resolution. The only method adequate to reveal total calcium data in small intracellular structures is electron energy loss spectroscopy, a technique that requires complex sample processing to be employed, and that therefore has been appropriately employed to deal with biological problems only in a few cases.

The $[Ca^{2+}]$ of another type of intracellular organelle, the mitochondrion, has been shown recently to follow closely the activity of the endoplasmic reticulum. Mitochondria do in fact accumulate in part the Ca^{2+} discharged by IP_3, working at the expense of their inner membrane potential. As a whole, mitochondria can thus play the role of an additional buffer with respect to large cytosolic $[Ca^{2+}]_i$ increases (Rizzuto et al., 1993). At the moment information about the physiological levels of mitochondrial total calcium is very limited. In contrast, mitochondria of damaged cells can pick up large amounts of Ca^{2+} that can ultimately react with phosphate and give rise to precipitates visible in the electron microscope.

Another organelle so far little known is the Golgi complex, which, however, is expected to contain high calcium because of the role of the cation in granule assembly (Chanat and Huttner, 1991). More, in contrast, is known about secretory organelles, synaptic (and synaptic-like) microvesicles and dense-core granules. Both are expected (and are in part known) to contain large amounts of calcium accumulated by means of exchange reactions with other ions, and released to the extracellular medium by exocytosis. These pools therefore appear to be distinct from those in rapid exchange with the cytosol described previously. The equilibrium between free and bound calcium fractions within secretory organelles is not yet known at the moment.

Conclusions

This rapid review about transmembrane signalling in neurosecretory cells is not intended to be complete. Indeed, large chapters, such as those on protein kinases and protein phosphatases, have been discussed only *en passant*, or completely overlooked. Our primary aim, in fact, was to provide information about areas which are directly investigated in our laboratory and which at the moment are in a phase of interesting development, with rapid acquisition not only of new information but also of new ideas. In these developments, the progress of techniques, such as single cell Ca^{2+} imaging and electron energy loss spectroscopy has been and will continue to be important. When properly applied, these techniques can open new vistas and lead to the reconsideration, from different standpoints, of pre-existing problems. Inevitably, and fortunately, progress keeps going on, and thus at least some of the points discussed here might become rapidly obsolete because other, new advanced vistas have been reached. This is one of the risks that inevitably exists when working in a hot field, a risk that can turn into enthusiasm inasmuch as the new vistas can emerge in part, and in some cases has indeed emerged, also from the work carried out in our laboratory.

Acknowledgements
The original experimental work reported in this paper has been supported by grants from the CNR (Target Project Biotechnology), AIRC (Italian Association for Cancer Research), and Telethon.

References

Bastianutto, C., Clementi, E., Codazzi, F., Podini, P., De Giorgi, F., Rizzuto, R., Meldolesi, J. and Pozzan, T. (1995) Overexpression of calreticulin increases the Ca^{2+} capacity of rapidly exchanging Ca^{2+} stores and reveals aspects of their lumenal microenvironment and function. *J. Cell Biol.* 130: 847-855.

Berridge, M.J. (1993) Inositol trisphosphate and calcium signalling. *Nature* 361: 315-325.

Blondel, O., Bell, G.I. and Seino, S. (1995) Inositol 1,4,5-trisphosphate receptors, secretory granules and secretion in endocrine and neuroendocrine cells. *Trends Neurosci.* 18: 157-161.

Chanat, E. and Huttner, W.B. (1991) Milieu-induced selective aggregation of regulated secretory proteins in the trans-Golgi network. *J. Cell Biol.* 115: 1505-1519.

Clementi, E., Racchetti, G., Zacchetti, D., Panzeri, M.C. and Meldolesi, J. (1992a) Differential expression of markers and activities in a group of PC12 nerve cell clones. *Eur. J. Neurosci.* 4: 944-953.

Clementi, E., Scheer, H., Zacchetti, D., Fasolato, C., Pozzan, T. and Meldolesi, J. (1992b) Receptor-activated Ca^{2+} influx. Two independently regulated mechanisms of influx stimulation coexist in neurosecretory PC12 cells. *J. Biol. Chem.* 267: 2164-2172.

Clementi, E., Martini, A., Stefani, G., Meldolesi, J. and Volpe, P. (1995) LU52396, an inhibitor of the store-dependent (capacitative) Ca^{2+} influx. *Eur. J. Pharmacol.* 289: 23-31.

Clementi, E., Sciorati, C., Riccio, M., Miloso, M, Meldolesi, J. and Nisticò, G. (1995) Nitric oxide action on growth factors-elicited signals. *J. Biol. Chem.*, in press.

D'Andrea, P., Zacchetti, D., Meldolesi, J. and Grohovaz, F. (1993) Mechanisms of $[Ca^{2+}]_i$ oscillations in rat chromaffin cells: complex Ca^{2+}-dependent regulation of a ryanodine-insensitive oscillator. *J. Biol. Chem.* 268: 15213-15220.

Fasolato, C., Pizzo, P. and Pozzan, T. (1990) Receptor mediated calcium influx in PC12 cells: ATP and bradykinin activate two independent pathways. *J. Biol. Chem.* 265: 20351-20355.

Fasolato, C., Zottini, M., Clementi, E., Zacchetti, D., Meldolesi, J. and Pozzan, T. (1991) Intracellular Ca^{2+} pools in PC12 cells. Three intracellular pools are distinguished by their turnover and mechanisms of Ca^{2+} accumulation, storage and release. *J. Biol. Chem.* 266: 20159-20167.

Gatti, G., Madeddu, L., Pandiella, A., Pozzan, T. and Meldolesi, J. (1988) Second-messenger generation in PC12 cells. *Biochem. J.* 255: 753-760.

Greene, L.A. and Tischler, A.S. (1976) Establishment of a noradrenergic clonal line of rat adrenal pheochromocytoma cells which respond to nerve growth factor. *Proc. Natl. Acad. Sci. USA* 73: 2424-2429.

Hirose, K. and Iino, M. (1994) Heterogeneity of channel density in inositol-1,4,5-trisphosphate-sensitive Ca^{2+} stores. *Nature* 372: 791-794.

Lechleiter, J., Girard, S., Peralta, E. and Clapham, D. (1991) Spiral calcium wave propagation and annihilation in *Xenopus laevis* oocytes. *Science* 252: 123-126.

López, M.G., Artalejo, A.R., García, A.G., Neher, E. and García-Sancho, J. (1995) Veratridine-induced oscillations of cytosolic calcium and membrane potential in bovine chromaffin cells. *J. Physiol.* 482: 15-27.

Lorenzon, P., Zacchetti, D., Codazzi, F., Fumagalli, G., Meldolesi, J. and Grohovaz, F. (1995) Ca^{2+} waves in PC12 neurites: a bidirectional, receptor-oriented form of Ca^{2+} signalling. *J. Cell Biol.* 129: 797-804.

Mark, M.D., Liu, Y., Wong, S.T., Hinds, T.R. and Storm, D.R. (1995) Stimulation of neurite outgrowth in PC12 cells by EGF and KCl depolarization: a Ca^{2+}-independent phenomenon. *J. Cell Biol.* 130: 701-710.

Meldolesi, J. and Pozzan, T. (1995) IP_3 receptors and secretory granules. *Trends Neurosci.* 18: 340-341.

Monkawa, T., Miyawaki, A. Sugiyama, T., Yoneshima, H., Yamamoto-Hino, M., Furuichi, T., Saruta, T., Hasegawa, M. and Mikoshiba, K. (1995) Heterotetrameric complex formation of inositol 1,4,5-trisphosphate receptor subunits. *J. Biol. Chem.* 270: 14700-14704.

Montero M., Brini, M., Marsault R., Alvarez J., Sitia, R., Pozzan, T. and Rizzuto, R. (1995) Monitoring dynamic changes in free Ca in the ER of intact cells. *EMBO J.*, in press.

Pozzan, T., Rizzuto, R., Volpe, P. and Meldolesi, J. (1994) Molecular and cellular physiology of intracellular Ca^{2+} stores. *Physiol. Rev.* 74: 595-637.

Rizzuto, R., Brini, M. and Pozzan, T. (1993) Microdomains with high Ca^{2+} close to IP_3-sensitive channels are sensed by neighboring mitochondria. *Science* 262: 744-746.

Togari, A., Baker, D., Dickens, G. and Guroff, G. (1983) The neurite-promoting effect of fibroblast growth factor on PC12 cells. *Biochem. Biophys. Res. Commun.* 114: 1189-1193.

Zacchetti, D., Clementi, E., Fasolato, C., Lorenzon, P., Zottini, M., Grohovaz, F., Fumagalli, G., Pozzan, T. and Meldolesi, J. (1991) Intracellular Ca^{2+} pools in PC12 cells. A unique, rapidly-exchanging pool is sensitive to both inositol 1,4,5-trisphosphate and caffeine-ryanodine. *J. Biol. Chem.* 266: 20152-20158.

The Peptidergic Neuron
B. Krisch and R. Mentlein (eds)
© 1996 Birkhäuser Verlag Basel/Switzerland

Multi-signal transduction of moth pheromone biosynthesis-activating neuropeptide (PBAN) and its modulation: Involvement of G-proteins?

A. Rafaeli and C. Gileadi

Department of Stored Products, ARO, The Volcani Center, Institute for Technology and Storage of Agricultural Products, P.O. Box 6, Bet Dagan, Israel

Summary. Pheromone production by many moths is regulated by the timely release of pheromone biosynthesis-activating neuropeptide (PBAN). Studies *in vitro* have resulted in the identification of the target cells for neuropeptide action. These cells are situated in the intersegmental tissue between the 8th and 9th abdominal segments of female moths. The involvement of G-proteins was implicated by the stimulatory action of NaF (1mM), which, like PBAN, stimulated both pheromone biosynthesis and intracellular cAMP levels. The adrenergic agonist, clonidine, inhibited the pheromonotropic actions (pheromone biosynthesis and intracellular cAMP elevations) due to PBAN and NaF. From our results we conclude that a negative regulation of pheromone biosynthesis occurs at the membrane receptor level by the interaction of an adrenergic receptor with the PBAN receptor.

Introduction

Females of several moth species exhibit a diel-periodicity of pheromone biosynthesis with a concomitant diel-periodicity of sexual behaviour. Regulation of the periodic biosynthesis of pheromone has been attributed to a neuropeptide, termed pheromone biosynthesis activating neuropeptide (PBAN), which is produced by neuroscretory cells of the suboesophageal ganglion (Raina et al., 1989). In our *in vitro* studies we have delineated the intersegmental membrane (situated between the 8th and 9th abdominal segments of the ovipositor tips of *Helicoverpa armigera* females) as the target tissue for PBAN stimulation (Rafaeli and Gileadi, 1995). Isolated intersegmental tissues are stimulated by PBAN to produce the main pheromone component of *Helicoverpa armigera* (Z-11 hexadecenal) (Rafaeli and Gileadi, 1995). The pheromonotropic response to PBAN has been shown to induce increases in intracellular cAMP levels in *Helicoverpa armigera* (Rafaeli and Soroker, 1989). Several pharmacological agents (forskolin, isobutylmethylxanthine and cAMP analogs) were shown to induce pheromone biosynthesis (Soroker and Rafaeli, 1989; Jurenka et al., 1991; Soroker and Rafaeli, 1995). The involvement of phosphatidylinositol breakdown was implicated by the stimulation of pheromone biosynthesis by phorbol esters and diacylglycerol analogs (Soroker and Rafaeli, 1995) although no inhibition was observed with staurosporine, a protein kinase-C inhibitor (Soroker and Rafaeli, 1995; Matsumoto et al., 1995).

The PBAN-induced pheromonotropic activity was found to be calcium-dependent (Jurenka et al., 1991; Fonagy et al., 1992; Soroker and Rafaeli, 1995; Ma and Roelofs, 1995). The ionophores, ionomycin (Soroker and Rafaeli, 1995) and thapsigargin (Rafaeli, unpublished) stimulated pheromone production and caused significant elevations of intracellular cAMP levels thereby suggesting that calcium activates adenylate cyclase during pheromone gland stimulation. Inhibition of both pheromone biosynthesis and intracellular cAMP formation was observed by the calcium-calmodulin inhibitor, W12 (Soroker and Rafaeli, 1995) and inhibition of pheromone biosynthesis was observed by the calcium-calmodulin inhibitors, W7 and trifluoperazine (Matsumoto et al., 1995). Omission of calcium from the incubation medium completely abolished the increase in intracellular cAMP levels due to PBAN stimulation (Soroker and Rafaeli, 1995). However, the stimulation, as a result of the cAMP analogs 8-bromo-cAMP (Jurenka et al., 1991) and adenosine 3′,5′- cyclic monophosphothioate-sp-isomer (Sp-cAMPS) (Soroker and Rafaeli, 1995), was not dependent on the presence of calcium in the incubation medium. These results suggested that calcium, entering the cell as a result of PBAN-receptor interaction, activates adenylate cyclase. In a recent study we demonstrated a pheromonostatic action, (inhibition of both pheromone biosynthesis and intracellular cAMP elevations) as a result of the presence of the adrenergic agonist, octopamine, on the response of intersegmental membrane cultures to PBAN (Rafaeli and Gileadi, 1995).

In this report the biochemical second messenger system during pheromonotropic and pheromonostatic activities was analysed and assessed.

Material and methods

The study was conducted on *Helicoverpa armigera,* reared as reported previously (Rafaeli and Soroker, 1989). We used the radiochemical bioassay as developed by us to monitor pheromone production by intersegmental membranes. The assay for cAMP production by intersegmental membranes was performed as reported previously (Soroker and Rafaeli, 1995; Rafaeli and Gileadi, 1995). The radio-immuno-assay for cAMP was performed as described previously (Rafaeli and Soroker, 1989). Synthetic Hez-PBAN (Hez, *Heliothis zea*) was purchased from Peninsula Lab. (Belmont CA, USA), other chemicals used were purchased from Sigma (St. Louis MO, USA).

Results

Fluoride ions, introduced as NaF, have often been used as evidence for the involvment of a G protein in a system (Bigay et al., 1987). The effect of NaF on pheromone biosynthesis was examined. A dose-response study (data not shown) revealed a stimulatory action at a concentration range of 1-2 mM (see stimulation at 1 mM, Table I). However, at higher NaF concentrations (> 4 mM), a much lower level of stimulation of pheromone biosynthesis was observed (see Table I, 10 mM NaF). In the presence of PBAN, high NaF levels inhibited its normal stimulatory action but low NaF levels showed an enhanced effect (Table I).

On examination of intracellular cAMP levels it was observed that NaF (at both 1 mM and 10 mM concentration ranges) stimulated intracellular cAMP levels (Table I). The α_2-adrenergic agonist, clonidine (10^{-4} M), inhibited both pheromone biosynthesis as well as intracellular cAMP levels which were induced by NaF (1 mM) (Table I). This inhibitory effect compared well with the inhibitory effect observed on PBAN-induced stimulation (Table I).

A. Rafaeli and C. Gileadi

Table I. The inhibitory effect of the adrenergic agonist, clonidine (10^{-4} M) on the pheromonotropic actions of both PBAN (0.05 µM) and NaF (1mM) showing effect on both pheromone production and intracellular cAMP production.

TREATMENT	PHEROMONE PRODUCTION (percentage stimulation above basal levels) means ± SEM (n)	PEAK INTRACELLULAR cAMP PRODUCTION (pmoles/ intersegment) means ± SEM (n)
Control	0 ± 0 (15) d	0.23 ± 0.14 (10) B
Clonidine	0 ± 0 (8) d	0.14 ± 0.07 (5) B
PBAN	643 ± 167 (15) a	1.43 ± 0.05 (8) A
PBAN + Clonidine	144 ± 68 (8) b	0.14 ± 0.047 (8) B
NaF (1 mM)	545 ± 122 (15) a	1.46 ± 0.046 (16) A
NaF (1 mM)+Clonidine	197 ± 138 (8) b	0.18 ± 0.094 (16) B
NaF (1 mM) + PBAN	2156 ± 670 (7) c	not tested
NaF (10 mM)	198 ± 50 (7) b	1.60 ± 0.032 (5) A
NaF (10 mM) + PBAN	254 ± 54 (16) b	not tested

The data are represented as the means ± SEM, numbers in brackets represent the number of replicates (n). Different letters indicate statistically significant differences (Fisher's protected LSD test $p < 0.05$), small letters for pheromone production, capital letters for cAMP production.

Discussion

Our results, showing a stimulatory action, at low NaF doses, indicates the involvement of G proteins (possibly G_S) in pheromonotropic activity. We observed an inhibitory effect on pheromone production by higher concentrations of NaF despite its stimulatory action on intracellular cAMP levels. The inhibitory action may be explained by negative feedback due to high intracellular cAMP levels which were observed under these conditions. Indeed, we have observed inhibitory effects on pheromone biosynthesis as a result of longterm exposure (2-3 h incubations) to isobutylmethyl-xanthine (IBMX, results not shown) which effectively increases intracellular cAMP levels due to its inhibitory action on cyclic-nucleotide phosphodiesterase. On the other hand, high NaF concentrations (50 mM) have also been shown to inhibit a variety of enzymatic systems such as phosphatases, phosphorylases, kinases and cytochrome oxidase (Ingebritsen et al., 1979). Indeed, an inhibitory effect on pheromone biosynthesis, which was attributed to inhibition of specific phosphatases, was observed in the moth *Bombyx mori* using 10 mM NaF (Matsumoto et al., 1995). In the present study the interaction of an α_2-adrenergic receptor in modulating pheromonotropic activity was shown. Its inhibitory effect on the G protein stimulated adenylate cyclase activity may indicate the involvement of a G_i protein linked adrenergic receptor. We cannot, however, explain the failure of NaF to stimulate G_i simultaneously but this may indicate that G_S linked receptors are more numerous in the intersegmental membrane cell. Recent cloning of the putative receptor for biogenic amines using *Drosophila* cDNA has shown that it contains seven transmembrane helices typical of a pattern associated with the family of cell-surface receptors that interact with G proteins (Saudou et al., 1990). These receptors have been shown to be negatively coupled to adenylate cyclase (Robb et al., 1991).

Conclusion

Our study has revealed that pheromone biosynthesis is induced by the interaction between a neuropeptide receptor and G protein which results in the stimulation of adenylate cyclase. We can conclude from our study that pheromone biosynthesis can be modulated at the membrane level by the interaction of an α-adrenergic receptor which results in the inhibition of adenylate cyclase. Pheromone biosynthesis may also be modulated at the cellular level by feedback inhibition due to increasing accumulation of intracellular cAMP levels.

Acknowledgements
This research was supported by a Binational Agricultural Research Development Grant (BARD Research Project No. IS 1645-89) to A. R. We thank Uzi Zisman for devoted attention to our insect culture.

References

Bigay, J., Deterre, P., Pfister, C. and Chabre, M. (1987) Fluoride complexes of aluminium or beryllium act on G-proteins as reversibly bound analogues of the γ phosphate of GTP. *EMBO J.* 6: 2907-2913.

Fonagy A., Matsumoto S., Uchiumi K. and Mitsui T. (1992) Role of calcium ion and cyclic nucleotides in pheromone production in *Bombyx mori. J. Pesticide Sci.* 17: 115-121.

Ingelbritsen, T.S., Geelen, M.J.H., Parker, R.A., Evenson, K.J. and Gibson, D.M. (1979) Modulation of hydroxymethylglutaryl-CoA reductase activity, reductase kinase activity, and cholesterol synthesis in rat hepatocytes in response to insulin and glucagon. *J. Biol. Chem.* 254: 9986-9989.

Jurenka R.A., Jacquin E. and Roelofs W.L. (1991) Stimulation of pheromone biosynthesis in the moth *Helicoverpa zea*: Action of a brain hormone on pheromone glands involves Ca^{2+} and cAMP as second messengers. *Proc. Natl. Acad. Sci. USA* 88: 8621-8625.

Ma, P.W.K. and Roelofs, W. (1995) Calcium involvement in the stimulation of sex pheromone production by PBAN in the European Corn Borer, *Ostrinia nubilalis* (Lepidoptera: Pyralidae). *Insect Biochem. Molec. Biol.* 25: 467-473.

Matsumoto, S., Ozawa, R., Nagamine, T., Kim, G-H., Uchiumi, K., Shono, T. and Mitsui, T. (1995) Intracellular transduction in the regulation of pheromone biosynthesis of the silkworm, *Bombyx mori*: Suggested involvement of calmodulin and phosphoprotein phosphatase. *Biosci. Biotech. Biochem.* 59: 560-562.

Raina A. K., Jaffe H., Kempe T. G., Keim P., Blacher R. W., Fales H. M., Riley C. T., Klun J. A., Ridgway R. and Hayes D. K. (1989) Identification of a neuropeptide hormone that regulates sex pheromone production in female moths. *Science* 244: 796-798.

Rafaeli A. and Soroker V. (1989) Cyclic-AMP mediation of the hormonal stimulation of ^{14}C-acetate incorporation by *Heliothis armigera* pheromone glands *in vitro. Mol. Cell. Endocrinol.* 65: 43-48.

Rafaeli, A. and Gileadi, C. (1995) Modulation of the PBAN-induced pheromonotropic activity in *Helicoverpa armigera. Insect Biochem. Molec. Biol.* 25: 827-834.

Robb, S., Cheek, T.R., Venter, J.C., Midgley, J.M. and Evans, P.D. (1991) The mode of action and pharmacology of a cloned *Drosophila* phentolamine receptor. *Neurotox '91 - An International Symposium,* pp. 369-371.

Saudou, F., Amlaiky, N., Plassat, J.-L., Borrelli, E. and Hen, R. (1990) Cloning and characterization of a *Drosophila* tyramine receptor. *EMBO J.* 9: 3611-3617.

Soroker V. and Rafaeli A. (1989) *In vitro* hormonal stimulation of acetate incorporation by *Heliothis armigera* pheromone glands. *Insect Biochem.* 67: 1- 5.

Soroker, V. and Rafaeli, A. (1995) Multi-signal transduction of the pheromonotropic response by pheromone gland incubations of *Helicoverpa armigera. Insect Biochem. Molec. Biol.* 25: 1-9.

Analysis of the effect of pituitary adenylate cyclase-activating polypeptide (PACAP) on growth hormone (GH) secretion in GH_3 cells

K. Koshimura, T. Miyake, Y. Murakami and Y. Kato

First Division, Department of Medicine, Shimane Medical University, Izumo 693, Japan

Summary. We examined the effects of PACAP on growth hormone (GH) secretion, intracellular Ca^{2+} concentrations ($[Ca^{2+}]i$) and ion channels in GH_3 cells. PACAP-38 and vasoactive intestinal polypeptide (VIP) ($10^{-9}M$) increased GH secretion. PACAP-induced increase in GH secretion was inhibited by PACAP(6-38), a PACAP antagonist. PACAP-38 ($10^{-10}M$) increased $[Ca^{2+}]_i$ in GH_3 cells. The increase in $[Ca^{2+}]_i$ was inhibited by 1) PACAP(6-38) ($10^{-6}M$), 2) nicardipine ($10^{-5}M$), a Ca^{2+} channel blocker, 3) depletion of extracellular Ca^{2+} or 4) tetrodotoxin ($3\times10^{-7}M$), a Na^+ channel blocker. Na^+ channel currents were increased by PACAP, whereas Ca^{2+} channel currents were unchanged. 8-bromo-cAMP ($10^{-3}M$) increased $[Ca^{2+}]_i$ and Rp-cAMPS ($10^{-3}M$), an inhibitor for protein kinase A, eliminated PACAP-induced increase in $[Ca^{2+}]_i$. Thus the present study suggests that PACAP activates Na^+ channels via cAMP-protein kinase A system coupled with its receptor to depolarize membrane potential, which in turn activates Ca^{2+} channels to induce an increase in $[Ca^{2+}]_i$, a trigger for GH secretion.

Introduction

Pituitary adenylate cyclase-activating polypeptide (PACAP) is a novel neuropeptide which is highly concentrated in the hypothalamus (Miyata et al., 1989, 1990). PACAP is known to stimulate adenylate cyclase in pituitary cells (Miyata et al., 1989) and to modulate the secretion of pituitary hormones (Murakami et al., 1995; Yamauchi et al., 1995). In the present study, we investigated the mechanism by which PACAP stimulates growth hormone (GH) secretion from the pituitary using GH_3 cells.

Material and methods

GH$_3$ cells were maintained and subcultured in Ham's F-10 medium supplemented with 10% horse serum, 5% fetal calf serum, penicillin (500 IU/ml) and streptomycin (500 mg/ml). GH secretion from superfused GH$_3$ cells was measured by radioimmunoassay (Murakami et al., 1995). $[Ca^{2+}]_i$ were measured with fura-2. Na^+ and Ca^{2+} channel currents were measured by whole-cell patch clamp recording.

Results

PACAP-38 (10^{-9} M) and VIP (10^{-9} M), which share binding to a PACAP receptor (10^{-9} M) increased GH secretion from GH$_3$ cells in the superfusion system (Fig. 1A). PACAP(6-38) (10^{-7} M), a PACAP receptor antagonist, blunted PACAP-induced increase in GH secretion (Fig. 1B). PACAP-38 (10^{-8} M) increased $[Ca^{2+}]_i$ in GH$_3$ cells (Fig. 2). The increase in $[Ca^{2+}]_i$ was blunted by PACAP(6-38) (10^{-6} M), removal of extracellular Ca^{2+} (calcium-free medium), nicardipine (10^{-5} M), a L-type Ca^{2+} channel blocker and tetrodotoxin (TTX, 3×10^{-7} M), a Na^+ channel blocker (Fig. 2), respectively. When protein kinase A activity was inhibited by Rp-cAMPs (10^{-3} M), PACAP did not increase $[Ca^{2+}]_i$ in GH$_3$ cells. The stimulating effect of PACAP on $[Ca^{2+}]_i$ was mimicked by 8-bromo-cAMP (10^{-3} M) (Fig. 2). PACAP-38 (10^{-9} M) increased Na^+ channel currents (Fig. 3A) but had little effect on Ca^{2+} channel currents in GH$_3$ cells (Fig. 3B).

Discussion

PACAP-38 and VIP increased GH secretion from superfused GH$_3$ cells. The stimulating effect of PACAP was blunted by a PACAP receptor antagonist, PACAP(6-38), indicating that PACAP stimulates GH secretion from GH$_3$ cells through a type-2 PACAP receptor. The stimulating effect of PACAP on $[Ca^{2+}]_i$ depends on extracellular Ca^{2+} and Ca^{2+}

channel activity, indicating that PACAP-induced increase in $[Ca^{2+}]_i$ is due to an increase in Ca^{2+} influx through Ca^{2+} channels. Na^+ channel current but not Ca^{2+} channel current was increased by PACAP. Furthermore, PACAP-induced increase in $[Ca^{2+}]_i$ was inhibited by a Na^+ channel blocker, tetrodotoxin (TTX). These data indicate that PACAP activates Na^+ channel to induce membrane depolarization, which in turn activates Ca^{2+} channel in GH_3 cells. PACAP-induced increase in $[Ca^{2+}]_i$ was also inhibited by a protein kinase A inhibitor, Rp-cAMPs and a cAMP analogue, 8-bromo-cAMP, mimicked the stimulating effect of PACAP on $[Ca^{2+}]_i$. These data suggest that PACAP activates Na^+ channel via cAMP-protein kinase A pathway.

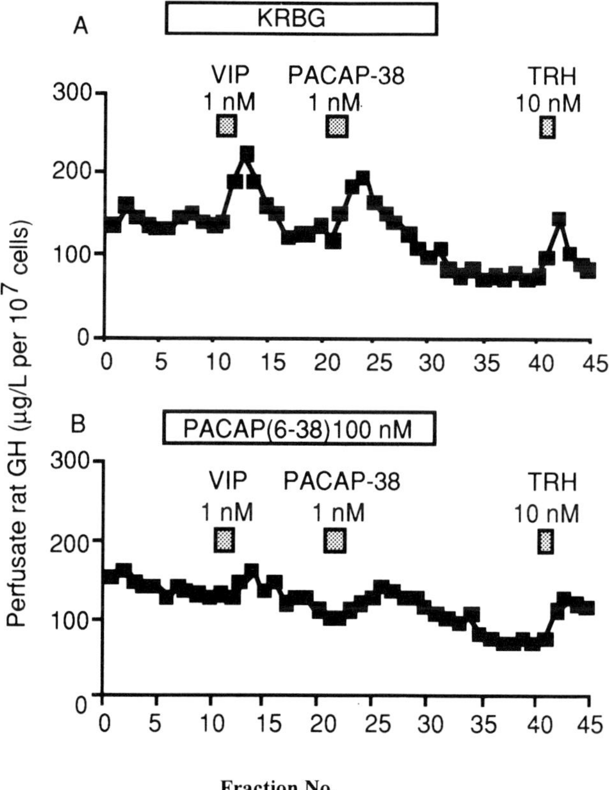

Fig. 1. Effect of PACAP and VIP on GH secretion from superfused GH_3 cells. GH_3 cells were superfused with Krebs Ringer bicarbonate buffer (KRBG, pH 7.4) and superfusate was fractionated every 5 min.

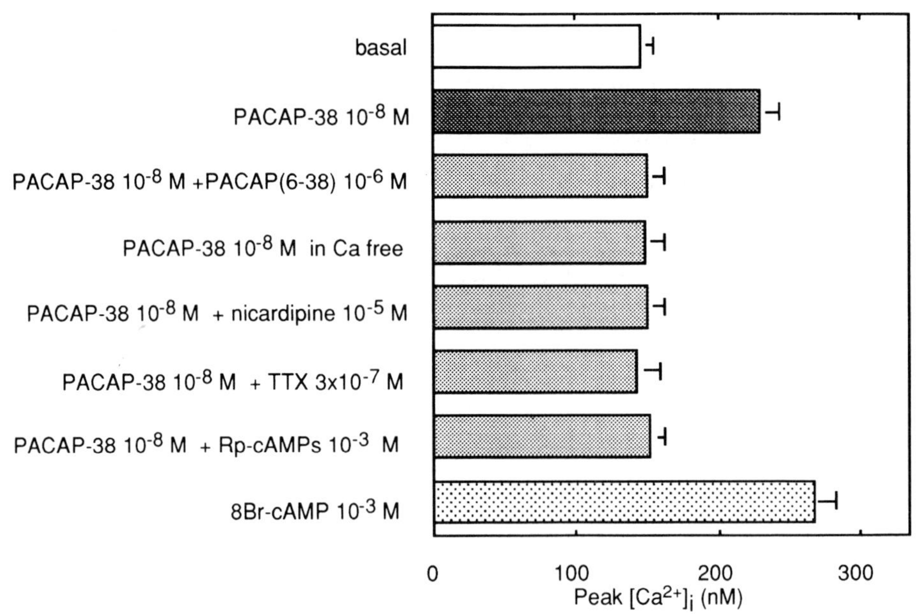

Fig. 2. Effect of PACAP and cAMP analogue on $[Ca^{2+}]i$ in GH_3 cells. Mean + SEM values are shown.

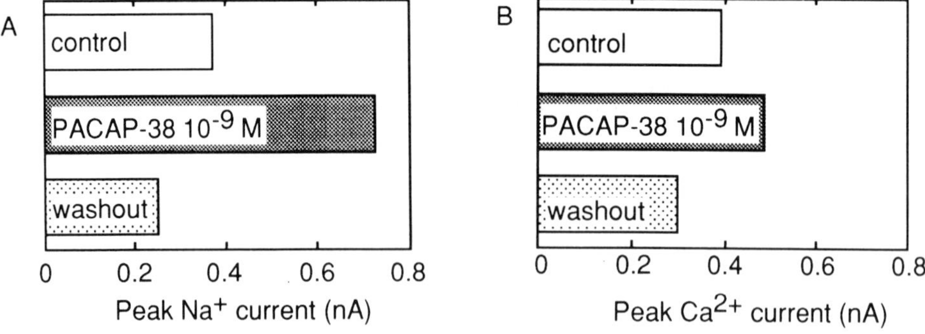

Fig. 3. Effect of PACAP-38 on Na^+ and Ca^{2+} channel currents in GH_3 cells. Bath solution was Ringer solution (pH 7.4) and pipette solution was 135 mM CsCl with 10 mM EGTA, 10 mM Hepes and 2 mM $MgCl_2$ (pH 7.4). Membrane potential was depolarized from - 80 mV to 10 mV.

Conclusion

Based on the present study, we hypothetize as follows: In GH_3 cells, PACAP activates Na^+ channels via cAMP-protein kinase A pathway coupled with a PACAP receptor, which induces membrane depolarization. As a result, a Ca^{2+} channel is activated to induce Ca^{2+} influx, a trigger for GH secretion. The effect on prolactin secretion, which should also be triggered by the mechanism proposed, remains to be studied.

Acknowledgments
We thank Miss Itoko Murakami and Mrs. Akemi Kageyama for excellent technical assistance.

References

Miyata, A., Arimura, A., Dahl, R.R., Minamino, N., Uehara, A., Jiang, L., Culler, M.D. and Coy, D.H. (1989) Isolation of a novel 38 residue-hypothalamic polypeptide which stimulates adenylate cyclase in pituitary cells. *Biochem. Biophys. Res. Commun.* 164: 567-574.

Miyata, A., Jiang, L., Dahl, R.D., Kitada, C., Kubo, K., Fujino, M., Minamino, N. and Arimura, A. (1990) Isolation of a neuropeptide corresponding to the N-terminal 27 residues of the pituitary adenylate cyclase activating polypeptide with 38 residues (PACAP38). *Biochem. Biophys. Res. Commun.* 170: 643-648.

Murakami, Y., Kato, Y., Shimatsu, A., Koshiyama, H., Hattori, N., Yanaihara, N. and Imura, H. (1989) Possible mechanisms involved in growth hormone secretion induced by galanin in the rat. *Endocrinology* 124: 1224-1229.

Murakami, Y., Koshimura, K., Yamauchi, K., Nishiki, M., Tanaka, J., Furuya, H., Miyake, T. and Kato, Y. (1995) Pituitary adenylate cyclase activating polypeptide (PACAP) stimulates growth hormone release from GH_3 cells through type II PACAP receptor. *Regul. Pept.* 56: 35-40.

Yamauchi, K., Murakami, Y., Nishiki, M., Tanaka, J., Koshimura, K. and Kato, Y. (1995) Possible involvement of vasoactive intestinal polypeptide in the central stimulating action of pituitary adenylate cyclase-activating polypeptide on prolactin secretion in the rat. *Neurosci. Lett.* 189: 131-134.

The Peptidergic Neuron
B. Krisch and R. Mentlein (eds)
© 1996 Birkhäuser Verlag Basel/Switzerland

Expression of vasotocin gene during metamorphosis in the bullfrog hypothalamus

S. Hyodo

Department of Biology, College of Arts and Sciences, University of Tokyo, Komaba, Meguro, Tokyo 153, Japan

Summary. To study physiological roles of vasotocin in anuran metamorphosis, expression of vasotocin gene was studied by Northern blot analysis in metamorphosing bullfrog tadpoles. Effects of osmotic stimulation on the vasotocin mRNA levels were also studied. The intensity of signal for vasotocin mRNA was gradually and consistently increased during prometamorphic development. Afterward, the vasotocin mRNA level was markedly increased at the metamorphic climax. The plasma osmolality and hematocrit values remained unchanged before metamorphosis, and increased after metamorphic climax. In the late metamorphic climax, the hematocrit value decreased to the levels of prometamorphic tadpoles. Immersion in 30% seawater for 3 days increased the plasma osmolality. Seawater treatment also increased the hematocrit values until early metamorphic climax, but did not alter those of tadpoles in the late metamorphic climax which stored a large volume of urine in their urinary bladder, as in adults. The urine osmolality of seawater-treated tadpoles was almost the same level with the plasma osmolality. Significant increase in the vasotocin mRNA levels by seawater-treatment was detected after preclimax stages. The present results suggest that vasotocin synthesis is increased by dehydration after metamorphic climax. Thus, vasotocin may have important roles in osmoregulatory and/or volume-regulatory mechanisms via osmoregulatory organs in metamorphosing bullfrogs, in relation to the adaptation to a semi-terrestrial habitat after metamorphosis.

Introduction

Neurohypophysial hormones have important osmoregulatory roles in vertebrates including amphibians. In adult bullfrogs and toads, hyperosmotic stimulation increased plasma vasotocin level (Pang, 1977). Injection of vasotocin induced water uptake at kidney, skin, and urinary bladder, resulting in body weight increase (Alvarado and Johnson, 1966; Shoemaker and Nagy, 1977). Since most anurans move from an aquatic to a semi-terrestrial habitat during metamorphosis, development of the hypothalamo-neurohypophysial system and its contribution to anuran osmoregulation from larval to adult stages are of interest.

Immunoreactive vasotocin neurons were virtually absent until Taylor-Kollros stage X (premetamorphic stage). Number of vasotocin neurons was increased from stage XII (Carr and Norris, 1990). Water retaining response to vasotocin injection is very small in young bullfrog tadpoles, but is obvious in the late stages of metamorphosis (Alvarado and Johnson, 1966). However, there has been no study on changes in activity of vasotocin producing neurons, such as regulation and activity of hormone synthetics, during metamorphosis of anuran tadpoles. Information on the hormone synthetics in normal and osmotically stimulated conditions is indispensable for clarifying the physiological roles of vasotocin in metamorphosing anurans.

In this study, expression level of vasotocin gene was studied by Northern blot analysis in metamorphosing bullfrog to assess the activity of vasotocin producing neurons. Effects of osmotic stimulation on the vasotocin mRNA levels were also studied.

Materials and methods

Animals. Adult male bullfrogs and tadpoles in various developmental stages ranging from V to XXV (based on stages for *Rana pipiens* by Taylor and Kollros, 1946) were obtained from a commercial supplier.

Experiment 1: Changes in mRNA levels during metamorphosis. Tadpoles were divided into 7 groups which represent stages V-X, XI-XIV, XV-XVI, XVII-XIX, XX, XXI-XXII and XXIII-XXV. After anaesthetization, hypothalami were removed and frozen in liquid nitrogen. Hematocrit value, plasma and urine osmolarities were measured.

Experiment 2: Effects of hyperosmotic stimulation. Since the time course study revealed that 3 days, but not 1 day, exposure to 30% diluted artificial seawater significantly increased the vasotocin mRNA levels in tadpoles of stage XXIII-XXV (data not shown), tadpoles and adult male bullfrogs were transferred from fresh water to 30% seawater, and were kept for 3 days. The hypothalami, plasma and urine samples were obtained and analyzed as described above.

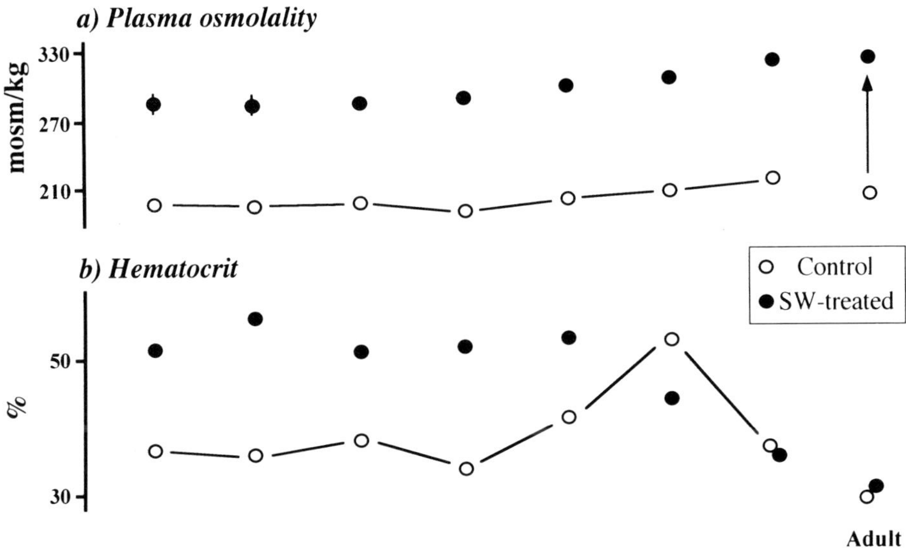

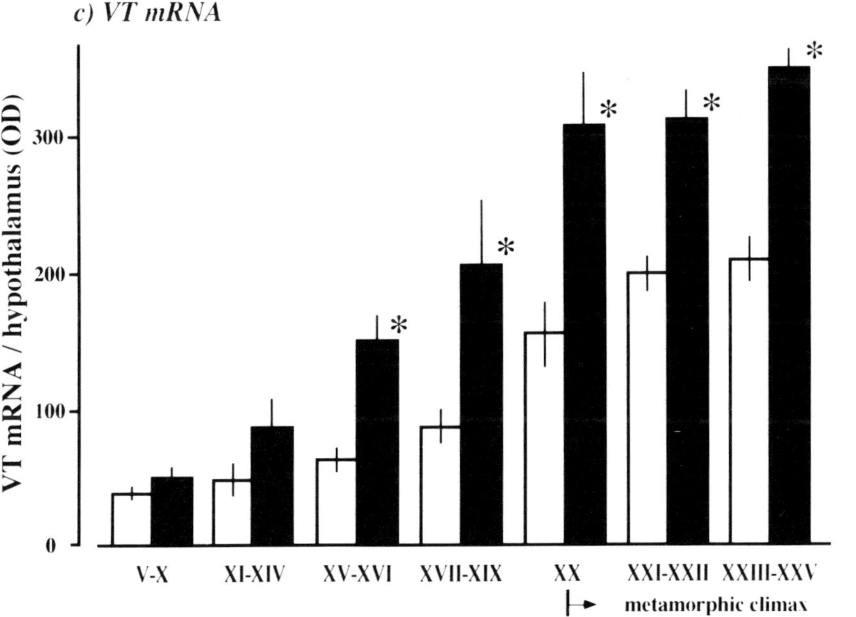

Fig. 1. Plasma osmolality (a), hematocrit values (b) and vasotocin mRNA levels (c) in freshwater(FW) (open bars) and seawater(SW)-treated (black bars) metamorphosing bullfrog tadpoles. OD, optical density.

Northern blot analysis. Extraction of total RNA and Northern blot analysis were performed as described previously (Hyodo et al., 1991). An oligonucleotide probe of fourty bases (5'-TGTTGTCTGTTGGCCATGTGCATGAGTCTGAGGAGCAGAT-3') corresponding to the region encoding glycopeptide moiety of vasotocin precursor was designed based on the nucleotide sequence of *Bufo japonicus* (Nojiri et al., 1987) and the amino acid sequence of *Rana esculenta* (Michel et al., 1987). Relative abundance of vasotocin mRNA was assessed as autoradiographic intensity per single hypothalamus.

Statistical analysis. The significance of differences in the measured values was examined by Duncan's new multiple range test and Cochran-Cox test, where appropriate.

Results

Developmental changes in the vasotocin mRNA levels during metamorphosis

The signal for vasotocin mRNA could be detected in the earliest premetamorphic group (stages V-X) examined in this study. Intensity of hybridization signal for vasotocin mRNA was gradually and consistently increased during prometamorphic development. Then, the vasotocin mRNA level was markedly increased at the metamorphic climax. The change from preclimax (stage XVII-XIX) to climax (stage XX) was statistically significant. At stage XX, frontlimbs of tadpoles are appeared, and tadpoles extend their habitation to a semi-terrestrial environment from this stage. Plasma osmolality and hematocrit value were unchanged before metamorphosis. After stage XX, plasma osmolality increased gradually. Hematocrit value also increased after stage XX, and attained at peak level at stage XXI-XXII (group 6). In the late metamorphic climax (stage XXIII-XXV; group 7), hematocrit value decreased to the levels of prometamorphic tadpoles. Most of animals in group 7 have urine in their urinary bladder.

Effects of osmotic stimulation in metamorphosing bullfrogs

At all stages examined, 3 days of seawater treatment increased plasma osmolality. Hematocrit values were also increased by the seawater treatment until stage XX. In tadpoles of group 7, osmotic stimulation did not change hematocrit value, as in adult bullfrogs. All seawater-treated animals in groups 6 and 7 (after stage XXI) had urine in their urinary bladder. Tadpoles of group 7 stored a large volume of urine (about 11% of body weight). The osmolality of urine of seawater-treated tadpoles was almost the same level with the plasma osmolality. The vasotocin mRNA levels were significantly increased by 3-day seawater treatment from stage XV (prometamorphic and climax stages). No significant change was observed before stage XV, though the vasotocin mRNA levels were slightly increased.

Discussion

The present study showed that expression of vasotocin gene in the bullfrog hypothalamus was observed from premetamorphic stages. The vasotocin mRNA level was increased gradually during prometamorphic stages, and elevated markedly at metamorphic climax. After stage XX, hematocrit value and plasma osmolality also increased, which were correlated well with the increases in vasotocin mRNA levels. Further, osmotic stimulation increased the vasotocin mRNA levels. Immersion in a 30% seawater significantly increased vasotocin mRNA level after stage XV, indicating that vasotocin producing neurons can respond to osmotic stimulation after stage XV. These results suggest that the increase in vasotocin mRNA level of control groups at metamorphic climax is, at least partly, due to the increases in plasma osmolality and/or dehydration.

In adult anurans, injection of vasotocin stimulates skin permeability to water and salt, antidiuresis in the kidney, and water and salt reabsorption from the urinary bladder to maintain blood volume and plasma osmolality. Injection of vasotocin retained water also in bullfrog tadpoles, particularly from stages just prior to metamorphic climax in a typical dose-response manner (Alvarado and Johnson, 1966). Although the mechanism underlying the effect of vasotocin on water and salt balance in tadpoles is still unclear, the skin may be one of the major target organs of vasotocin in tadpoles. The isolated ventral abdominal skin of metamorphosing bullfrogs have been shown to develop a short-circuit current and display a transepithelial potential difference after stage XX (Taylor and Barker, 1965).

In addition to the skin, the urinary bladder may be another physiological target organ of vasotocin in tadpoles of the late metamorphic climax. The urinary bladder was apparent after metamorphic climax. Tadpoles of control groups in the late metamorphic climax (all animals of stage XXIII-XXV and some animals of stage XXII) had urine in their urinary bladder. Immersion in diluted seawater increased the volume of urine (about 11% of body weight) in the late metamorphic tadpoles. The urine of the seawater-treated tadpoles was iso-osmotic to their plasma, as in adults. These results suggest that the late metamorphic tadpoles may be able to store urine in their urinary bladder, and reabsorb water from urine. In fact, osmotic stimulation increased hematocrit values of tadpoles except for those in the late metamorphic stages, indicating that water absorption in the skin and reabsorption in the kidney and urinary bladder prevent the increase in hematocrit value caused by dehydration in the late metamorphic climax. Although plasma levels of vasotocin in tadpoles are not clear yet, the present and previous results suggest that vasotocin has important roles in osmoregulatory and/or volume-regulatory mechanisms of metamorphosing bullfrogs through the osmoregulatory organs. Thus, vasotocin may be important for translocation from aquatic to semi-terrestrial habitation following metamorphosis.

Acknowledgments
The author is grateful to Prof. A. Urano, Hokkaido University, for critical reading of the manuscript. This study was supported in part by grants from the Ministry of Education, Japan, and Sasakawa Foundation.

References

Alvarado, R.H. and Johnson, S.R. (1966) The effects of neurohypophysial hormones on water and sodium balance in larval and adult bullfrogs. *Comp. Biochem. Physiol.* 18: 549-561.

Carr, J.A. and Norris, D.O. (1990) Immunohistochemical localization of corticotropin-releasing factor- and arginine vasotocin-like immunoreactivities in the brain and pituitary of the American bullfrog (*Rana catesbeiana*) during development and metamorphosis. *Gen. Comp. Endocrinol.* 78:180-188.

Hyodo, S., Kato, Y., Ono, M. and Urano, A. (1991) Cloning and sequence analysis of cDNAs encoding vasotocin and isotocin precursors of chum salmon: evolutionary relationships of neurohypophysial hormone precursors. *J. Comp. Physiol., B* 160: 601-608.

Michel, G., Chauvet, J., Chauvet, M-T. and Acher, R. (1987) One-step processing of the amphibian vasotocin precursor: structure of a frog "big" neurophysin. *Biochem. Biophys. Res. Commun.* 149: 538-544.

Nojiri, H., Ishida, I., Miyashita, E., Sato, M., Urano, A. and Deguchi, T. (1987) Cloning and sequence analysis of cDNAs for neurohypophysial hormones vasotocin and mesotocin for the hypothalamus of toad, *Bufo japonicus. Proc. Natl. Acad. Sci. USA,* 84: 3043-3046.

Pang, P.K.T. (1977) Osmoregulatory functions of neurohypophysial hormones in fishes and amphibians. *Amer. Zoologists* 17: 739-749.

Taylor, R.E. and Barker, S.B. (1965) Transepidermal potential difference: development in anuran larvae. *Science* 148: 1612-1613.

Taylor, A. and Kollros, J.J. (1946) Stages in the normal development of *Rana pipiens* larvae. *Anat. Records* 94: 7-23.

Shoemaker, V.H. and Nagy, K.A. (1977) Osmoregulation in amphibians and reptiles. *Ann. Rev. Physiol.* 39: 449-471.

Basal-medial hypothalamus conducts the development of GnRH neurons

S. Daikoku and I. Koide

Tokushima Research Institute, Otsuka Pharamaceutical Co. Ltd., Kawauchi-Cho, 771-01 Tokushima, Japan

Summary. In rats, gonadoliberin- (GnRH-) neurons deriving from the medial nasal placode (NAP) penetrate into the forebrain vesicle and medial septum, and migrate further to various brain parts concerned with hypophysio-gonadal activities or with sexual behaviour. Such heterotaxic migration occurred in a spacio-temporal fashion. In the brain of 21.5-day-old rat (E 21.5) embryo whose nasal cavity had been scratched unilaterally at E 16.5, GnRH neurons were diminished in number in the ipsilateral septo-preoptic area. The presence of the heterotaxic attitude of the neurons was further examined *in vitro*. In co-culture of E 1.5 nasal placodes with various brain parts of E 14.5 embryos, GnRH neurons emerging from the nasal placodes migrated into all the brain tissue, most frequently, however, into the median eminence-arcurate complex (ME-Arc). Together with our previous findings (Daikoku et al., 1995), it is suggested that the developing medial basal hypothalamus has certain attractions for intracerebral migration and development of GnRH neurons in rats.

Introduction

Since its discovery by Schwanzel-Fukuda and Pfaff (1989), the hypothesis that the septo-preoptic and olfactory gonadoliberin neuron systems originate from a unique source, the olfactory placode, has generally been accepted in mammals. In other animal species, however, for example in fishes or in birds, at least two origins have been proposed (Muske, 1993). Even in mammals, two different origins/sources of (1) neurons containing hypophysiotrophic GnRH and (2) neurons containing the peptide acting as neurotransmitter have long been assumed from their heterogeneous localization and projections in the brain. Aim of the present study was to obtain insight into the underlying mechanism of such heterotaxic migration of GnRH neurons in developing rats.

Materials and methods

Sprague-Dawley strain rats were used. Embryonic days were determined as indicated before (Daikoku et al., 1995).

In vivo study: For histological observation, embryos were obtained on each day from E 12.5 to 22.5 (birthday) after sacrificing their mothers by cervical dislocation. Embryonic heads were fixed in Bouin's fluid for immunohistochemical demonstration of GnRH or in 4% paraformaldehyde fixative to demonstrate cell adhesion molecules (CAMs). Immuno-histochemical staining procedures have been described previously (Daikoku-Ishido et al., 1990; Koide and Daikoku, 1995). Antibodies for CAMs were kindly supplied by Prof. Y. Arai and Dr. T. Seki (School of Medicine Juntendo University, Tokyo) [antibody 12ES, NCAM(A)], Dr. Gerald M. Edelmann, (Scripps Res. Inst., La Jolla) [NCAM(E)] and Prof. K. Uemura and Dr. H. Aso, (School of Medicine, Keio University, Tokyo) [L1(FN)]. The staining specificities of the antibodies have been presented elsewhere (Murakami et al., 1991; Miura et al., 1992; Schwanzel-Fukuda et al., 1992). The embryonic heads were cut serially at 7μm in the frontal, sagittal or horizontal plane. Paraformaldehyde-fixed sections were double-stained with anti-CAMs and anti-rat GnRH associated peptide (rGAP).

In vitro study: E 12.5 and 14.5 embryos were used. The nasal placodes were removed from E 12.5 embryos and cultured with the anterior ventromedial portion of the forebrain vesicle of E 12.5 embryos. Further, nasal placodes were cultured with the Rathke's pouch, the median eminence-arcuate nucleus complex (ME-Arc), the ventral wall of the mesencephalon (MC) or the ventral wall of the medulla oblongata (Med) of E 14.5 embryos. Additionally, the vomeronasal organ of E 14.5 embryos was cultured with the ME-Arc, MC or Med of the same age. All cultures were kept for 7 days in a basal culture medium supplemented with 1% water extract of stalk-medial basal hypothalamus of E 18.5 rat embryos (Daikoku et al., 1995) at 37°C in 95% CO_2 moist atmosphere. The cultures were fixed in Bouin's fixative or paraformaldehyde and examined by whole mount immunohistochemistry (Koide and Daikoku, 1995).

Nasal scratching: Under sodium pentobarbital anesthesia, laparotomy was performed in animals pregnant for 16.5 days and nasal scratching was done in the embryos as previously described (Daikoku, 1966). In brief, when exposing the uterine horns embryonal nose is easily recognized through the wall of the transparent antimesometrical wall of the uterine horn. With a small needle with silk thread (ME-24, Kowa, Tokyo) 4 stitches are placed in the wall like a string to tie the opening of a pouch surrounding an area (3 mm in diameter) within which the apex of the nose is visualized. The ends of the thread are lightly tied to make a circle. A small incision is made at the center of the circled uterine wall area to expose the apex of the nose. Either one of the nasal cavities is lesioned by inserting an emery-powder grain concreted with an adhesive agent at the tip of a thin needle. The nose is replaced in the uterine cavity, and the wound of the uterine wall is closed by tying the thread. The operation is carried out on two or three embryos in each uterine horn. The uterine horns are replaced in the abdominal cavity, and the abdominal wall is sutured. The animals are kept under normal conditions, and killed 6 days later (E 21.5). The operated embryos are removed, weighed and fixed in Bouin's fixative. The brain and the nose are paraffin-embedded, cut serially into 7 μm-thickness in the horizontal plane, and processed for immunostaining with anti-rGAP.

Results and discussion

As shown in a previous paper (Daikoku-Ishodo et al., 1990), the vomeronasal organ develops from the medial evagination of the medial wall of the nasal placode on E 13.5. At this stage, GnRH neurons are first recognized beneath the epithelium of the organ. During the following days the cells migrate toward the anterior tip of the forebrain vesicle (E 14.5) and after E 15.5, they have invaded the brain parenchyma along the fibers expressing cell surface adhesion molecules (CAMs). Penetrating further into the brain, the GnRH-cells spread in the neuropile in posterior, dorsal and lateral direction and are evidenced in the olfactory tubercule, piriform cortex, prehippocampal area, and medial septum.

By E 15.5-16.5, the anterior tip of the forebrain vesicle has developed a slight elongation to form a rudimental olfactory bulb. Concomitantly, olfactory axons increase in number under the olfactory bulb forming a dense fiber plexus. As shown in rats or in guinea pigs by Schwanzel-Fukuda et al. (1985) and Schwanzel-Fukuda and Silverman (1980), GnRH neurons move through the pia-arachnoid membrane to the caudal end of the olfactory bulb along the ventro-medial surface of the olfactory bulb and forebrain vesicle after E 16.5. Subsequently, the cells, as singles or accompanied by capillaries, infiltrate the brain parenchyma, mainly the medial septum and the preoptic area. After E 17.5, perikarya and fibres are found in the piriform cortex, medial septum, preoptic area, diagonal band, dorsal aspect of the telencephalon, hippocampal area, thalamus, subfornical organ, stria terminalis, dorsal aspect of the main and accessory olfactory bulb, medial forebrain bundle, and amygdalo-hippocampal region.

Thus, the migration of GnRH-neurons developing after E 16.5 should be disturbed if the nasal epithelium including the vomeronasal organ was destroyed at this stage. Therefore, we unilaterally scratched the epithelium of the nasal cavity at E 16.5. As a result, the number of GnRH neurons diminished remarkably in the ipsilateral side of the medial septum and accessory olfactory bulb, but only slightly in the ipsilateral side of the diagonal band and median forebrain bundle. GnRH-fibres were diminished in the ipsilateral part of the median eminence and organum vasculosum of the lamina terminalis.

These findings suggest that the major part of GnRH-neurons deriving from the olfactory placode before E 16.5 may not be involved in hypophysio-gonadal activity. In terms of functional properties, such a spaciotemporal development of GnRH-neurons corresponds with results on other hypothalamic peptidergic neurons deriving from the neural tube (Okamura et al., 1991).

If this is the fact, it is of interest to know what directs the migration of GnRH neurons into different brain targets. At present, however, we have not yet obtained enough knowledge on the mechanisms which induce GnRH neurons to invade throughout the brain parenchyma. Although we have demonstrated GnRH neurons emerging from the medial basal hypothalamus *in vitro* (Koide and Daikoku, 1995) and moving retrogradely along NCAM(A)-expressing fibres, we were unable to demonstrate CAMs structures along which

GnRH neurons traverse to the medial basal hypothalamus *in vivo*. Therefore, we examined *in vitro* whether certain brain parts have distinct chemical attractants in guiding GnRH-neurons and axon projections.

We co-cultured nasal placodes or vomeronasal organs obtained from 12.5 or 14.5 day-old embryonic rats together with the anterior ventromedial part of the forebrain vesicle, the median eminence-arcuate nucleus comlex, the ventral wall of the mesencephalon, of the medulla oblongata or together with Rathke's pouch obtained from E 14.5 embryos. After 7 days in culture, we found GnRH neurons from the nasal placode invading all the co-cultured tissues. However, the number of GnRH neurons penetrating into the co-cultured tissues was highest in the median eminence-arcuate nucleus complex and in Rathke's pouch preparations. These findings indicate that nasal placode-derived GnRH neurons can immigrate and project axons into various brain regions, preferentially into the median eminence-arcuate nucleus complex. Thus, it is highly probable that the medial basal hypothalamus may contain certain substances which have attractive effects on nasal placode-derived GnRH neurons.

Recently we found a 75 kD molecule which promotes GnRH neuron growth and survival (Daikoku et al., 1995). Together with those findings, we conclude that the medial basal hypothalamus conducts the development of GnRH neurons in ontogenic processes.

References

Daikoku, S. (1966) A method of diencephalon in the fetal rat. *Okajimas Fol. Anat. Jap.* 42: 39-49.

Daikoku, S., Koide, I., Yoshinaka, Y., Oka, T. and Natori, Y. (1995) How the developing septo-preoptic medial basal hypothalamus stimulates the development of placode-derived LHRH neurons. *Arch. Histol. Cytol.* 58: 77-95.

Daikoku-Ishido, H., Okamura, Y., Yanaihara, N. and Daikoku, S. (1990) Development of the hypothalamic luteinizing hormone-releasing hormone-containing neuron system in the rat: *In vivo* and in transplantation studies. *Dev. Biol.* 140: 374-387.

Koide, I. and Daikoku, S. (1995) *In vitro* analysis of the centripetal migration mechanisms of developing LHRH neurons. *Arch. Histol. Cytol.* 58: 265-283.

Miura, M., Asou, H., Kobayashi, M. and Uyemura, K. (1992) Functional expression of a full-length cDNA coding for rat neural cell adhesion molecule L1 mediates homophilic intercellular adhesion and migration of cerebellar neurons. *J. Biol. Chem.* 267: 10752-10758.

Murakami, S., Seki, T., Wakabayashi, K. and Arai, Y. (1991) The ontogeny of luteinizing hormone-releasing hormone (LHRH) producing neurons in the chick embryo: possible evidence for migrating LHRH neurons from the olfactory epithelium expressing a highly polysialated neural cell adhesion molecule. *Neurosci. Res.* 12: 421-431.

Muske, L.E. (1993) Evolution of gonadotropin-releasing hormone (GnRH) neuronal systems. *Brain Behav. Evol.* 42: 215-230.

Okamura, Y., Kawano, H. and Daikoku, S. (1991) Spatial-temporal appearance of developing immuno-reactive TRH neurons in the neuroepithelial wall of the diencephalon. *Dev. Brain Res.* 63: 21-31.

Schwanzel-Fukada, M. and Silverman, A.J. (1980) The nervus terminalis of the guinea pig: A new luteinizing hormone-releasing hormone (LHRH) neuronal system. *J. Comp. Neurol.* 191: 213-225.

Schwanzel-Fukada, M., Morrell, J.I. and Pfaff, D.W. (1985) Ontogenesis of neurons producing luteinizing hormone-releasing hormone (LHRH) in the nervus terminalis of the rat. *J. Comp. Neurol.* 238: 348-364.

Schwanzel-Fukada, M. and Pfaff, D.W. (1989) Origin of luteinizing hormone releasing hormone neurons. *Nature* 338: 161-164.

Schwanzel-Fukada, M., Abraham, S., Crossin, K.L., Edelman, G.M. and Pfaff, D.W. (1992) Immuno-histochemical demonstration of neural cell adhesion molecule (NCAM) along the migration rout of luteinizing hormone-releasing hormone (LHRH) neurons in mice. *J. Comp. Neurol.* 321: 1-18.

VI. Behavioural effects of neuropeptides

Behavioural effects of neuropeptides: central and peripheral mechanisms of action of vasopressin

B. Bohus

Department of Animal Physiology and Graduate School of Behavioural and Cognitive Neurosciences, University of Groningen, P.O.Box 14, 9750 AA Haren, The Netherlands

Summary. It has become well established that vasopressin is not only a peripheral hormone with antidiuretic and pressor activities, but is also a neurotransmitter/neuromodulator with numerous functions, found in many parts of the central nervous system. The first behavioural effect of vasopressin to be recognized was its influence on learning and memory. Other studies on neuropeptides have emphasized the role of oxytocin as an amnesic substance. This paper focusses on two major aspects of the manner in which vasopressin and oxytocin affect the central nervous system. The role of peripheral factors, particularly that of adrenal epinephrine, in memory functions influenced by vasopressin is described. The involvement of vasopressinergic and oxytocinergic mechanisms in the central nucleus of the amygdala during behavioural and physiological coping is discussed.

Introduction

It was thirty years ago that the effect of vasopressin on cognitive processes was first described. Removal of the posterior pituitary gland, an intervention that disturbed pituitary-adrenal responses to emotional stress, impaired avoidance behaviour in rats. The deficit in the behavioural response was restored by treating the animals with pitressin, an extract of the posterior pituitary with high levels of pressor and antidiuretic activities (De Wied, 1965). Administration of pitressin to intact rats caused long term retention of active avoidance behaviour, even well beyond the point at which the exogenous peptide could no longer be detected in the animals (De Wied and Bohus, 1966). Subsequent studies showed that vasopressin was the active peptide which caused the long lasting behavioural change (Bohus, 1971; De Wied, 1971). A hypothesis was formulated stating that vasopressin affects long term memory processing. In the nineteen-seventies, it became clear that the structurally related neurosecretory peptide oxytocin, although it frequently mimics the effects of vasopressin, has its own effect on behaviour opposite to that of vasopressin. Oxytocin can be classified as a neuropeptide with amnesic properties. Crucial experiments with intracerebro-ventricular administration of both vasopressin and oxytocin in rats which had learned a simple task showed that these neuropeptides change behavioural performance by affecting memory processing in the consolidation as well as in the retrieval phase (Bohus et al., 1978). The initial reports were followed by hundreds of others that were extensively and frequently reviewed (e.g. Strupp and Levitsky, 1985; De Wied et al., 1993). Although the interpretations of the data differed, it was generally recognized that vasopressin and oxytocin are involved in cognitive processes. However, both peptides have profound behavioural (and physiological) effects that cannot be interpreted as mnemonic (see De Wied et al., 1993). The structural basis for the multiplicity of central nervous functions attributed to these peptides can be seen in the fact that the 'classical' hypothalamo-hypophyseal neurosecretory system is not the only source of vasopressin, and that hypothalamic paraventricular neurosecretory cells also project to the limbic forebrain, to lower brain structures and to the spinal cord (e.g. Buijs, 1982). In addition, a distinct receptor system for vasopressin and oxytocin has been discovered in neural tissues (Jard et al., 1987). The presence of this system, in conjunction with manifold peptidergic neuronal networks, makes the multiplicity of actions understandable.

This contribution focusses on several novel aspects of the behavioural effects of vasopressin and oxytocin, with particular reference to peripheral factors influencing behaviour. A large portion of our findings point to the central amygdala as a key structure in peptidergic modulation of behavioural and physiological processes. The discussion is therefore centered on the involvement of vasopressinergic and oxytocinergic mechanisms in the central amygdala in coping with environmental challenge.

Adrenomedullary hormones and vasopressin-induced modulation of memory

Studies with intraventricular and circumscribed intracerebral administration of vasopressin and oxytocin have supported a view that central peptidergic networks are involved in brain processes serving memory (Bohus et al., 1978; Kovacs et al., 1979; De Wied et al., 1993). However, more or less from the very first findings, it was clear that systemically administered peptides affect learning and memory processes in a manner similar to that of centrally available peptides (e.g. vasopressin: Bohus, 1982; Koob et al., 1991). Although the precise mechanisms by which a peripheral vasopressinergic message reaches the brain is not yet well understood, there is data suggesting that peripheral hormonal mechanisms, particularly of adrenomedullary origin, are important for the appearance of certain behaviour (Bohus et al., 1993). In a series of elegant experiments McGaugh and colleagues showed that circulating epinephrine is an important modulator of memory consolidation (McGaugh, 1989). Administration of epinephrine in low and moderate doses enhanced memory in rats, whereas high doses impaired memory consolidation. Adrenergic mechanisms in the brain have been proposed to mediate the effect of peripheral catecholamines on memory. Other studies of our own showed that removal of the adrenal glands or adrenomedullectomy resulted in amnesia (i.e., memory deficit) toward an averse experience. Systemic administration of low and moderate doses of epinephrine reinstated memory (Borrell et al., 1983). Centrally available vasopressin enhanced memory provided that noradrenergic transmission in the brain was intact (Kovacs et al., 1979). Finally, administration of vasopressin both systemically and centrally was shown to reverse experimentally induced loss of memory (see Bohus, 1982). Taken together, these data suggest that there is an interaction between catecholamine and vasopressin that affects memory functions.

Indeed, in adrenalectomized or adrenomedullectomized rats, neither systemically nor intracerebroventricularly administered vasopressin was able to restore memory of a painful event such as receiving footshock in a certain environment (Borrell et al., 1984). These findings suggested that a properly functioning adrenal medulla is essential for the mnemonic action of vasopressin. Two alternative hypotheses were developed. First, the release of epinephrine by vasopressin in intact rats may cause changes in memory functions. Second, peripherally available epinephrine may act permissively - e.g. by activating central noradrenergic mechanisms - allowing vasopressin to facilitate memory.

As far as the first hypothesis is concerned, we have shown that systemically administered vasopressin did not raise catecholamine concentrations in the bloodstream, either in resting or under new stress conditions: epinephrine levels remained unaltered, whereas norepinephrine levels dropped following injections of peptide (Buwalda et al., 1993). Microinjections of vasopressin into the central nucleus of the amygdala also failed to alter peripheral levels of catecholamines in stress-free rats (Roozendaal et al., 1993). The central amygdala is evidently an important site in which vasopressin acts on memory by affecting local catecholaminergic transmission (Kovacs et al., 1979). Taken together, these results suggest that the first hypothesis can be rejected.

The results of numerous experiments with both adrenalectomized and adreno-medullectomized rats which had received different doses of vasopressin and catecholamine, either systemically or centrally, then favored the hypothesis that circulating epinephrine allows vasopressin to facilitate memory (Borrell et al., 1984; Bohus et al., 1993). Vasopressin evidently causes a dose-dependent facilitation of memory for a behavioural task called a *passive (inhibitory) avoidance response* in adrenalectomized and adrenomedullectomized rats provided that catecholamine was administered in low or moderate doses shortly before the peptide became available. It was suggested that circulating epinephrine and probably a momentary occupation of peripheral ß-adrenergic receptors were essential for long-term influencing of memory by vasopressin. This rather straight forward interaction is probably operable under physiological conditions, and the cooperative action between vasopressin and epinephrine appears to be mutual. This suggestion is supported by studies using Brattleboro rats, homozygous or heterozygous for hereditary hypothalamic diabetes insipidus (Borrell et al., 1985; Bohus et al., 1993). The homozygous variant shows serious deficits in the synthesis of hypothalamic vasopressin. The heterozygous variant has only slight deficits in vasopressin release from the posterior pituitary.

Systemical administration of vasopressin facilitated memory in a dose-dependent manner in both homo- and heterozygous intact Brattleboro rats. The effective doses of vasopressin did not facilitate memory in adrenalectomized Brattleboro rats supplemented with physiological amounts of corticosterone. Accordingly, as in Wistar rats, vasopressin failed to restore adrenalectomy-induced memory deficits in Brattleboro rats. Epinephrine treatment of heterozygous adrenalectomized Brattleboro rats, however, improved memory in a dose-dependent manner. The same doses of the catecholamine were not effective in the adrenalectomized homozygous variant (Bohus et al., 1993). Taken together, the results of these studies suggest that sufficient levels of circulating vasopressin are essential for eliciting a mnemonic action by epinephrine. Accordingly, the cooperative action of the two hormones is mutual.

The central nucleus of the amygdala and the modulation of memory by vasopressin

As mentioned in the Introduction, the brain has several anatomically distinct vasopressinergic and oxytocinergic systems. This multiplicity is magnified by the multiplicity of receptor mechanisms for these peptide hormones. The diversity of behavioural and physiological actions, as summarized by De Wied et al. (1993), fits the morphological and functional multiplicity. Of the many sites in the brain which appear to mediate memory-related activity of vasopressin (Kovacs et al., 1979), our interest was especially focussed on the central nucleus of the amygdala. This nucleus apparently receives vasopressinergic innervation from the bed nucleus of the stria terminalis. An oxytocinergic pathway is known to arise from the hypothalamic paraventricular nucleus (De Vries and Buijs, 1983). It contains a high density of vasopressin V_{1A} and oxytocin receptors as shown by autoradiographic binding (Tribollet et al., 1988) and mRNA transcript distribution (Ostrowski et al., 1992). The central amygdala is a functionally heterogeneous area involved in the integration of external and internal influences necessary for coping with environmental challenges. The nucleus has been suggested to play a central role in initiating (acquisition) and consolidating the processing of environmental information (see Aggleton, 1992). Studies of the past decade have emphasized its role in mnemonic processes that can be designated as emotional memory (LeDoux, 1993; Davis, 1994). In McGaugh's view (McGaugh et al., 1992), the amygdala modulates the storage of information in other parts of the brain by activating physiological systems by means of affective stimulation. This view implies that the amygdala is the site in which hormones interact with neurotransmitters and are thus able to modulate memory. As with other neuropeptides and adrenal cortical hormones, vasopressin and oxytocin, interacting with peripheral and brain amines, also fitted into this picture (Bohus, 1994). The role of vasopressin and oxytocin in behavioural and physiological regulation was studied in stress-free Wistar rats by microinfusing these peptides into the central amygdala.

A low dose of vasopressin (20 pg) caused long-lasting bradycardia, associated with immobility behaviour. A rise in plasma corticosterone levels was also observed, but no changes occurred in plasma catecholamine levels (epinephrine and norepinephrine). Higher doses of vasopressin (200 pg and 2 ng) induced opposite changes in heart rate and behaviour. Dose-dependent tachycardia became apparent and reached maximum levels after termination of the infusion. It was associated with increased exploratory activity. Pretreatment with the antagonist of the oxytocin receptor, dPTyr(Me)OVT, blocked the tachycardic effects of even the highest dose of vasopressin. Similar doses of oxytocin (20 and 200 pg, 2 ng) induced tachycardia and increased exploration. In addition, both vasopressin and oxytocin increased plasma corticosterone levels, whereas plasma epinephrine and norepinephrine concentrations remained unaffected (Roozendaal et al., 1993). Taken together, these findings suggest that a vasopressin receptor mechanism in the central amygdala activates parasympathetic (vagal) output to the heart and induces immobility behaviour. Oxytocin receptors probably mediate the effects of high doses of vasopressin and all doses of oxytocin, resulting in vagal inhibition, rather than sympathetic activation. These suggestions are supported by the fact that there was no rise in plasma catecholamines. Finally, the activation of the pituitary-adrenal system is also related to an oxytocin receptor-mediated process with remarkably high sensitivity for both vasopressin and oxytocin.

Observations in genetically inbred rats subjected to conditioned emotional stress (i.e. prior to inescapable footshock), suggest that vasopressinergic and oxytocinergic control of the central amygdala may be of physiological significance. Rats selected for *low active avoidance behaviour* (Roman Low Avoidance rats: RLA; Driscoll and Bättig, 1982) display 'freezing' and very marked bradycardia in response to the emotional stressor. The low dose of vasopressin (20 pg) infused into the central amygdala intensified both the cardiac and behavioural reactions to stress. A high dose of vasopressin (2 ng) or oxytocin (200 pg) prevented 'freezing' and bradycardia. Both peptides were ineffective in rats selected for *high active avoidance behaviour* (Roman High Avoidance rats: RHA). The RHA rats reacted with behavioural activation and tachycardia to the same conditioned emotional stress (Roozendaal et al., 1992). In summary, both vasopressin and oxytocin receptor-mediated processes may be involved in behavioural and physiological coping with stressful stimuli. The activation of vagus output and induction of freezing behaviour is probably related to the V_{1A} vasopressin receptor. Oxytocin receptors are likely to mediate behavioural activation via disinhibition, and to cause tachycardia through vagal inhibition. These relationships are most probable because the effects of the peptides are absent in RHA rats. These animals are by definition very active, and their sympathetic nervous systems are also highly active. The elevation of corticosterone levels in the circulating blood (i.e., activation of the hypothalamo-pituitary-adrenal axis via the paraventricular hypothalamic nuclei), brought about by both vasopressin and oxytocin, was elicited through a high affinity oxytocin receptor. It is interesting to note that central facilitation of memory by vasopressin appeared to be mediated by oxytocin receptors in the brain for which vasopressin was an agonist and oxytocin an inverse agonist (see De Wied et al., 1993). In contrast, the effect of peripheral vasopressin on learning and memory was blocked by the V_{1A} antagonist dPTTyr(Me)AVP (Koob et al., 1991). Accordingly, the effects of vasopressin on memory processes may be induced not only by mechanisms that arouse the sympathetic system, but also through mechanisms that cause vagal activation. That a distinction between peripheral and central is not an essential factor in the involvement of different receptor mechanisms is supported by the observation that cardiac vagal activation and behavioural inhibition (freezing) may also be elicited by systemically administered vasopressin (Bohus et al., 1990; Buwalda et al., 1992). These effects were observed during both conditioned emotional stress and mild stress situations. The vagal control of heart rate response to stressors was absent in aged rats but was restored by vasopressin. The fact that amphetamine, but not apomorphine, was also effective in aged rats suggests that noradrenergic rather than dopaminergic mechanisms are involved (Buwalda et al., 1992).

Vasopressin: a central nervous and peripheral peptide involved in passive coping

Manipulation of the vasopressinergic system, whether genetically, during the perinatal period, or in adult and aged rats, resulted in altered learning and memory processing. Vasopressin enhances passivity and parasympathetic activation. Impairment of the vasopressinergic function results in more active behaviour with memory deficit and sympathetic activation. Coping with environmental challenge needs more than a single strategy. Animals and human beings might adopt either an active or a passive strategy. The active strategy is characterized by rapid learning of active responses such as fight or flight toward or from a social or non-social environment. A low tendency to be attentive or to 'freeze', and high sympathetic activity is typical for this kind of adaptive change. By contrast, a passive strategy is defined as a kind of conservative/withdrawal response with a high level of freezing and attention, and a low level of fight or flight. A high level of vagal and pituitary-adrenal activation is the physiological basis of passive coping (see Bohus et al., 1990). Because of its behavioural and physiological activity profile, mediated by a V_{1A} receptor, vasopressin may be considered to be an important mediator of passive coping. Oxytocin, via the oxytocin receptor, may transform coping strategy from passive to active. As far as memory processes are concerned, the action of vasopressin in the central nucleus of the amygdala may represent the memory component of passive coping. Activation of the central amygdala, followed by physiological and neuroendocrine responses, is an essential mechanism for acquiring new information and consolidating behavioural patterns. Sympatho-adrenal activation is an important feature of the stress response that accompanies learning (Roozendaal et al., 1991). This sympathetic involvement may underlie the interaction between vasopressin and epinephrine which influences memory processes.

The observations in Roman Low and High Avoidance (RLA and RHA) rats suggest that preference for a certain coping strategy and the role of the vasopressin/oxytocin system is a genetically established mechanism. Although direct data on the Roman lines are not yet available, findings in genetically selected mouse lines support this view. The system of vasopressinergic fibres in the lateral septum is more dense and there are more vasopressin-immunoreactive cell bodies in the bed nucleus of the stria terminalis in the passivly coping, non-aggressive mouse line than in the aggressive, active copers (Compaan et al., 1993).

Conclusions

The opening hypothesis of this paper was that vasopressin affects learning and memory. As reviewed recently, there is a wide variety of behavioural effects of this neuropeptide, among which those requiring memory in avoidance and approach situations are well established (De Wied et al., 1993). The data summarized here indicate that vasopressin serves mnemonic processes related to passive coping. One may wonder, however, how the long-term effects of vasopressin on the maintenance of active avoidance behaviour can be reconciled with the hypothesis on passive coping. Recent research does not provide a simple answer to this question. The long-lasting effect of the peptide may reflect a kind of pathological hypermnesia, an inextinquishable memory. Whether this hypermnesic state might be the basis of compulsive or phobic behaviour, remains to be shown.

Acknowledgements
Parts of this contribution were supported by a special grant from the University of Groningen, Faculty of Sciences, for the promotion of research on ageing; and by the Netherlands Organization for Scientific Research and the Dutch Heart Foundation.

References

Aggleton, J.P. (1992) *The Amygdala. Neurobiological Aspects of Emotion, Memory, and Mental Dysfunction.* Wiley-Liss, New York.
Bohus, B. (1971) Effect of hypophyseal peptides on memory functions in rats. *In:* G. Adam and J. Szentagothai (eds.): *The Biology of Memory,* Akademiai Kiado, Budapest, pp. 93-100.
Bohus, B. (1982) Neuropeptides and memory. *In:* R.L. Isacson and N.E. Spear (eds.): *The Expression of Knowledge,* Plenum Press, New York, pp. 141-177.
Bohus, B. (1994) Humoral modulation of learning and memory processes: Physiological significance of brain and peripheral mechanisms. *In:* J. Delacour (ed.): *The Memory System of the Brain,* World Scientific, Singapore, pp. 337-364.
Bohus, B., Kovacs, G.L. and De Wied, D. (1978) Oxytocin, vasopressin and memory processes: Opposite effects on consolidation and retrieval processes. *Brain Res.* 157: 414-417.

Bohus, B., Koolhaas, J.M., Nyakas, C., Luiten, P.G.M., Versteeg, C.A.M., Korte, S.M., Timmerman, W. and Eisenga, W. (1990) Neuropeptides and behavioural and physiological stress response: The role of vasopressin and related peptides. *In:* S. Puglisi-Allegra and A. Oliverio (eds.): *Psychobiology of Stress,* Kluwer, Dordrecht, pp. 103-123.

Bohus, B., Borrell, J., Koolhaas, J.M., Nyakas, C., Buwalda, B., Compaan, J.C. and Roozendaal, B. (1993) The neurohypophyseal peptides, learning, and memory processing. *Ann. N. Y. Acad. Sci.* 689: 285-299.

Borrell, J., De Kloet, E.R., Versteeg, D.H.G. and Bohus, B. (1983) Inhibitory avoidance deficit following short-term adrenalectomy in the rat. *Behav. Neural Biol.* 39:241-258.

Borrell, J., De Kloet, E.R., Versteeg, D.H.G., Bohus, B. and De Wied, D. (1984) Neuropeptides and memory: Interactions with peripheral catecholamines. *In:* E. Usdin, R. Kvetnansky and J. Axelrod (eds.): *Stress. The Role of Catecholamines and Other Transmitters,* Gordon and Breach, New York, pp. 391-402.

Borrell, J., Del Cerro, S., Guaza, C., Zubiaur, M., De Wied, D. and Bohus, B. (1985) Interactions between adrenaline and neuropeptides in modulation of memory processes. *In:* J.L. McGaugh (ed.): *Contemporary Psychology: Biological Processes and Theoretical Issues,* North-Holland, Amsterdam, pp. 17-36.

Buijs, R.M. (1982) Vasopressin and oxytocin - their role in neurotransmission. *Pharmacol. Ther.* 22: 127-141.

Buwalda, B., Koolhaas, J.M. and Bohus, B. (1992) Behavioural and cardiac responses to mild stress in young and aged rats: effects of amphetamine and vasopressin. *Physiol. Behav.* 51:211-216.

Buwalda, B., Nyakas, C., Koolhaas, J.M. and Bohus, B. (1993) Neuroendocrine and behavioural effects of vasopressin in resting and mild stress conditions. *Physiol. Behav.* 54: 947-853.

Compaan, J.C., Buijs, R.M., Pool, C.W., De Ruiter, A.J.H. and Koolhaas, J.M. (1993) Differential lateral septum vasopressin innervation in aggressive and non-aggressive male mice. *Brain Res. Bull.* 30:1-6.

Davis, M. (1994) The role of the amygdala in emotional learning. *Int. Rev. Neurobiol.* 36: 225-266.

De Vries, G.J. and Buijs, R.M. (1983) The origin of vasopressinergic and oxytocinergic innervation of the rat brain with special reference to the lateral septum. *Brain Res.* 273: 307-317.

De Wied, D. (1965) The influence of the posterior and intermediate lobe of the pituitary and pituitary peptides on the maintenance of a conditioned avoidance response in rats. *Int. J. Neuropharmacol.* 4: 157-167.

De Wied, D. (1971) Long-term effect of vasopressin on the maintenance of a conditioned avoidance response in rats. *Nature* 232: 58-60.

De Wied, D. and Bohus, B. (1966) Long-term and short-term effects on retention of a conditioned avoidance response in rats by treatment with long acting pitressin and α-MSH. *Nature* 212: 1484-1486.

De Wied, D., Diamant, M. and Fodor, M. (1993) Central nervous effects of the neurohypophyseal hormones and related peptides. *Frontiers Neuroendocrinology* 14: 251-302.

Driscoll, P. and Bättig, K. (1982) Behavioural, emotional and neurochemical profiles of rats selected for extreme differences in active, two-way avoidance performance. *In:* I. Lieblich (ed.): *Genetics of the Brain,* Elsevier, Amsterdam, pp. 95-123.

Jard, S., Barberis, C., Audiger, S. and Tribollet, E. (1987) Neurohypophyseal hormone receptor systems in brain and periphery. *Progr. Brain Res.* 72: 173-187.

Koob, G.F., Lebrun, C., Bluthe, R-M., Dantzer, R., Dorsa, D.M. and Le Moal, M. (1991) Vasopressin and learning: Peripheral and central mechanism. *In:* R.C.A. Fredrickson, J.L. McGaugh and D.L. Felten (eds.): *Peripheral Signalling of the Brain,* Hogrefe and Huber, Toronto, pp. 351-363.

Kovacs, G.L., Bohus, B. and Versteeg, D.H.G. (1979) The effects of vasopressin on memory processes: Role of noradrenergic transmission. *Neuroscience* 4: 1529-1537.

LeDoux, J.E. (1993) Emotional memory systems in the brain. *Behav. Brain Res.* 58: 69-79.

McGaugh, J.L. (1989) Involvement of hormonal and neuromodulatory systems in the regulation of memory storage. *Ann. Rev. Neurosci.* 12: 255-287.

McGaugh, J.L., Introini-Collison, I.B., Cahill, L., Kim, M. and Liang, K.C. (1992) Involvement of the amygdala in neuromodulatory influences on memory storage. *In:* J.P. Aggleton (ed.): *The Amygdala. Neurobiological Aspects of Emotion, Memory and Mental Dysfunction,* Wiley-Liss, New York, pp. 431-451.

Ostrowski, N.L., Lolait, S.J., Bradley, D.J., O'Carroll, A.M., Brownstein, M.J. and Young, W.S. III (1992) Distribution of V$_{1A}$ and V$_2$ vasopressin receptor messenger ribonucleic acids in rat liver, kidney, pituitary and brain. *Endocrinology* 131: 533-535.

Roozendaal, B., Koolhaas, J.M. and Bohus, B. (1991) Attenuated cardiovascular, neuroendocrine, and behavioural responses after a single footshock in central amygdaloid lesioned male rats. *Physiol. Behav.* 50: 771-775.

Roozendaal, B., Wiersma, A., Driscoll, P., Koolhaas, J.M. and Bohus, B. (1992) Vasopressinergic modulation of stress responses in the central amygdala of the Roman high-avoidance and low-avoidance rat. *Brain Res,* 596: 35-40.

Roozendaal, B., Schoorlemmer, G.H.M., Koolhaas, J.M. and Bohus, B. (1993) Cardiac, neuroendocrine and behavioural effects of central amygdaloid vasopressinergic and oxytocinergic mechanisms under stress-free conditions in rats. *Brain Res. Bull.* 32: 573-579.

Strupp, B.J. and Levitsky, D. (1985) A mnemonic role for vasopressin: The evidence for and against. *Neurosci. Biobehav. Rev.* 9: 399-411.

Tribollet, E., Barberis, C., Jard, S., Dubois-Dauphin, M. and Dreifuss, J.J. (1988) Localization and pharmacological characterization of high affinity binding sites for vasopressin and oxytocin in the rat brain by light microscopic autoradiography. *Brain Res.* 442: 105-118.

Effects of tank colour and stress on melanin-concentrating hormone gene expression in the rainbow trout

M. Suzuki, B. I. Baker and A. Levy[1]

University of Bath, School of Biology and Biochemistry, Claverton Down, Bath BA2 7AY, United Kingdom
[1] Bristol Royal Infirmary, Dept. of Medicine, Bristol BS2 8HW, United Kingdom

Summary. Using *in situ* hybridization histochemistry, we have studied the influence of environmental colour and stress on the melanin-concentrating hormone (MCH) gene expression in the rainbow trout. Although MCH is encoded by two homologous genes (MCH1 and 2) in this fish, only oligoprobes against MCH2 hybridized to brain sections. The hybridization signal was four-fold greater in white-reared trout than in their black counterparts. This observation applied equally to magnocellular neurons in the nucleus lateralis tuberis (NLT), and parvocellular neurons above the lateral ventricular recess (LVR). Repeated moderate stress (5 min low water, 3 times daily) had no effect after 1 day, but enhanced MCH gene expression throughout the nucleus lateralis tuberis after 5 days (P<0.05). MCH neurons in the lateral ventricular recess showed no response to this stress. During early development, the response of the MHC neurons in the nucleus lateralis tuberis to tank colour was observed within 1 week after hatching, but MCH mRNA in these cells was elevated by a chronic stress (confinement and 1 min iced water daily) only 3 weeks posthatch. On the other hand, MHC neurons in the lateral ventricular recess did not respond to tank colour until 4 weeks posthatch, but MCH signal was enhanced by the stress from 2 weeks after hatching onwards. These results indicate a difference in sensitivity of the two groups of MCH neurons to background colour and stress.

Introduction

The melanin-concentrating hormone (MCH) is a ubiquitous neuropeptide, found in the brains of all vertebrates. Its physiological roles are still poorly understood. One of its effects, in fish as in mammals, is to depress the release of adrenocorticotropin during a mild stress by acting at some central site (Baker, 1991; Bluet-Pajot et al., 1995). Conversely, the release of MCH itself, at least in fish, is enhanced during repeated stress, a response that is prevented by glucocorticoids (Green and Baker, 1991). In fish, MCH has evolved a second function as an adaptive colour change hormone, aiding skin pallor (Baker, 1994). In this capacity, MCH neurons project to the neural lobe of the pituitary gland from magnocellular neurons located on the floor of the hypothalamus, around the pituitary stalk (nucleus lateralis tuberis, NLT). MCH is released from the pituitary into the circulation when the fish is placed in a pale-coloured environment.

The MCH neurons in the trout brain are located in two paired nuclei: the nucleus lateralis tuberis contains numerous magnocellular MCH neurons whose axons project to the pituitary, as already mentioned, and also into the brain. A second group of relatively fewer, parvocellular MCH neurons are located more dorsally, above the lateral ventricular recesses (LVR). Their axons project mainly into the thalamus and do not appear to contribute to the neurohypophysial tracts (Baker et al., 1995). It is not yet known whether the ability to respond to background colour and to stress is shown by all MCH neurons or by separate populations. To investigate this question, we have examined the influence of tank colour and stress on MCH mRNA expression in MCH neurons in both the nucleus lateralis tuberis and above the lateral ventricular recess, both in adult trout and in trout fry from the earliest stages after hatching.

Materials and methods

Fish. Rainbow trout were reared from eggs in either black or white tanks at 11°C. Photoperiod was 16.5 h light: 7.5 h dark. Young fry were sampled from either tank at 14:00 h on days 7, 14, 21 and 28 after hatching. They were decapitated under anaesthesia and their heads were fixed in 4% paraformaldehyde in phosphate buffer overnight, embedded in paraffin wax and sectioned at 10 μm for hybridization. During weeks 1, 2, 3 and 4, some white-reared fry were stressed by being confined over a 5 day period in a small cage of white plastic netting (10 x 10 x 2.5 cm) and exposed daily to ice-cold water for 1 min. They were processed as described above. Adult trout of 18 months, reared and maintained in either white or black tanks, were killed at about 16:00 h and their brains fixed in 4% paraformaldehyde. Other white-reared adult trout were stressed by being transferred once or thrice daily to a tank with a low (2.5 cm) water level. They were killed after either 24 h or 5 days of such stress, and their brains processed as described.

In situ hybridization. Paraformaldehyde- fixed brains were embedded in wax and 10 μm sections were hybridized with either of two ^{35}S-labelled 28mer oligonucleotide probes, specific for one or other of the two MCH genes (MCH1 and 2), as described in detail previously (Baker et al., 1995; Suzuki et al., 1995). Hybridized sections were apposed to autoradiography film, together with ^{14}C-standards (microscales, Amersham). The intensity of the hybridization signals over background was quantified by reference to a standard curve, constructed from the standards by the curve-fitting function of the computer, which takes into consideration the non-linear response of the film to β-irradiation (Suzuki et al., 1995). The intensity of the hybridization signals is expressed as arbitrary units that are assumed to be proportional to mRNA. Student's t-test or the Mann-Whitney U test were used for statistical analyses.

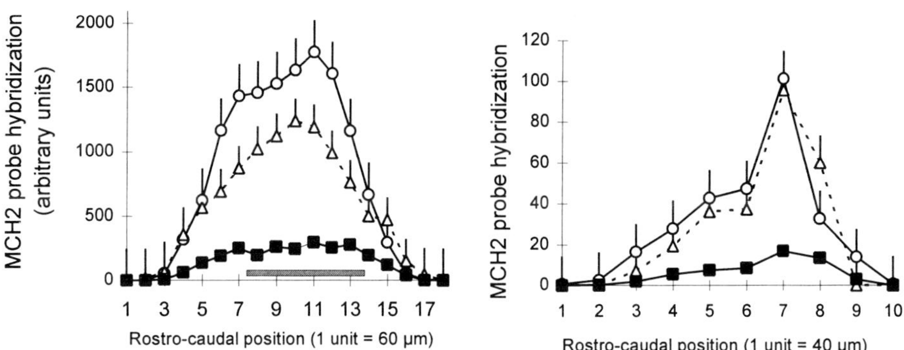

Fig. 1. MCH2 hybridization signal in the nucleus lateralis tuberis (left graph) and above the lateral ventricular recess (right graph) of adult white-reared trout (Δ), white-reared trout stressed by three daily transfers to low water for 5 min each time, for 5 days (O), and black-reared trout (■). Hatched bar indicates the position of the pituitary stalk. Rostro-caudal position 1 of the lateral ventricular recess sections is located at approximately rostro-caudal position 14 of the nucleus lateralis tuberis graph.

Results and discussion

Background effect. In adult trout, MCH neurons throughout the nucleus lateralis tuberis and also those above the lateral ventricular recess respond to background colour (Fig. 1a, b). In both sites, MCH mRNA levels were about 4-fold (P<0.01) greater in fish maintained in pale-coloured tanks than those from black tanks. This responsiveness to background colour was seen within the first week after hatching (Table I), at which time, MCH gene transcripts were clearly detectable in the nucleus lateralis tuberis and were significantly (166%, P<0.05) greater in the white-reared fish. Total MCH mRNA levels were much lower in neurons over the lateral ventricular recess and only showed a significant increase in response to a white-coloured background by the end of week 4 after hatching. Thus, although both groups of MCH neurons are affected by the colour of the tank, the more immediate response of neurons in the nucleus lateralis tuberis and their axonal connection with the pituitary gland probably indicates their greater involvement with pigmentation.

Stress effects. MCH gene expression was also affected by stress in both adult and developing trout. In the developing fry, repeated cold shock resulted in a consistent, marked increase of MCH mRNA levels in neurons over the lateral ventricular recess in the 2nd, 3rd and 4th weeks after hatching; that is, these MCH neurons responded to stress earlier than they responded to background. They also showed enhanced MCH mRNA levels in response to stress,which was evident in the 2nd. and 3rd. weeks after hatching, although this increase was statistically significant only in week three.

Table I. Effect of background colour and stress on MCH gene expression in the nucleus lateralis tuberis and above the lateral ventricular recess of developing trout.

NLT-MCH2 probe hybridization units

	WEEK 1	WEEK 2	WEEK 3	WEEK 4
Black	*186±25	468±44	*545±305	*1870±361
White	310±40	677±88	1989±214	3499±342
Stressed white	252±16	951±126	*3291±214	4153±380

LVR-MCH2 probe hybridization units

	WEEK 1	WEEK 2	WEEK 3	WEEK 4
Black	10.5±5.6	13.7±3.3	29.7±18.2	***65.5±5.0
White	5.6±2.4	13.7±3.3	43.8±6.9	108±6.9
Stressed white	9.7±2.7	*34.1±7.1	*68.6±8.3	*158±14.2

Stressed fry were kept in a perforated white container (10 x 10 cm) and exposed to ice-cold water for 1 min daily for 5 days. Values are the means ± SEM. *P<0.05, ***P<0.01compared with white-reared fish.

In adult trout, stressed either once or 3 times daily by transfer to low water levels the MCH neurons in the nucleus lateralis tuberis exhibited significantly increased levels of mRNA after 5 days of daily stress, (Fig. 1a, showing response to 3-times daily stress). This increase was not observed after only 24 h. MCH neurons over the lateral ventricular recess did not respond to these moderate stresses. However, in other experiments, in which adult trout were stressed by different manoeuvres, such as maintenance in dirty, static water, neurons in the lateral ventricular recess showed an upregulation of MCH gene expression. Thus, MCH cells in either location will respond to prolonged, stressful situations by an increase in MCH gene transcription and cells in either group may be involved with the physiological response to stress. As observed with background colour changes, no regional difference in response was evident within either the nucleus lateralis tuberis or above the lateral ventricular recess. It is not yet clear why one or other group of cells respond more or less intensely to a particular stress. The sites in the brain to which the MCH axons project and which are affected by this neuropeptide are still unknown. It has been found that MCH depresses the release of corticotropin-releasing hormone (CRH) from hypothalamic fragments *in vitro* (Green and Baker, 1991) but whether this is achieved by a direct or an indirect pathway to the CRH neurons has not been examined.

Two oligoprobes to MCH were used throughout this work, specific for MCH1 and MCH2 genes. Only that against the gene termed MCH2 (Baker et al., 1995) hybridized with the sections, and it is concluded that the MCH1 gene is not expressed under any of the experimental conditions examined here.

Acknowledgements
This work was supported by a research grant from the BBSRC.

References

Baker, B. I. (1991) MCH: a general vertebrate neuropeptide. *Int. Rev. Cytol.* 126: 1-47.
Baker, B. I. (1994) Melanin-concentrating hormone updated-functional considerations. *Trends Endocrinol. Metabolism.* 5: 120-126.
Baker, B.I., Levy, A., Hall, L. and Lightman, S. (1995) Cloning and expression of melanin-concentrating hormone genes in the rainbow trout brain. *Neuroendocrinology* 61: 67-76.

Bluet-Pajot, M.-T., Presse, F., Voko, Z., Hoeger, C., Mounier, F., Epelbaum, J. and Nahon., J. -L. (1995) Neuropeptide-E-I antagonizes the action of melanin-concentrating hormone on stress-induced release of adrenocorticotropin in the rat. *J. Neuroendocrinol.* 7: 297-303.

Green, J.A. and Baker, B.I. (1991) The influence of repeated stress on the release of melanin-concentrating hormone in the rainbow trout. *J. Endocrinol.* 128: 261-266.

Suzuki, M., Narnaware, Y.K., Baker, B.I. and Andy, A. (1995) Influence of environmental colour and diurnal phase on MCH gene expression in the trout. *J. Neuroendocrinol.* 7: 319-328.

The Peptidergic Neuron
B. Krisch and R. Mentlein (eds)
© 1996 Birkhäuser Verlag Basel/Switzerland

Endocrine influence on vasopressin-enhanced retrieval of a passive avoidance response

H. Schwarzberg, J. Onnasch and M. Pross

Otto von Guericke University, Institute of Neurophysiology, Leipziger Straße 44, D-39120 Magdeburg, Germany

Summary. There is ample evidence that peptide-peptide interactions and similar mechanisms between peptidergic and other hormonal systems exist in regulation of behaviour. Therefore, rats were pretreated with oxytocin or tri-iodothyronine, and vasopressin-enhanced retrieval of a passive avoidance response was measured. When oxytocin and vasopressin or tri-iodothyronine and vasopressin were applied successively, the effect of vasopressin was not observed. In rats made hypothyroid by thiouracil, vasopressin caused reduced retrieval. The present findings support the hypothesis that vasopressin needs regulated levels of different hormones to develop its effects on behaviour.

Introduction

Different effects of arginine-vasopressin (AVP) on behaviour of rats are well documented. The peptide is known to influence open-field behaviour, learning and memory processes, electrical self-stimulation, and drug self-administration. In similar experiments, oxytocin (OXT) developed opposite effects (for review see Kovacs and Telegdy, 1985). Hormones of the thyroid gland are also involved in the regulation of avoidance behaviour. Passive avoidance behaviour was attenuated in rats with depressed thyroxine levels caused by iodine-deficient diet (Overstreet et al. 1984).

The problem of peptide-peptide interactions has been discussed in the literature. Pretreatment with OXT reduced AVP-induced barrel rotation convulsions of rats (Abood et al., 1980), and a combined application of somatostatin and AVP resulted in a cumulative effect (Balaban et al., 1988). A mixture of proctolin, adrenocorticotropin-(4-10) and AVP,

Results

In experiment 1, the intravenous administration of 1.5 IU vasopressin significantly increased plasma vasopressin levels compared to placebo, 0.025 IU vasopressin intravenous, and 20 IU vasopressin intranasal ($p < 0.01$, respectively). Plasma vasopressin levels following 0.025 IU vasopressin intravenous and following 20 IU vasopressin intranasal were nearly identical (Fig. 1). The change in P3 amplitude was not correlated with the increase in plasma vasopressin. P3 amplitude was significantly increased following administration of 20 IU vasopressin intranasally compared to the effects of all other treatments, i.e., to the administration of 1.5 IU vasopressin intravenously ($p < 0.05$), of 0.025 IU vasopressin intravenously ($p < 0.05$) and of placebo ($p < 0.05$; Fig. 1).

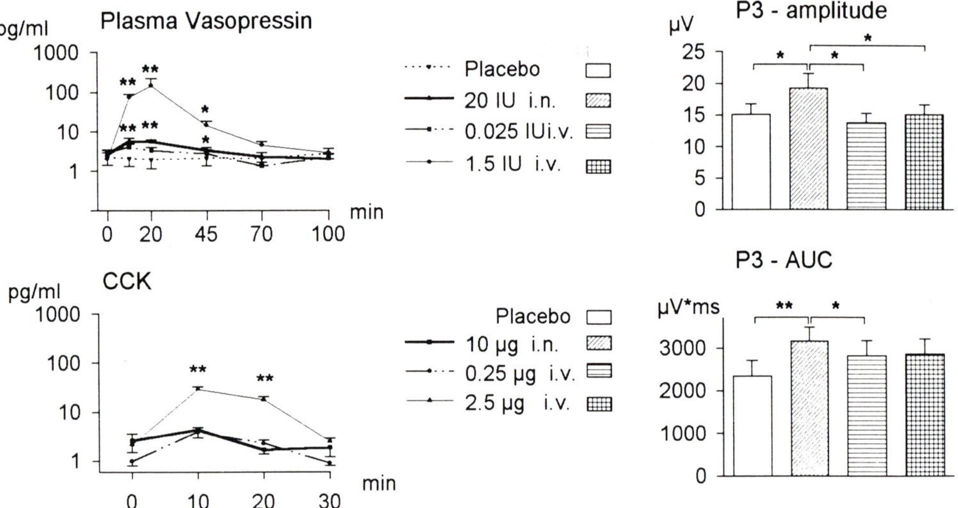

Fig. 1. Top left: Plasma vasopressin (AVP) levels following 20 IU AVP intranasal (i.n.), 0.025 IU AVP intravenous (i.v.), 1.5 IU AVP i.v., and placebo at 0, 10, 20, 30, 45, 70, and 100 min following administration. Right: Peak amplitude of the P3 component following placebo and i.n. and i.v. AVP administrations. **Bottom** left: Plasma CCK levels following 10 µg CCK i.n., 0.25 µg CCK i.v., and 2.5 µg CCK i.v. Right: Area under the curve (AUC) of the P3 amplitude following placebo and i.n. and i.v. CCK-8 administration, * $p < 0.05$, ** $p < 0.01$.

In experiment 2, CCK plasma concentrations were distinctly higher following the 2.5 µg CCK intravenous administration compared to the other treatments (p<0.01). The plasma CCK concentrations following intranasal administration of 10 µg CCK and following intravenous administration of 0.25 µg CCK were nearly identical (Fig. 1). P3 area under the curve was significantly increased following the intranasal administration of CCK compared to placebo (p<0.01) and to the low dose (0.25 µg) intravenous CCK (p<0.05; Fig. 1).

Discussion

The results show central nervous effects of peptides administered intranasally or intravenously in doses leading to comparable plasma levels, selectively following the intranasal route of administration. This excludes that the central nervous effects of these peptides when given intranasally emerged after entrance of the peptide into systemic blood. Results suggest a direct pathway from the nose to the brain for peptides, as was previously shown for other substances (e.g. Balin et al., 1986; Barthold, 1988; Barnett and Perlman, 1993).

While the lower dose of peptides administered intravenously (with similar plasma levels to the intranasal administration) did not affect event-related potentials, moderate effects were observed after high doses of intravenously admininistered CCK. This observation indicates that peptides circulating in the peripheral blood are not completely ineffective with regard to central nervous actions. However, the mechanisms transferring effects of blood borne peptides across the blood-brain barrier on the brain are obviously different than those mediating brain effects of intranasally administered peptides.

Several mechanisms have been considered transporting molecules from the nose to the brain. Peptides could be taken up into olfactory nerves and transported intracellularly (Barthold, 1988; Perlman et al., 1990). Also, they may pass through intercellular clefts in the olfactory epithelium to diffuse extracellularly to further distant sites of the central nervous system expressing peptide receptors (Balin et al., 1986). Finally, substances may influence the central nervous system not only by a spread along the olfactory system, but also along the accessory olfactory system, i.e., the vomeronasal organ, which is a chemosensory organ differing in morphology and neuroanatomical connections from the olfacory system (Monti-Bloch et al., 1994; Meredith and Fernandez- Fewell, 1994).

Conclusions

The present findings suggest that a direct nose-brain pathway exists for peptides. However, the results provide only functional evidence. Thus, it can not be decided whether peptides are transported into the brain following intranasal administration or whether their actions result from binding to receptors at peripheral sites of the olfactory systems. Nevertheless, since the actions of intranasally administered peptides do not depend on resorption into systemic blood, the nose-brain pathway may be of relevance for a more direct treatment of central nervous system diseases.

References

Balin, B.J., Broadwell, R.D., Salcman, M. and El-Kalliny, M. (1986) Avenues for entry of peripherally administered protein to the central nervous system in mouse, rat, and squirrel monkey. *J. Comp. Neurol.* 251: 260-280.
Barnett, E.W. and Perlman, S. (1993) The olfactory nerve and not the trigeminal nerve is the major site of CNS entry for mouse hepatitis virus, strain JHM. *Virology* 194: 185-191.
Barthold, S.W. (1988) Olfactory neural pathway in mouse hepatitis virus nasoencephalitis. *Acta Neuropathol.* 76: 502-506.
Eseri, M.M. and Tomlinson, A.H. (1984) Herpes simplex encephalitis. *J. Neurol. Sci.* 64: 213-217.
Fehm-Wolfsdorf, G. and Born, J. (1991) Behavioural effects of neurohypophyseal peptides in healthy volunteers: 10 years of research. *Peptides* 12: 1399-1406.
Meredith, M. and Fernandez-Fewell, G. (1994) Vomeronasal system, LHRH, and sex behaviour. *Psychoneuroendocrinology* 19: 657-672.
Monti-Bloch, L., Jennings-White, C., Dolberg, D.S. and Berliner, D.L. (1994) The human vomeronasal system. *Psychoneuroendocrinology* 19: 673-686.
Perlman, S., Evans, G. and Afifi, A. (1990) Effects of olfactory bulb ablation on spread of a neurotropic coronavirus into the mouse brain. *J. Exp. Med.* 172: 1127-1132.

C-type natriuretic peptide (CNP) in the mammalian pineal gland: An endogenous autocrine peptide ?

J. Olcese, R. Middendorff[1], E. Maronde, H.J. Paust and M.S. Davidoff[1]

Institute for Hormone and Fertility Research at the University of Hamburg, Grandweg 64, 22529 Hamburg
[1] *Institute of Anatomy, University of Hamburg, Martinistrasse 52, 20246 Hamburg, Germany*

Summary. Immunohistochemical data at the electron-microscopical level support the idea that C-type natriuretic peptide (CNP) is present in secretory vesicles in the bovine pineal gland. Polymerase chain reaction amplification of reverse-transcribed mRNA from bovine and rat pineal glands revealed the presence of pro-CNP transcripts. These data, together with previous evidence demonstrating the expression of functional natriuretic peptide receptors in pinealocytes, suggest that CNP may serve to regulate cyclic guanosine monophosphate (cGMP) pathways in the pineal gland via an autocrine feedback loop.

Introduction

We have shown previously (Olcese et al., 1994) that membrane-bound guanylyl cyclases type A (GC-A) and type B (GC-B) are expressed in the rat pineal gland and mediate increases in cyclic GMP production upon activation by natriuretic peptides. The cellular localization of natriuretic peptides in this tissue has not been clarified. In the present studies, we examined whether atrial natriuretic peptide (ANP) and c-type natriuretic peptide (CNP) could be visualized immunocytochemically, and whether pro-CNP transcripts are present in the pineal gland. In addition, we attempted to confirm the expression of functional GC-A and GC-B receptors in the bovine pineal gland.

                        J. Olcese et al.

Material and methods

For ultrastructural studies ten fresh bovine glands were fixed in 4% paraformaldehyde (PFA) with 0.08% glutaraldehyde in 0.1 M phosphate buffer for 3 h, followed by incubation overnight in PFA. After rinsing in 0.5 M phosphate buffer, sections (50 µm) were cut on a Vibratome, and processed according to the combination peroxidase-antiperoxidase / avidin-biotin-peroxidase method of Davidoff and Schulze (1990). Incubation with the primary antiserum (1:50) was carried out at 4°C for 48 h. Controls involved the use of preabsorbed anti-CNP serum, normal rabbit serum, or omission of primary, secondary or tertiary antibodies. Upon completion of the procedure, the sections were horizontally embedded in Epon, sectioned on an Ultratome and viewed under a Philips 300 electron microscope.

For detection of pro-CNP transcripts, reverse-transcriptase polymerase chain reaction (PCR) was performed on 5 µg of bovine total RNA using established protocols. Oligonucleotide primers encompassing different exons of the pro-CNP gene were commercially synthesized (MWG, Göttingen). These primers are equivalent to nucleotides 121-142 (5') and 462-485 (3') of the rat proCNP coding region (Kojima et al., 1990). Following agarose gel electrophoresis and ethidium bromide staining, the amplified PCR product was visualized under UV light. GC-A and GC-B transcripts were detected by Southern blot hybridization using previously described methods (Olcese et al., 1994).

Results

Both ANP- and CNP-immunopositive pinealocytes have been observed in tissue sections as well as in cultures of bovine cells (Middendorff et al., submitted). Ultrastructural examination showed the presence of CNP-immunoreactivity in many secretory vesicles having an average diameter of 250 nm (Fig. 1). Visualization of amplified cDNAs from several tissues revealed the presence in bovine pineal tissue of a PCR fragment having the expected size (364 bp) for the pro-CNP cDNA (Fig. 2). Similar to these results from rat pineal glands (Olcese et al., 1994), both GC-A and GC-B cDNAs have been identified in bovine pineal gland extracts following Southern blot hybridization (Middendorff et al., submitted). Furthermore, addition of ANP, brain natriuretic peptide (BNP) or CNP to cultured bovine pinealocytes resulted in dose-dependent increases in cGMP production with CNP having a potency (EC_{50}) of 5 nM.

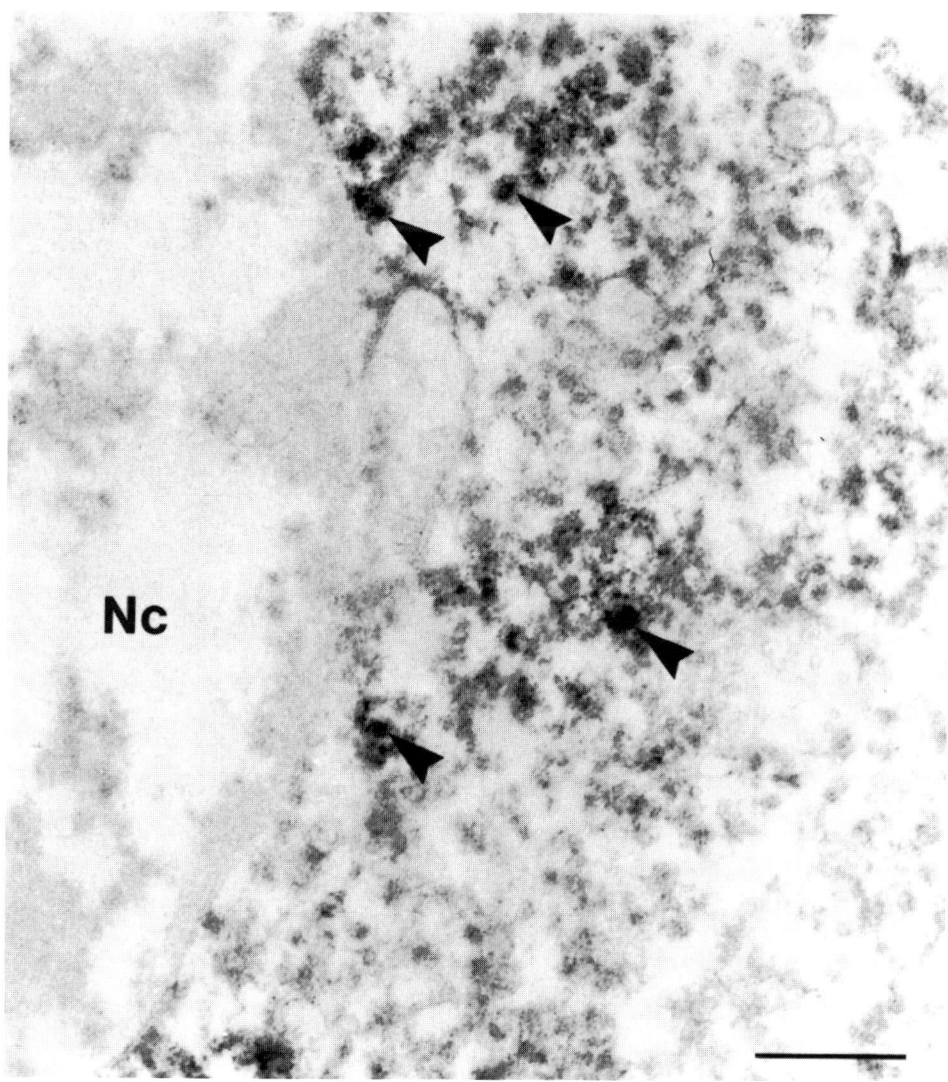

Fig. 1. Ultrastructural demonstration of CNP-immunoreactivity in a pinealocyte. Reaction product is unevenly distributed in the cytoplasm, while nucleus (Nc) is negative. Arrows point to secretory vesicles. Bar = 0.5 mm.

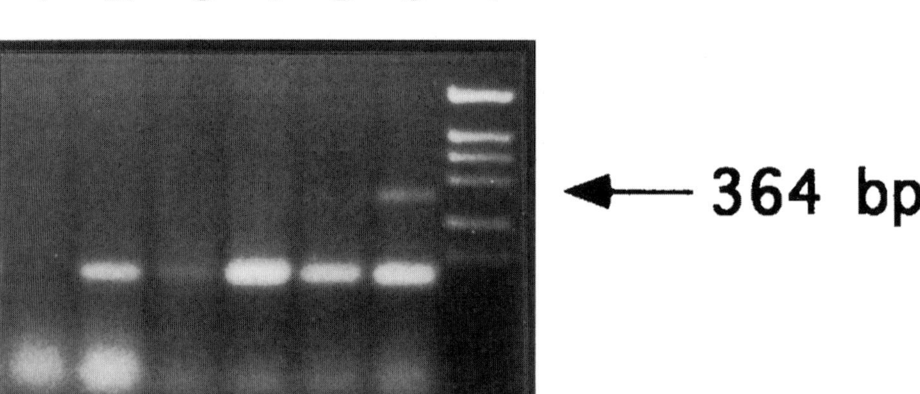

Fig. 2. PCR fragments generated with pineal cDNA (lane 6) reveal fragments corresponding in size (364 bp) to pro-CNP (arrow). Lane 1 is control (water) while lane 7 shows size markers (● type VI; Boehringer). Other cDNAs (lanes 2-5) were generated from bovine liver, lung, heart and cortex, respectively.

Discussion

The mammalian pineal contains both the cytosolic (NO-responsive) and membrane forms of GC (Spessart et al., 1993; Olcese et al., 1994). Whereas the regulation of cyclic GMP production by the former has been well-investigated, the role and the localization of natriuretic peptides in the pineal has received little attention. We have recently reported (Olcese et al., 1994) that both forms of natriuretic peptide receptors (GC-A and GC-B) are present and activated by their natural ligands in the rat pineal gland. In addition, the presence of a related "orphan" receptor (GC-E) in the rat pineal has also been demonstrated recently (Yang et al., 1995). We have also confirmed that both GC-A and GC-B are common transcripts in the bovine pineal gland, and that they can be activated by nanomolar conentrations of natriuretic peptides, with CNP being the most potent (Middendorff et al., submitted). Hence, several signalling pathways appear to be involved in generating cGMP in the pineal, suggesting that the intra- or extracellular targets for this second messenger molecule are important generally for pineal function. Although a role for cGMP in the mammalian pineal gland remains uncertain, recent reports indicate the presence of cGMP-gated cationic channels (Distler et al., 1994; Schaad et al., 1995) and a cGMP-phosphodiesterase (Carcamo et al., 1995).

Our present findings that 1) both ANP as well as CNP are present within a subset of pinealocytes, 2) CNP-immunoreactivity appears to localize to secretory vesicles, and 3) proCNP transcripts are clearly present, supports the view that natriuretic peptides may represent endogenous pineal products with autocrine or paracrine activities in the mammalian pineal gland. Although we have as yet been unable to obtain clear evidence of CNP secretion from pinealocytes, these data raise the possibility that pinealocytes may communicate information via peptidergic signals. Potential cell targets may be pinealocytes themselves, sympathetic nerve fibres innervating the gland, or vascular smooth muscle cells. A specific change in cGMP levels due to activation of natriuretic peptide receptors at any of these sites could be anticipated to have an impact, directly or indirectly, on pineal activity or on regional blood flow, as has been demonstrated in other tissues (Babinski et al., 1995; Drewett et al., 1995).

Conclusion

In addition to the presence of functional GC-A and GC-B receptors, the mammalian pineal also expresses pro-CNP transcripts, and CNP appears localized within cytoplasmic vesicles. This suggests that CNP may have an autocrine or paracrine role within the pineal.

Acknowledgements
We gratefully acknowledge the technical assistance provided by A. Samalecos. These studies were carried out with financial support provided by the Deutsche Forschungsgesellschaft (#Ol 45/4-1).

References

Babinski, K., Haddad, P., Vallerand, D., McNicoll, N., DeLean A. and Ong H. (1995) Natriuretic peptides inhibit nicotine-induced whole-cell currents and catecholamine secretion in bovine chromaffin cells: Evidence for the involvement of the atrial natriuretic factor R2 receptors. *J. Neurochem.* 64: 1080-1087.

Carcamo, B., Hurwitz, M.Y., Craft, C.M. and Hurwitz, R.L. (1995) The mammalian pineal expresses the cone but not the rod cyclic GMP phosphodiesterase. *J. Neurochem.* 65: 1085-1092.

Davidoff, M. and Schulze, W. (1990) Combination of the peroxidase-antiperoxidase (PAP)- and the avidin-biotin-peroxidase complex (ABC)-techniques: An amplification alternative in immunocytochemical staining. *Histochemistry* 93: 531-536.

Distler, M., Biel, M., Flockerzi, V. and Hofmann, F. (1994) Expression of cyclic nucleotide-gated cation channels in non-sensory tissues and cells. *Neuropharmacology* 33: 1275-82.

Drewett, J.G., Fendly, B.M., Garbers, D.L. and Lowe, D.G. (1995) Natriuretic peptide receptor-B (guanylyl cyclase-B) mediates C-type natriuretic peptide relaxation of precontracted aorta. *J. Biol. Chem.* 270: 4668-4674.

Kojima, M., Minamino, N., Kangawa, K. and Matsuo, H. (1990) Cloning and sequence analysis of a cDNA encoding a precursor for rat C-type natriuretic peptide (CNP). *FEBS Lett.* 276: 209-213.

Middendorff, R., Maronde, E., Paust, H.-J., Müller, D., Davidoff, M. and Olcese, J. (1996) Expression of C-type natriuretic peptide in the bovine pineal gland. *J. Neurochem.; in press.*

Olcese, J. , Müller, D., Münker, M. and Schmidt, C. (1994) Natriuretic peptides elevate 3',5'-guanosine monophosphate levels in cultured rat pinealocytes: Evidence for guanylate cyclase-linked membrane receptors. *Mol. Cell. Endocrinol.* 103: 95-100.

Rüppel, R. and Olcese, J. (1991) Bovine pinealocytes in monolayer culture: Studies on the adrenergic regulation of melatonin secretion. *Endocrinology* 129: 2655-2662.

Schaad, N.C., Vanecek, J., Rodriguez, I.R., Klein, D.C., Holtzclaw, L. and Russell, J.T. (1995) Vasoactive intestinal peptide elevates pinealocyte intracellular calcium concentrations by enhancing influx: Evidence for involvement of a cyclic GMP-dependent mechanism. *Mol. Pharmacol.* 47: 923-933.

Spessart, R., Layes, E. and Vollrath, L. (1993) Adrenergic regulation of cyclic GMP formation requires NO-dependent activation of cytosolic guanylate cyclase in rat pinealocytes. *J. Neurochem.* 61: 138-144.

Yang, R.-B., Foster, D.C., Garbers, D.L. and Fülle, H.-J. (1995) Two membrane forms of guanylyl cyclase found in the eye. *Proc. Natl. Acad. Sci. USA* 92: 602-606.

VII. Neuropeptides in immune defense and pathology

The Peptidergic Neuron
B. Krisch and R. Mentlein (eds)
© 1996 Birkhäuser Verlag Basel/Switzerland

The role of neuropeptides in immunoregulatory processes

G. B. Stefano

Multidisciplinary Center for the Study of Aging, Old Westbury Neuroscience Research Institute, State University of New York, College at Old Westbury, Old Westbury, N.Y. 11568-0210, USA

Summary. Among the various non-neuronal cell types known to produce and make use of neuropeptides, those of the immune system have received much attention in recent years. Comparative studies in vertebrates and invertebrates have shown that some of these compounds, especially opioid peptides, are engaged in receptor-mediated autoregulating processes of the immune system as well as the exchange of information with the neuroendocrine system. Most of those activities observed are immunostimulatory, as determined by their effects on conformational changes indicative of immunocyte activation, cellular motility, and phagocytosis. In co-operation with cytokines, opiate alkaloids, and certain regulatory enzymes (neutral endopeptidase 24.11), they form an effective network of messenger molecules. The immunostimulatory effects observed in this system are balanced (counteracted) by the primarily inhibitory influence of morphine and related opiates. These signals are mediated by several types of receptors with different degrees of selectivity. Among them the recently identified, opioid insensitive μ_3 receptor deserves attention on account of its specificity for opiate alkaloids.

Introduction

An important development in the study of neurosecretion is the growing realization that neuropeptides, formerly thought to be produced by a selected group of neurons, are found in a variety of non-neuronal cells. Comparative studies in vertebrates and invertebrates (see Stefano and Scharrer, 1994) have shown that among them are immunoactive cells which make use of neuropeptides for autoregulatory control as well as the bidirectional exchange of information between the immune system and the neuroendocrine system. Along with cyto-kines, long considered to be the primary messenger molecules of the immune system, and with endogenous opioid peptides, these neuropeptides form an effective network of communication.

Previous studies demonstrated the presence of Met-enkephalin, Leu-enkephalin and Met-enkephalin-Arg-Phe in the nervous and immune system of both invertebrates and vertebrates along with highly selective opioid receptors (see Stefano, 1982; Leung and Stefano, 1987; Stefano, 1989; Stefano and Scharrer, 1994).

Activities associated with immunoinflammatory reactions that have been analyzed in detail in higher invertebrates as well as vertebrates are cellular adherence, locomotory activity and conformational changes of immunocompetent cells. Adherence of mammalian neutrophils and invertebrate immunocytes is enhanced by opioid neuropeptides (see Scharrer and Stefano, 1994; Stefano, 1994). In human neutrophils substance P (10^{-11} M) modulates the expression of two cell-surface adhesion molecules, Mo 1 and LAM-1 (Shipp et al., 1991). Polymorphonuclear leukocytes, lymphocytes, and monocytes as well as invertebrate immunocytes in addition to random movements (chemokinesis), show migratory behaviour directed toward sites of inflammation or antigenic challenge (chemotaxis). In an *in vitro* study, T-lymphocytes exhibited chemotaxis in the presence of a concentration gradient of Met-enkephalin or β-endorphin. Synthetic enkephalin analogues, including DADLE (D-Ala^2-D-Leu^5-enkephalin), DPDPE (D-Pen^2-D-Pen^5-enkephalin; Pen, L-penicillamine or β,β-dimethyl-L-cysteine), and DAG (D-Ala^2-MePhe^4-Gly^5-enkephalin) stimulated the T-cells to a lesser extent, a finding that will be discussed below.

In vitro tests with immunocytes of *Mytilus* and *Leucophaea* showed chemotactic movements and the formation of large cellular clumps after opioid peptide exposure along with the same poor reaction to DADLE (see Stefano, 1989). Furthermore, changes in the activity of human and invertebrate immunocytes are preceded by characteristic conformational alterations (see Scharrer and Stefano, 1994). Prior to the onset of locomotory behaviour, stimulated by opioid peptides, indications of cellular activation are observed: mammalian and invertebrate cells in the inactive condition are more or less rounded, upon activation show an increase in cellular size and surface area and/or the formation of pseudopodia. D-Ala^2-Met^5-enkephalinamide (DAMA) is most effective in inducing these changes. DADLE, Leu-enkephalin analogue most closely related to DAMA (a Met-enkephalin analogue), is not as potent in mammalian and invertebrate immunocytes (see Stefano, 1989). The distinctly lower effectiveness of DADLE in both human and invertebrate immune reactions is in contrast to the situation in the mammalian nervous system where no discrepancy in the binding potency of Met-enkephalin and Leu-enkephalin has been observed (see Scharrer and Stefano, 1994). Taken together, these studies demonstrate that opioid peptides exhibit, in general, immunostimulatory actions.

In human and invertebrate immunocytes a possible immunoregulatory function for Met-enkephalin-Arg^6-Phe^7 involving has been reported as well (Stefano et al., 1991a). Met-enkephalin-Arg^6-Phe^7 exhibits stimulatory effects comparable to those of Met-enkephalin. Furthermore, phosphoramidon, a potent inhibitor of neutral endopeptidase 24.11, potentiates the effects of the heptapeptide in inducing conformational change in both human and invertebrate immunocytes (Shipp et al., 1990). The major metabolic products of neutral endopeptidase activity, Tyr-Gly-Gly and especially Phe-Met-Arg-Phe, appear to be themselves potent antagonists of this enzyme (Stefano et al., 1991a). Thus, protease cleavage is another naturally occurring mechanism to diminish the efficiency of the heptapeptide.

Immunocyte opioid receptors

Deltorphin I, a naturally occurring opioid peptide isolated from amphibian skin, has the ability to modulate both human and invertebrate immunoregulatory activities in a manner quite similar to Met-enkephalin . Its binding and pharmacological studies also have provided evidence for a special subtype of delta opioid receptor δ_2, sensitive to naltrindole antagonism (Stefano and Scharrer, 1994), on human and invertebrate immune cells (Stefano et al., 1992b). The results obtained with deltorphin I support the view that the special role played by endogenous Met-enkephalin in immunobiological activities of vertebrates and invertebrates is mediated by a special subtype of delta opioid receptor, δ_2. It is also of interest to note that both the invertebrate immunocytes and human granulocytes thus have δ_1 and δ_2 receptors (Stefano et al., 1992b).

Opiate alkaloids

While opiate alkaloids, e.g., morphine, are not opioid peptides they do deserve special attention within the context of this review for several reasons. First, in addition to antinociceptive mechanisms, opiate alkaloids and opioid peptides initiate different immunocyte behaviours. As noted above, opioid peptides may be generally regarded as immunocyte stimulatory and/or activating ligands whereas morphine, noted first by Wybran et al. (1979), is inhibitory. Secondly, confusion exists in the scientific literature as to the proper terminology for these ligands, e.g., opioid alkaloid and opiate peptide. Thirdly, a novel opiate alkaloid and opioid peptide insensitive receptor, namely μ_3, has been demonstrated which does not recognize μ-opioid ligands. Lastly, opiate alkaloids appear to be naturally occurring substances found both in mammals and invertebrates (see Stefano et al., 1993; Stefano and Scharrer, 1994).

Recently, as in mammals, morphine-like and codeine-like substances were demonstrated in the pedal ganglia, hemolymph and mantle tissues of the mollusc *Mytilus edulis* (Stefano et al., 1993). The pharmacological activities of the endogenous morphine-like material resemble those of authentic morphine. Both substances were found to counteract, in a dose dependent manner, the stimulatory effect of tumour necrosis factor α (TNF-α) or interleukin-1α on human monocytes and *Mytilus* immunocytes. The immunosuppressive effect of this opiate material expresses itself in a lowering of chemotactic activity, cellular velocity and adherence as well as making active immunocytes inactive (rounded). These pharmacological effects of morphine on immunocytes are consistent with those actions attributed to opiates reported in the literature. Indeed, it has been surmised that transmission by morphins may regulate the downregulation of immune activation (see Stefano and Scharrer, 1994).

Along with the opiate substances found in animal tissues came the recent discovery of a specific high-affinity and novel receptor site (μ_3) for opiate alkaloids on human monocytes as well as *Mytilus* immunocytes. A variety of opioid peptides, tested by two methods, were found to be ineffective in displacing specifically bound 3-dihydromorphine. By contrast, the opiate alkaloid μ ligands were potent and κ ligands dynorphin 1-17 and ethyl-keto-cyclazocine were weak competitors. Based on this novel displacement information we assigned this opioid peptide insensitive and opiate alkaloid sensitive site the name μ_3 (Stefano et al., 1993). Recent studies demonstrate that human granulocytes also contain the μ_3 subtype opiate receptor mediating inhibition by morphine and other opiates of cytokine-induced activation and chemotaxis (Makman et al., 1995). Furthermore, in the presence of

NaCl (50 mM) plus the GTP analogue guanylyl-5′-imidodiphosphate (GppNHp; 100 μM), there was a significant decrease in specific high-affinity binding of the agonist ligand ^{3}H-morphine (Makman et al., 1995). The influence of the GTP analogue GppNHp on binding indicated that the granulocyte receptor was linked to a G protein. The discovery of this receptor site mediating opiate effects was first found in an invertebrate and then in man, again demonstrating the value of the comparative approach (Stefano et al., 1993).

Recently, this opiate alkaloid sensitive receptor has also been found in established macrophage cell lines (Makman et al., 1995). Furthermore, its presence on invertebrate and vertebrate microglial cells as well as in invertebrate neurons demonstrates its widespread occurrence (Sonetti et al., 1994; Stefano and Scharrer, 1995). It is worth noting that these cell lines have been utilized previously as model systems for study of actions of interferons, interleukins, colony stimulating factor-1 as well as other regulators of macrophage and immune functions.

It is important to note that the cloning of δ-, μ- and κ- receptors has now been accomplished (Evans et al., 1992; Kieffer et al., 1992; Yasuda et al., 1993). As a result of these and other studies now published, it will be possible to study individual receptors regarding their effector coupling, pharmacological characteristics, regulation of expression as well as their regional distributions. Important information will also become available regarding their evolvement.

Opioid degradation mechanisms in hemolymph

A major concern that relates to the presence of neuropeptides in hemolymph and plasma is, how they are degraded or their action terminated. In this regard, in mammals CD10 (CALLA, common acute lymphoblastic leukemia antigen / neutral endopeptidase 24.11; NEP, "enkephalinase") hydrolyzes a number of naturally occurring peptides including the endogenous opioid pentapeptides Met- and Leu-enkephalin (see Kenny, pp. 87 - 101). This enzyme in the mammalian brain has been termed "enkephalinase". *Mytilus edulis* hemocytes also express a CD10/NEP related enzyme and the abrogation of CD10/NEP enzymatic activity reduces the amount of Met-enkephalin required for hemocyte activation by five orders of magnitude (Shipp et al., 1990). Human CD10-positive polymorphonuclear leukocytes have similar responsiveness. Thus, in vertebrates and invertebrates a precise mammalian-like mechanism for degrading peptides in the animals immune/defense system is present as well as a mechanism that functions to control responsiveness to Met-enkephalin and other peptides.

Opioid-cytokine link

Opioid induction of an interleukin-1-like substance in *Mytilus edulis* pedal ganglia and immunocytes has been demonstrated (see Stefano, 1992a). Both the immune and nervous system of *Mytilus* contain an interleukin-1-like molecule. In nervous tissue it is localized in microglial cells (Paemen et al., 1992) thus adding a new signal system to invertebrate neurobiology as well as the potential for invertebrate microglial cells to alter neural signaling. In *Mytilus*, recombinant human interleukin-6, although not activating cells directly, potentiated interleukin-1 activation of immunocytes (Hughes et al., 1991). Furthermore an immunoreactive interleukin-6 appears to be present in *Mytilus edulis* and the insect *Leucophaea* hemolymph (0.82 ng/ml; Hughes et al., 1991). It was also found that immunoreactive interleukin-6 is produced in pedal ganglia in response to the pharmacological challenge of the Met-enkephalin analogue DAMA. These data demonstrate that immune signal molecules may have functions that transcend immunomodulation.

Corticotropin-releasing factor, adrenocorticotropin and melanotropin-like substances

Adrenocorticotropin (ACTH)-like immunoreactivity is present in representatives of most phyla (see Smith et al., 1991). This compound is found in neuroendocrine as well as immunoactive structures (Hansen et al., 1986). Ottaviani and colleagues have done extensive comparative studies (Ottaviani et al., 1990; Sonetti et al., 1994) and demonstrated ACTH and ß-endorphin in the hemolymph and hemocytes of the freshwater snail *Planorbarius corneus*. The ACTH-derived peptide α-melanocyte stimulating hormone (α-MSH) has also been demonstrated in *Mytilus* ganglia and hemolymph (Stefano and Martin, 1983; Smith et al., 1991). The role of ACTH in invertebrate hemolymph may be immunoregulation in addition to some other signal function. When ACTH (1-39) and MSH were assessed for the ability to alter immunocyte conformation and motility in *Mytilus edulis*, results similar to those in human granulocytes were obtained (Stefano et al., 1991b; Smith et al., 1992a). ACTH itself did not appear to alter the activity, i.e., the conformational state of inactive or active hemocytes in short-term assays, whereas both α- and ß-MSH induced active ameboid hemocytes to become rounded and inactive in a dose-dependent fashion. α-MSH also inactivated ameboid hemocytes which had been stimulated with ß-endorphin or TNF (Stefano et al., 1991b). As in human granulocytes, ACTH inactivated the immunocytes after four hours of incubation whereas the effect of α-MSH occurred in minutes (Smith et al., 1992a). The effect of ACTH was blocked by phosphoramidon. Therefore, it is likely that the major inhibitory activity on *Mytilus* immunocytes is due to MSH following ACTH cleavage by neutral endopeptidase (Smith et al., 1992a; Duvaux-Miret et al., 1992a,b). Furthermore, at extremely low doses ACTH has been reported to enhance immunocyte migration, phagocytic activity of and the release of biogenic amines from immunocytes of the snail *Planorbarius corneus* (Ottaviani et al., 1990, 1991). Recently, Genedani et al. (1993) demonstrated that different ACTH fragments induce different activities in the same cell. ACTH fragments -(1-24), -(1-4), -(4-9), -(1-13), -(1-17), -(11-24) significantly stimulate molluscan hemocyte migration, whereas the whole sequence -(1-39) and the fragment -(4-11) have an inhibitory effect.

In invertebrates, corticotropin-releasing factor (CRF)-like molecules were first identified both in hemocytes and serum of *Planobarius corneus* (Ottaviani et al., 1990) by radioimmunoassay. Recently, CRF was shown to mimic the effects of α-MSH while exhibiting a longer duration of action (Smith et al., 1992b). α-Helical CRF, a specific inhibitor of CRF, antagonized CRF-induced cellular immunosuppression (ameboid conformation and chemotaxis) but was ineffective in altering MSH induced immunosuppression. Both human and *Mytilus* immunocytes appear to have specific CRF receptors (Smith et al., 1992b).

Biomedical significance

The biomedical importance of a well balanced immunoregulatory system is illustrated by the consequence of interference with its normal operation (see Jankovic and Maric, 1994). Recent studies have shown that immunosuppression effected by neuropeptides may determine the course of certain diseases caused by parasites or viral infection (Duvaux-Miret et al., 1992a,b; Smith et al., 1992a). There is experimental evidence supporting the concept that in schistosomiasis the parasite escapes detection and an effective immune reaction in the host by using the same signal molecules operating in the human immune and autoimmunoregulatory system (Duvaux-Miret et al., 1992a,b). The release of ACTH by the adult parasite, and its conversion to α-MSH by neutral endopeptidase on human polymorphnuclear leukocytes, inactivates specific defense cells and thus interferes with proper surveillance. Furthermore, the human immunodeficiency virus appears to have the ability to stimulate the production of ACTH by human immune cells (Smith et al., 1992a), thus creating a scenario similar to that described for the parasitic worm. It is becoming quite clear that these peptides play important immunoregulatory roles, actions that include neuroimmune as well as autoimmunoregulatory mechanisms.

Conclusion

I surmise that we are just scratching the surface of the involvement of neuropeptides in immune regulation. This review has mainly emphasized the roles of opioid and related peptides, clearly leaving out many other types of peptidergic signaling compounds. For the most part, it is the opioid/opiate and ACTH "story" that has emerged in recent years. Thus, we will undoubtedly look forward to the activities and presence of other peptidergic signaling molecules being used both in autoimmunoregulation and neuroimmunoregulation. Given the presence of many of these signaling molecules in neuroendocrine structures the field of neurosecretion will grow to include, if it hasn't already done so, neuroimmunology.

Acknowledgements
I acknowledge the following grant support: NIMH-NIDA COR 17138, NIH Fogarty International Center 00045, NIDA-09010 and the Research Foundation/SUNY (GBS).

References

Duvaux-Miret, O., Stefano, G.B., Smith, E.M., Dissous, C. and Capron, A. (1992a) Immunosuppression in the definitive and intermediate hosts of the human parasite *Schistosoma mansoni* by release of immunoactive neuropeptides. *Proc. Natl. Acad. Sci. USA* 89: 778-781.

Duvaux-Miret, O., Stefano, G.B., Smith, E.M., Mallozzi, L.A. and Capron, A. (1992b) Proopiomelanocortin-derived peptides as tools of immune evasion for the human trematode *Schistosoma mansoni. Acta Biol. Hungaria* 43: 281-286.

Evans, C., Keith, D., Morrison, H., Magendzo, K. and Edwards, R. (1992) Cloning of a delta opioid receptor by functional expression. *Science* 258: 1952-1955.

Genedani, S., Bernardi, M., Ottaviani, E., Franceschi, C., Leung, M.K. and Stefano, G.B. (1993) Differential modulation of invertebrate hemocyte motility by CRF, ACTH, and its fragments. *Peptides* 15: 203-206.

Hansen, B.L., Hansen, B.N. and Scharrer, B. (1986) Immunocytochemical demonstration of a material resembling vertebrate ACTH and MSH in the corpus cardiacum - corpus allatum complex of the insect *Leucophaea maderae. In:* G.B. Stefano (ed.): *Handbook of Comparative Aspects of Opioid and Related Neuropeptide Mechanisms*, Vol 1., CRC Press, Boca Raton, pp. 213-222.

Hughes, T.K., Smith, E.M. and Stefano, G.B. (1991) Detection of immunoreactive interleukin-6 in invertebrate hemolymph and nervous tissue. *Prog. NeuroImmuneEndocrin.* 4: 234-239.

Jankovic, B.D. and Maric, D. (1994) Enkephalins as regulators of inflammatory immune reactions. *In:* B. Scharrer, E.M. Smith and G.B. Stefano (eds.): *Neuropeptides and Immunoregulation,* Springer-Verlag, Heidelberg, pp. 76-100.

Kieffer, B.L., Befort, K., Gaveraux-Ruff, C. and Hirth, C.G. (1992) The delta opioid receptor: Isolation of cDNA by expression cloning and pharmacological characterization. *Proc. Natl. Acad. Sci. USA* 89: 12048-12052.

Leung, M.K. and Stefano, G.B. (1987) Comparative neurobiology of opioids in invertebrates with special attention to senescent alterations. *Prog. Neurobiol.* 28: 131-159.

Makman, M.H., Bilfinger, T.V. and Stefano, G.B. (1995) Human granulocytes contain an opiate receptor mediating inhibition of cytokine-induced activation and chemotaxis. *J. Immunol.* 154: 1323-1330.

Ottaviani, E., Petraglia, F., Montagnani, G., Cossarizza, A., Monti, D. and Franceschi, C. (1990) Presence of ACTH and β-endorphin immunoreactive molecules in the freshwater snail *Planorbarius corneus* (L.) (Gastropoda, Pulmonata) and their possible role in phagocytosis. *Regul. Pept.* 27: 1-9.

Ottaviani, E., Caselgrandi, E., Bondi, M., Cossarizza, A., Monti, D. and Franceschi, C. (1991) The "immune-mobile brain": evolutionary evidence. *Adv. Neuroimmunol.* 1: 27-39.

Paemen, L.R., Porchet-Hennere, E., Masson, M., Leung, M.K., Hughes, T.K. and Stefano, G.B. (1992) Glial localization of interleukin-1α in invertebrate ganglia. *Cell. Mol. Neurobiol.* 12: 463-472.

Scharrer, B. and Stefano, G.B. (1994) Neuropeptides and autoregulatory immune processes *In:* B. Scharrer, E.M. Smith and G. B. Stefano (eds.): *Neuropeptides and Immunoregulation,* Springer-Verlag, Heidelberg, pp. 1-18.

Shipp, M.A., Stefano, G.B., D'Adamio, L., Switzer, S.N., Howard,F.D., Sinisterra, J., Scharrer, B. and Reinherz, E. (1990) Downregulation of enkephalin-mediated inflammatory responses by CD10/neutral endopeptidase 24.11. *Nature* 347: 394-396.

Shipp, M.A., Stefano, G.B., Switzer, S.N., Griffin, J.D. and Reinherz, E.L. (1991) CD10 (CALLA)/ Neutral endopeptidase 24.11 modulates inflammatory peptide-induced changes in neutrophil morphology, migration, and adhesion proteins and is itself regulated by neutrophil activation. *Blood* 78: 1834-1841.

Smith, E. M., Hughes, T.K., Leung, M.K. and Stefano, G.B. (1991) The production and action of ACTH-related peptides in invertebrate hemocytes. *Adv. Neuroimmunol.* 1: 7-16.

Smith, E.M., Hughes, T.K., Hashemi, F. and Stefano, G.B. (1992a) Immunosuppressive effects of ACTH and MSH and their possible significance in human immunodeficiency virus-infection. *Proc. Natl. Acad. Sci. USA* 89: 782-786.

Smith, E.M., Hughes, T. K., Cadet, P. and Stefano, G.B. (1992b) CRF induced immunosuppression in human and invertebrate immunocytes. *Cell. Mol. Neurobiol.* 12: 473-482.

Sonetti, D., Ottaviani, E., Bianchi, F, Rodriguez, M., Stefano, M.L., Scharrer, B. and Stefano, G.B. (1995) Microglia in invertebrate ganglia. *Proc. Natl. Acad. Sci USA*, in press.

Stefano, G.B. (1982) Comparative Aspects of Opioid-Dopamine Interaction. *Cell. Mol. Neurobiol.* 2: 167-178.

Stefano, G. B. (1989) Role of opioid neuropeptides in immunoregulation. *Prog. Neurobiol.* 33: 149-159.

Stefano, G.B. (1992) Invertebrate and vertebrate immune and nervous system signal molecule commonalties. *Cell. Mol. Neurobiol.* 12: 357-366.

Stefano, G. B. (1994) Pharmacological and binding evidence for opioid receptors on vertebrate and invertebrate blood cells. *In:* B. Scharrer, E.M. Smith and G. B. Stefano (eds.): *Neuropeptides and Immunoregulation,* Springer-Verlag, Heidelberg, pp. 139-151.

Stefano, G.B. and Martin, R. (1983) Enkephalin-like immunoreactivity in the pedal ganglion of *Mytilus edulis* (bivalvia) and its proximity to dopamine containing structures. *Cell Tissue Res.* 230: 147-153.

Stefano, G. B., Shipp, M.A. and Scharrer, B. (1991a) A possible immunoregulatory function for Met-enkephalin-Arg6-Phe7 involving human and invertebrate granulocytes. *J. Neuroimmunol.* 31: 97-103.

Stefano, G.B., Smith, D.M., Smith, E.M. and Hughes, T.K. (1991b) MSH can deactivate both TNF stimulated and spontaneously active immunocytes. *In:* K.S. Kits, H.H. Boer and J. Joosse (eds.): *Molluscan Neurobiology,* North Holland Publishing Co., Amsterdam, pp. 206-209.

Stefano, G.B., Kimura, T., Stefano, J.M., Finn III, J.P., Leung, M.K., Smith, E.M., Mallozzi, L., Pryor, S. and Hughes Jr., T.K. (1992a). Autoimmunomodulation: Age-related opioid differences in vertebrate and invertebrate immune systems. *Ann. New York Acad. Sci.* 663: 396-402.

Stefano, G.B., Melchiorri, P., Negri, L., Hughes, T.K. and Scharrer, B. (1992b) (D-Ala2)-Deltorphin I binding and pharmacological evidence for a special subtype of delta opioid receptor on human and invertebrate immune cells. *Proc. Natl. Acad. Sci. USA* 89: 9316-9320.

Stefano, G.B., Digenis, A., Spector, S., Leung, M.K., Bilfinger, T.V., Makman, M.H., Scharrer, B. and Abumrad, N.N. (1993) Opiate - like substances in an invertebrate, a novel opiate receptor on invertebrate and human immunocytes, and a role in immunosuppression. *Proc. Natl. Acad. Sci. USA* 90: 11099-11103.

Stefano, G.B. and Scharrer, B. (1994) Endogenous morphine and related opiates, a new class of chemical messengers. *Adv. Neuroimmunol.* 4: 57-68.

Stefano, G.B. and Scharrer, B. (1995) The presence of the μ3 opiate receptor in invertebrate neural tissues. *Comp. Biochem. Physiol.; in press.*

Wybran, J., Appelboom, T., Famaey, J.P. and Govaerts, A. (1979) Suggestive evidence for receptors for morphine and methionine enkephalin on normal human blood T- lymphocytes. *J. Immunol.* 123: 1068-1070.

Yasuda, K., Raynor, K., Kong, H., Breder, C., Takeda, J., Reisine, T. and Bell G.I. (1993) Cloning and functional comparison of κ and δ receptors from mouse brain. *Proc. Natl. Acad. Sci. USA* 90: 6736-6740.

Neuropeptides in hypothalamic pathology

D.F. Swaab

Graduate School Neurosciences Amsterdam, Netherlands Institute for Brain Research, Meibergdreef 33, 1105 AZ Amsterdam ZO, The Netherlands

Summary. The human hypothalamus is involved in a wide range of functions in the developing, adult and aging subject and disorders and degenerative changes in this brain region are responsible for a large number of symptoms of neuroendocrine, neurological and psychiatric diseases. The present paper gives only a few examples of such neuropathological changes in the suprachiasmatic nucleus (SCN) and paraventricular nucleus (PVN).

The SCN shows seasonal and circadian fluctuations in the number of vasopressin neurons in young subjects. During normal aging the seasonal and circadian fluctuations in vasopressin neurons disappear after the age of 50. The number of vasopressin expressing neurons decreases after the age of 80 and even more so and at an earlier age in Alzheimer's disease. The SCN is affected in Alzheimer's disease, which may lead to wandering, agitation and sleep disorders, that can be treated with light therapy.

The vasopressin and oxytocin cells of the supraoptic (SON) and paraventricular nucleus project to the neurohypophysis where these peptides are released as neurohormones that are involved in water metabolism, sexual arousal, ejaculation, labor and lactation. One year after hypophysectomy 80% of the SON and PVN neurons is lost. Animal experiments have shown that oxytocin neurons that project to the brain stem inhibit eating behaviour and are the putative satiety neurons of the brain. This idea is reinforced by our observations in Prader-Willi-syndrome patients that are characterized by gross obesity and insatiable hunger. In these patients we found that the PVN total cell number was 38% lower and the PVN oxytocin neuron number 42% lower than in controls. Diabetes insipidus may have different hypothalamic causes. Apart from trauma, ischemia, hemorrhage, inflammation and surgical manipulations, familial hypothalamic diabetes insipidus can be present, based upon a point mutation in the vasopressin-neurophysin-glycopeptide gene. Urine production may amount to up to 20 liters per day. Neuronal death in the SON and PVN has been reported in this disorder. In addition, an autoimmune form of diabetes insipidus exists with circulating antibodies against the vasopressin neuron cell surface.

Parvicellular corticotropin-releasing hormone (CRH)-containing neurons in the PVN are moderately activated during the course of normal aging, slightly more in Alzheimer's disease and very strongly in depression and multiple sclerosis. Activation of CRH neurons has been established on the basis of an increase in the number of neurons expressing this peptide, an increased colocalization of vasopressin in CRH neurons and an increase in the amount of CRH mRNA. However, the pattern in which CRH neurons are activated differs in different disorders. CRH cells are not only involved in the regulation of the hypothalamo-pituitary adrenal axis, but project centrally as well. These centrally projecting CRH neurons may be responsible for mood changes, such as depression. The activation of vasopressin and oxytocin neurons in the PVN of depressed patients might potentiate the effects of CRH activation, while oxytocin hyperactivity might contribute to the inhibition of food intake in depression.

The human hypothalamus is involved in a wide range of functions in the developing, adult and aging subject, as well as in various diseases of different etiologies. Alterations in hypothalamic structures and functions are thought to be operative in signs and symptoms of diseases such as anorexia nervosa, bulimia, depression, Cushing's disease, diabetes insipidus, Wolfram's syndrome, Prader-Willi syndrome, polycystic ovaries syndrome and the malignant neuroleptic syndrome, as well as in disturbances in sleep and temperature regulation. Alterations in the hypothalamus have also been found in Sudden-Infant-Death-syndrome. In addition, the hypothalamus is affected in neurodegenerative diseases, which may lead to particular symptoms in, e.g., Alzheimer's, Parkinson's, and Huntington's disease, and of Multiple Sclerosis. Moreover, this brain region is presumed to alter as a result of endocrine effects on brain development in the adrenogenital syndrome, due to hormones administered during development (e.g. diethylstibestrol), as well as in transsexuality, in Adrenogenital, Turner's, Klinefelter's, and Kallmann's syndrome. Attention is now paid to the relationship between the structural development of the human hypothalamus, gender and sexual orientation (Swaab and Hofman, 1995). The present paper deals only with a few peptidergic changes in the human suprachiasmatic and paraventricular nucleus. (For a more extensive review see Swaab, 1996).

Suprachiasmatic nucleus

The suprachiasmatic nucleus (SCN) is a small structure that is considered to be the major circadian pacemaker of the mammalian brain and to coordinate hormonal and behavioural circadian and circannual rhythms (Hofman and Swaab, 1993). The vasopressin subnucleus of the SCN has a volume of 0.25 mm^3 on each side (Swaab et al., 1985). A lesion in the suprachiasmatic region of the anterior hypothalamus, e.g. as the result of a tumour, indeed results in disturbed circadian rhythms in human beings (Schwartz et al., 1986; Cohen and Albers, 1991). The SCN itself generates biological rhythms with a period of approximately 24 h. The endogenous SCN rhythm is normally synchronized for its period and phase to the environmental light-dark cycle. This process is called "entraining". It is performed by a direct neuronal pathway from the retina to the SCN that exists also in human as was shown by staining degenerating neurons in patients with incurred prior optic nerve damage (Sadun et al., 1984). The retinohypothalamic tract is the principal pathway mediating the entraining effects of light on the circadian pacemaker, the SCN. Totally blind people often lack the entraining effects of light and may show free-running temperature, cortisol and melatonin rhythms. They may also suffer from sleep disturbances (Sack et al., 1992). Surprisingly, some blind people maintain circadian entrainment and show light-induced suppression of melatonin secretion, despite the apparently total lack of pupillary light reflexes and with no conscious perception of light (Czeisler et al., 1995). It has been proposed that in these patients the retinohypothalamic pathway that passes through the SCN would still be intact (Czeisler et al., 1995). It might be of practical importance to distinguish these patients since enucleation might in this group cause recurring insomnia and other symptoms associated with the loss of circadian rhythms.

Circadian and circannual rhythms in aging and Alzheimer's disease

Age-related changes have been found in many circadian rhythms in man (Touitou, 1995). An example of age-related changes is the fragmented sleep-wake pattern which occurs in senescence but which is even more pronounced in Alzheimer's disease (Mirmiran et al., 1988; Witting et al., 1990; Prinz and Vitiello, 1993). In Alzheimer's disease the disruptions of the circadian rhythms are often so severe that they are even thought to contribute to mental decline (Moe et al., 1995). Moreover, disruption of the sleep of the caregiver due to nocturnal problems of the patient is a more important reason for placement in a nursing home than cognitive impairment (Pollak and Perlick, 1991). Because of the disruption of circadian rhythms the number of vasopressin-expressing cells in the SCN was determined during aging and in Alzheimer's disease. A marked decrease was found in the number of vasopressin-expressing neurons in the SCN in subjects of 80 to 100 years of age, while in Alzheimer's disease these changes occurred even earlier and were more dramatic (Swaab et al., 1985). The number of vasoactive intestinal peptide (VIP)-expressing neurons in the

                                              D.F. Swaab

SCN of women did not show any age-related change as opposed to the neurons expressing vasopressin, whereas in men a complex pattern of changes was observed of VIP-expressing neurons with advancing age. Between 10 and 40 years the male SCN contained twice as many VIP neurons as the female one, but a subsequent decrease in the number of male VIP neurons between 40 and 65 years of age resulted in fewer VIP neurons in men than in women. After 65 years of age the sex difference remained just short of significance (Zhou et al., 1995). In a group of young subjects (6 to 47 years of age) we observed a significant fluctuation in the number of vasopressin-expressing neurons over the 24-hour period. During the daytime the SCN contained 1.8 times as many vasopressin neurons as during the nighttime, with peak values in vasopressin cell number occurring in early morning (Hofman and Swaab, 1993). The number of vasopressin-containing neurons in the SCN was also found to fluctuate over the year with values being two to three times higher in the autumn than in the summer (Hofman and Swaab, 1992, 1993). Photoperiod seems to be the major Zeitgeber for the observed annual variations in the SCN (Hofman et al., 1993). The circadian and circannual fluctuations in vasopressin-expressing neuron numbers in the SCN decrease during aging. The marked diurnal oscillation in the number of vasopressin-expressing neurons in the SCN of young subjects, i.e. low vasopressin neurons numbers during the night and peak values during the early morning, disappeared in subjects over the age of 50 (Hofman and Swaab, 1994). Whereas in young subjects low vasopressin neuron numbers were found during the summer, and peak values in autumn, the SCN of people over 50 years of age showed a disruption of the annual cycle with a reduced amplitude (Hofman and Swaab, 1995).

The SCN is affected by Alzheimer's disease since the typical cytoskeletal alterations have also been found in the SCN of Alzheimer patients (Swaab et al., 1992; Van de Nes et al., 1993). With respect to the occurrence of degenerative changes of the SCN it may be important to note that a decreased light input to the SCN is present. Both the retina and the optic nerve, which provide direct and indirect light input to the SCN, show degenerative changes in Alzheimer's disease (Hinton et al., 1986; Trick et al., 1989; Katz et al., 1989). In the macula of Alzheimer patients retinal cell degeneration has been observed without neurofibrillary tangles, neuritic plaques or amyloid angiopathy present in the retina or optic nerves (Blanks et al., 1989). In addition to degenerative changes in the visual system, Alzheimer patients are generally exposed to less environmental light than their age-matched controls (Campbell et al., 1988). Apparently both the input of the visual system to the SCN and the SCN itself may be seriously affected in Alzheimer's disease. The more demented the Alzheimer patients, the more fragmented their sleep. Increased wandering at night and more aggressive behaviour during the day are associated with the use of sedative-hypnotics and with going to bed early (Ancoli-Israel et al., 1994). Regression analysis showed that rest-activity rhythm disturbances are influenced by daytime activity and light (Van Someren et al., 1995) and that sleep/wake variables were highly correlated with and explained significant variance in cognitive and functional measures (Moe et al., 1995). Indeed, following exposure to extra amounts of bright light behavioural disorders such as wandering, agitation or delirium almost disappeared, and sleep-wake rhythm disorders improved in Alzheimer patients (Hozumi et al., 1990; Okawa et al., 1991; Satlin et al., 1992; Mishima et al., 1994) Enforcement of social interaction with nurses was also effective (Okawa et al., 1991). These observations indicate that stimulation of the circadian system may have important therapeutic consequences for Alzheimer patients.

Supraoptic and paraventricular nucleus (SON, PVN)

The SON and PVN and their axons running to the neurohypophysis form the classical neuroendocrine system. In order to establish the proportion of SON and PVN cells that project to the neurohypophysis, Morton (1961) determined neuronal numbers in these nuclei for a period of 12-45 months following hypophysectomy as a palliative measure in the treatment of hormone-dependent metastatic mammary carcinoma. After hypophysectomy there was an average loss of neurons from both the SON and PVN of over 80%. From this observation it was concluded that most neurons of the SON and PVN project to the neurohypophysis.

The SON is subdivided in three parts. The largest part, the dorsolateral SON, has a volume of 3 mm^3 (Goudsmit et al., 1990) and contains 53,000 neurons, 90% of which contain vasopressin and 10% oxytocin (Fliers et al., 1985). The PVN has a volume of 6 mm^3 (Goudsmit et al., 1990) and was estimated to consist of about 56,000 neurons (Morton, 1961) of which some 25,000 contain oxytocin and 21,000 express vasopressin (Wierda et al., 1991; Purba et al., 1993; Van der Woude et al., 1995).

The famous case of a man who in 1912 received a bullit wound in the sella-turcica which destroyed the posterior lobe and caused diabetes insipidus (Brooks, 1988) revealed the antidiuretic function of the neurohypophysis. However, the concept of "neurosecretion", based, e.g., upon the large neurons of the human supraoptic and paraventricular nucleus, was not proposed until 1939, by the Scharrers (Scharrer and Scharrer, 1940; Brooks, 1988). According to the critics of those days, this concept was based on "nothing more than signs of pathological processes, postmortem changes or fixation artifacts". In the 1940s "practically everybody vigorously" or even "viciously" rejected the concept that a neuron could have a glandular function (B. Scharrer, pers. comm.; Scharrer, 1933). This initially highly charged reception of the neurosecretion concept was followed by acceptance only when Bargmann (1949) demonstrated the same Gomori-positive material in the neurohypophysis and in the neurons of the SON and PVN.

Oxytocin, food intake and Prader-Willi syndrome

Animal experiments have shown that the parvocellular oxytocin neurons of the hypothalamic PVN are crucial for the regulation of food intake. In the rat these oxytocin neurons project to brain stem nuclei, for example the nucleus of the solitary tract and the dorsal motor nucleus of the nervus vagus. Small lesions in the rat PVN are responsible for overeating and obesity (Leibowitz et al., 1981), suggesting that the PVN usually has an inhibitory effect on eating and body weight. Central administration of oxytocin or oxytocin agonists inhibits food intake, whereas these effects are prevented by an oxytocin receptor antagonist (Olson et al., 1991a,b).

We recently investigated whether a disorder of the PVN, or, more particularly, of its putative satiety neurons - the oxytocin neurons - might be the basis of the insatiable hunger and obesity in the most common type of human genetic obesity, the Prader-(Labhart)Willi syndrome (PWS). Apart from gross obesity and problems during the process of birth (Wharton and Bresman, 1989), this syndrome is characterized by diminished fetal motor activity, severe infantile hypotonia, mental retardation, hypogonadism and hypogenitalism (Prader et al., 1956).

The thionin-stained volume of the PVN is 28% smaller in PWS patients and the total cell number of the PVN is 38% lower than in controls (Swaab, 1995). Following immunocytochemistry the immunoreactivities for oxytocin and vasopressin are decreased in PWS patients, although the variation within the groups was high. A large and highly significant decrease (42%) in the number of oxytocin-expressing neurons was found in all five PWS patients. The volume of the PVN containing the oxytocin-expressing neurons is 54% lower in PWS. The number of vasopressin-expressing neurons in the PVN did not change significantly. The finding that volume and total cell number and oxytocin cell number was so much lower in PWS patients points to a developmental hypothalamic disorder. Consequently oxytocin neurons of the PVN may be good candidates for a physiological role as "satiety neurons" in ingestive behaviour, also in the human brain (Swaab, 1995).

The supraoptic and paraventricular nucleus in diabetes insipidus

Familial hypothalamic diabetes insipidus is transmitted as an autosomal dominant gene. Affected individuals have low or undetectable levels of circulating vasopressin and suffer from polydipsia and polyuria, but they respond to substitution therapy with exogenous vasopressin or analogues. Urine production may amount to some 20 litres per day. Members of a Dutch family suffering from this disease appeared to have a point mutation in one allele of the affected family members, based upon a G to T transversion at position 17 of the neurophysin encoding exon B on chromosome 20 (Bahnsen et al., 1992). In a Japanese diabetes insipidus family a G to A transition has been described in the same exon (Ito et al., 1991). The few postmortem histological observations in other families with hereditary hypothalamic diabetes insipidus point to severe neuronal death in the SON and PVN in the case of familiar hypothalamic diabetes insipidus associated with a loss of nerve fibers in the posterior pituitary (Braverman et al., 1965; Nagai et al., 1984; Bergeron et al., 1991) suggesting that the mutated product might be toxic to the neurosecretory cell. Such toxicity would also explain the variable age at onset of the disease (Schmale et al., 1993) e.g. our observation that diabetes insipidus did not strike until an individual reached the age of approximately 9 years (Bahnsen et al., 1992). More data from age-related studies and more postmortem observations are needed to establish such an effect definitively.

In addition, diabetes insipidus has been observed following closed-head trauma, ischemia, hemorrhage, inflammation and surgical manipulations affecting either the SON and PVN or, more frequently, the hypothalamo-neurohypophysial tract (Rudelli and Deck, 1979).

Recently an autoimmune form of hypothalamic diabetes insipidus has been described with circulating autoantibodies against the vasopressin cell surface. Such autoantibodies could not be demonstrated in hereditary forms of diabetes insipidus. It has not yet been established whether the autoantibodies observed in diabetes insipidus are indeed cytotoxic and might destruct the vasopressin cell bodies (Scherbaum, 1992). Yet it seems certainly worthwhile to look for autoimmune processes that may be directed towards other hypothalamic neurons and might be an explanation for hypothalamic symptoms in neurological, psychiatric or neuroendocrine diseases.

Corticotropin-releasing hormone (CRH) neurons in the paraventricular nucleus

CRH is a crucial neuropeptide in the regulation of the hypothalamo-pituitary-adrenal axis. CRH immunoreactivity is present in the human hypothalamus only in parvicellular neurons of the PVN, and in their fibers that run to, e.g., the median eminence. The CRH-expressing neurons are not located in a distinct subnucleus of the PVN, as it is in the rat, but to be scattered throughout the PVN, with only relatively few cells in the rostral part (Raadsheer et al., 1993).

The total number of CRH-expressing neurons in the human PVN increases with age in controls and Alzheimer's disease brains to the same degree (Raadsheer et al., 1994a). The age-dependent increase in the absolute number of neurons expressing CRH within the PVN of both control and Alzheimer's disease patients is interpreted as a sign that CRH neurons become increasingly active with age. Parvicellular neurons containing both CRH and vasopressin were found in subjects ranging between 43 and 91 years of age, whereas the CRH neurons in the PVN of younger subjects (23-27 years of age) did not contain vasopressin. The colocalization of vasopressin in CRH neurons is also an index of the activity of CRH neurons. This index was much the same in controls and Alzheimer's disease patients. In both groups a similar increase with age is present in the number of CRH neurons that colocalize vasopressin (Raadsheer et al., 1994b). The third parameter for activity of CRH neurons measured in this material was the total amount of CRH-mRNA as determined by quantitative *in situ* hybridization histochemistry. In contrast to the two parameters mentioned above, CRH-mRNA is higher in Alzheimer's disease patients than in age-matched controls (Raadsheer et al., 1995). In conclusion, CRH neurons in Alzheimer's disease patients were moderately activated as compared to normal controls merely due to a difference in CRH-mRNA.

Depressed patients showed a much stronger CRH neuron activation than Alzheimer's disease patients on the basis of the total number of cells expressing CRH the total number of CRH neurons showing vasopressin colocalization and the amount of CRH-mRNA in the PVN (Raadsheer et al., 1994c, 1995).

From the immunocytochemical and *in situ* hybridization studies it appears that activated CRH neurons show different activation patterns in aging, Alzheimer's disease and depression. The process of aging is accompanied by increased CRH cell numbers and an increased fraction of CRH neurons showing vasopressin colocalization. Alzheimer's disease goes together with only an extra increased production of CRH per neuron and in depression the numbers of CRH neurons, the number of vasopressin-coexpressing neurons and the total amount of CRH-mRNA in the PVN are increased, but not the amount of CRH-mRNA per neuron (Raadsheer et al., 1994a-c; 1995).

The observation that the number of non-vasopressin-expressing CRH neurons increased more in depression than the number of vasopressin-expressing CRH neurons (Raadsheer et al., 1994c) could mean that different phenotypic subtypes of CRH neurons are present in humans and that these subtypes are activated differentially in depressed patients (Raadsheer et al., 1995). This view is supported by the finding of two subtypes of CRH neurons in the PVN of experimental animals (Whitnall et al., 1993). One type colocalized vasopressin and projects to the median eminence, whereas the other type does not coproduce vasopressin and projects to the brain stem and spinal cord (Sawchenko and Swanson, 1982). Although in the rat these non-neuroendocrine neurons represent only a minor subpopulation of the CRH neurons in the PVN (Swanson et al., 1983; Mezey et al., 1984), our data indicate that this fraction may be considerably larger in humans. Furthermore this fraction of CRH neurons that does not colocalize vasopressin seems to be strongly activated in depression.

At present there are arguments suggesting that CRH might be a causal factor in the development of depression. Firstly, there is a strong increase of CRH activity in major depression (Raadsheer et al., 1994c, 1995) - the total number of CRH neurons of the major depressed patients was 4 times higher than in controls (Raadsheer et al., 1994c) - and CRH-mRNA as determined by quantitative *in situ* hybridization was strongly activated in major depression (Raadsheer et al., 1995). Secondly, the symptoms resembling depression, e.g. decreased food intake, decreased sexual activity, disturbed sleep and motor behaviour and increased anxiety can be induced in experimental animals by intracerebroventricular injection of CRH (Holsboer et al., 1992). Thirdly, antidepressant drugs attenuate the synthesis of CRH (Fischer et al., 1990; Brady et al., 1991; 1992; Delbende et al., 1991) and the CRH concentrations in CSF in healthy volunteers (Veith et al., 1993) and lastly, a transgenic mouse model which has an overproduction of CRH appeared to have symptoms that are usually related with major depression which can be counteracted by injection of CRH antagonist (Stenzel-Poore et al., 1994). The sum of these arguments leads us to propose a CRH-hypothesis of depression, i.e., that the hyperactivity of the hypothalamic-pituitary-adrenal axis may contribute to the development of symptoms of depression rather than to the pathogenesis of Alzheimer's disease and that especially the subgroup of CRH neurons that does not project to the median eminence may be activated in depression.

The effects of the activated CRH neurons in depressed patients may be potentiated by the hyperactive vasopressin and oxytocin neurons in the PVN in this disorder. We observed in the PVN an increase in the total number of vasopressin and oxytocin expressing neurons of 56% and 23% respectively (Purba et al., 1995). Not only vasopressin, but also oxytocin may potentiate the effects of CRH release. In addition, CRH and vasopressin may act synergistically on behaviour. Our observations confirm the postulation of Bardeleben and Holsboer (1989) that the action of CRH in depression is enhanced by vasopressin. The activation of oxytocin neurons might also contribute to the inhibition of food intake in depression.

Acknowledgements
I want to thank Ms W.T.P. Verweij and Ms T. Eikelboom for their secretarial assistance.
Hypothalamic tissue was obtained from the Netherlands Brain Bank (coordinator Dr. R. Ravid).

References

Ancoli-Israel, S., Klauber, M.R., Gillin, J.C., Campbell, S.S. and Hofstetter, C.R. (1994) Sleep in non-institutionalized Alzheimer's disease patients. *Aging Clin. Exp. Res.* 6; *in press.*

Bahnsen, U., Oosting, P., Swaab, D.F., Nahke, P., Richter, D. and Schmale, H. (1992) A missense mutation in the vasopressin-neurophysin precursor gene cosegregates with human autosomal dominant neurohypophyseal diabetes insipidus. *EMBO J.* 11: 19-23.

Bardeleben, U. and Holsboer, F. (1989) Corticol response to a combined dexamethasone-human corticotropin-releasing hormone challenge in patients with depression. *J. Neuroendocrinol.* 1: 485-488.

Bargmann, W. (1949) Über die neurosekretorische Verknüpfung von Hypothalamus und Neurohypophyse. *Z. Zellforsch.* 34: 610-634.

Bergeron, C., Kovacs, K., Ezrin, C. and Mizzen, C. (1991) Hereditary diabetes insipidus: an immunohistochemical study of the hypothalamus and pituitary gland. *Acta Neuropathol.* 81: 345-348.

Blanks, J.C., Hinton, D.R., Sadun, A.A. and Miller, C.A. (1989) Retinal ganglion cell degeneration in Alzheimer's disease. *Brain Res.* 501: 364-372.

Brady, L.S., Whitfield, H.J.Jr., Fox, R.J., Gold, P.W. and Herkenham, M. (1991) Long-term antidepressant administration alters corticotropin-releasing hormone, tyrosine hydroxylase, and mineralcorticoid receptor gene expression in rat brain. *J. Clin. Invest.* 87: 831-837.

Brady, L.S., Gold, P.W., Herkenham, M., Lynn, A.B. and Whitfield, H.J. Jr. (1992) The antidepressants fluoxetine, idazoxan and phenylzine alter corticotropin-releasing hormone and tyrosine hydroxylase mRNA levels in rat brain: therapeutic implication. *Brain Res.* 572: 117-125.

Braverman, L.E., Mancini, J.P. and McGoldrick, D.M. (1965) Hereditary idiopathic diabetes insipidus. A case report with autopsy findings. *Ann. Intern. Med.* 63: 503-508.

Brooks, C.M.C. (1988) The history of thought concerning the hypothalamus and its functions. *Brain Res. Bull.* 20: 657-667.

Campbell, S.S., Kripke, D.F., Gillin, J.C. and Hrubovcak, J.C. (1988) Exposure to light in healthy elder subjects and Alzheimer patients. *Physiol. Behav.* 42: 141-144.

Cohen, R.A. and Albers, H.E. (1991) Disruption of human circadian and cognitive regulation following a discrete hypothalamic lesion: a case study. *Neurology* 41: 726-729.

Czeisler, C.A., Shanahan, T.L., Klerman, E.B., Martens, H., Brotman, D.J., Emens, J.S., Klein, T. and Rizzo, J.F. (1995) Suppression of melatonin secretion in some blind patients by exposure to bright light. *N. Engl. J. Med.* 332: 6-11.

Delbende, C., Contesse, V., Mocaër, E., Kamoun, A. and Vaudry, H. (1991) The novel antidepressant tianeptine reduces stress-evoked stimulation of the hypothalamo-pituitary-adrenal axis. *Eur. J. Pharmacol.* 202: 391-392.

Fischer, P., Simanyi, M. and Danielczyk, W. (1990) Depression in dementia of the Alzheimer type and in multi-infarct dementia. *Am. J. Psychiatry* 147: 1484-1487.

Fliers, E., Swaab, D.F., Pool, C.W. and Verwer, R.W.H. (1985) The vasopressin and oxytocin neurons in the human supraoptic and paraventricular nucleus: Changes with aging and in senile dementia. *Brain Res.* 342: 45-53.

Goudsmit, E., Hofman, M.A., Fliers, E. and Swaab, D.F. (1990) The supraoptic and paraventricular nuclei of the human hypothalamus in relation to sex, age and Alzheimer's disease. *Neurobiol. Aging* 11: 529-536.

Hinton, D.R., Sadun, A.A., Blanks, J.C. and Miller, C.A. (1986) Optic nerve degeneration in Alzheimer's disease. *N. Engl. J. Med.* 315: 485-487.

Hofman, M.A. and Swaab, D.F. (1992) Seasonal changes in the suprachiasmatic nucleus of man. *Neurosci. Lett.* 139: 257-260.

Hofman, M.A. and Swaab, D.F. (1993) Diurnal and seasonal rhythms of neuronal activity in the suprachiasmatic nucleus of humans. *J. Biol. Rhythms* 8(4): 283-295.

Hofman, M.A. and Swaab, D.F. (1994) Alterations in circadian rhythmicity of the vasopressin-producing neurons of the human suprachiasmatic nucleus (SCN) with aging. *Brain Res.* 651: 134-142.

Hofman, M.A. and Swaab, D.F. (1995) Influence of aging on the seasonal rhythm of the human suprachiasmatic nucleus (SCN) *Neurobiol. Aging; in press.*

Hofman, M.A., Purba, J.S. and Swaab, D.F. (1993) Annual variations in the vasopressin neuron population of the human suprachiasmatic nucleus. *Neuroscience* 53: 1103-1112.

Holsboer, F., Spengler, D. and Heuser, I. (1992) The role of corticotropin-releasing hormone in the pathogenesis of Cushing's disease, anorexia nervosa, alcoholism, affective disorders and dementia. *In:* D.F. Swaab, M.A. Hofman, M. Mirmiran, R. Ravid and F.W. Van Leeuwen (eds.): *The Human Hypothalamus in Health and Disease. Progress in Brain Research, Vol. 93*, Elsevier, Amsterdam, pp. 385-417.

Hozumi, S., Okawa, M., Mishima, K., Hishikawa, Y., Hori, H. and Takahashi. K. (1990) Phototherapy for elderly patients with dementia and sleep-wake rhythm disorders - a comparison between morning and evening light exposure. *Jpn. J. Psychiatry Neurol.* 44: 813-814.

Ito, M., Mori, Y., Oiso, Y. and Saito, H. (1991) A single base substitution in the coding region for neurophysin II associated with familial central diabetes insipidus. *J. Clin. Invest.* 87: 725-728.

Katz, B., Rimmer, S., Iragui, V. and Katzman, R. (1989) Abnormal pattern electroretinogram in Alzheimer's disease: evidence for retinal ganglion cell degeneration? *Ann. Neurol.* 26: 221-225.

Leibowitz, S.F., Hammer, N.J. and Chang, K. (1981) Hypothalamic paraventricular nucleus lesions produce overeating and obesity in the rat. *Physiol. Behav.* 27: 1031-1040.

Mezey, E., Kiss, J.Z., Skirboll, L.R., Goldstein, M. and Axelrod, J. (1984) Increase of corticotropin releasing factor staining in the rat paraventricular nucleus staining by depletion of hypothalamic adrenaline. *Nature* 310: 140-141.

Mirmiran, M., Overdijk, J., Witting, W., Klop, A. and Swaab, D.F. (1988) A simple method for recording and analyzing circadian rhythms in man. *J. Neurosci. Meth.* 25: 209-214.

Mishima, K., Okawa, M., Hishikawa, Y., Hozumi, S., Hori, H. and Takahashi, K. (1994) Morning bright light therapy for sleep and behaviour disorders in elderly patients with dementia. *Acta Psychiatr. Scand.* 89: 1-7.

Moe, K.E., Vitiello, M.V., Larsen, L.H. and Prinz, P.N. (1995) Sleep/wake patterns in Alzheimer's disease: relationships with cognition and function. *J. Sleep Res.* 4: 15-20.

Morton, A. (1961) A quantitative analysis of the normal neuron population of the hypothalamic magnocellular nuclei in man and of their projections to the neurohypophysis. *J. Comp. Neurol.* 136: 143-158.

Nagai, L., Li, C.H., Hsieh, S.M., Kizaki, T. and Urano, Y. (1984) Two cases of hereditary diabetes insipidus, with an autopsy finding in one. *Acta Endocrinol.* 105: 318-323.

Okawa, M., Mishima, K., Hishikawa, Y., Hozumi, S., Hori, H. and Takahashi, K. (1991) Circadian rhythm disorders in sleep-waking and body temperature in elderly patients with dementia and their treatment. *Sleep* 14: 478-485.

Olson, B.R., Drutarosky, M.D., Stricker, E.M. and Verbalis, J.G. (1991a) Brain oxytocin receptor antagonism blunts the effects of anorexigenic treatments in rats: evidence for central oxytocin inhibition of food intake. *Endocrinology* 129: 785-791.

Olson, B.R., Drutarosky, M.D., Chow, M.S., Hruby, V.J., Stricker, E.M. and Verbalis, J.G. (1991b) Oxytocin and an oxytocin agonist administered centrally decrease food intake in rats. *Peptides* 12: 113-118.

Pollak, C.P. and Perlick, D. (1991) Sleep problems and institutionalization of the elderly. *J. Geriatr. Psychiatric Neurol.* 4: 204-210.

Prader, A., Labhart, A. and Willi, H. (1956) Ein Syndrom von Adipositas, Kleinwuchs, Krytorchismus und Oligophrenie nach Myotonieartigem Zustand in Neugeborenalter. *Schweiz. Med. Wochenschr.* 86: 1260-1261.

Prinz, P.N. and Vitiello, M.V. (1993) Sleep in Alzheimer's disease. *In:* J.L. Alberrede, J.E. Marley, T. Roth and B.J. Vellas (eds.): *Sleep Disorders and Insomnia in the Elderly. Facts and Research in Gerontology,* Vol.7.

Purba, J.S., Hofman, M.A., Portegies, P., Troost, D. and Swaab, D.F. (1993) Decreased number of oxytocin neurons in the paraventricular nucleus of the human hypothalamus in AIDS. *Brain* 116: 795-809.

Purba, J.S., Hoogendijk, W.J.G., Hofman, M.A. and Swaab, D.F. (1995) Increased number of vasopressin and oxytocin expressing neurons in the paraventricular nucleus of the human hypothalamus in depression. *Arch. Psych.; in press.*

Raadsheer, F.C., Sluiter, A.A., Ravid, R., Tilders, F.J.H. and Swaab, D.F. (1993) Localization of corticotropin-releasing hormone (CRH) neurons in the paraventricular nucleus of the human hypothalamus; age-dependent colocalization with vasopressin. *Brain Res.* 615: 50-62.

Raadsheer, F.C., Oorschot, D.E., Verwer, R.W.H., Tilders, F.J.H. and Swaab, D.F. (1994a) Age-related increase in the total number of corticotropin-releasing hormone neurons in the human paraventricular nucleus in controls and Alzheimer's disease: comparison of the disector with an unfolding method. *J. Comp. Neurol.* 399: 447-457.

Raadsheer, F.C., Tilders, F.J.H. and Swaab, D.F. (1994b) Similar age related increase of vasopressin colocalization in paraventricular corticoptropin-releasing hormone neurons in controls and Alzheimer patients. *J. Neuroendocrinology* 6: 131-133.

Raadsheer, F.C., Hoogendijk, W.J.G., Stam, F.C., Tilders, F.J.H. and Swaab, D.F. (1994c) Increased numbers of corticotropin-releasing hormone expressing neurons in the hypothalamic paraventricular nucleus of depressed patients. *Neuroendocrinology* 60: 436-444.

Raadsheer, F.C., Van Heerikhuize, J.J., Lucassen, P.J., Hoogendijk, W.J.G., Tilders, F.J.H. and Swaab, D.F. (1995) Increased corticotropin-releasing hormone (CRF)-mRNA in the paraventricular nucleus of patients with Alzheimer's disease or depression. *Am. J. Psych.; in press.*

Rudelli, R. and Deck, J.H.N. (1979) Selective traumatic infarction of the human anterior hypothalamus. Clinical anatomical correlation. *J. Neurosurg.* 50: 645-654.

Sack, R.L., Lewy, A.J., Blood, M.L., Keith, L.D. and Nakagawa, H. (1992) Circadian rhythm abnormalities in totally blind people: incidence and clinical significance. *J. Clin. Endocrin. Metab.* 75: 127-134.

Sadun, A.A., Schaechter, J.D. and Smith, L.E.H. (1984) A retinohypothalamic pathway in man: light mediation of circadian rhythms. *Brain Res.* 302: 371-377.

Satlin, A., Volicer, L., Ross, V., Herz, L. and Campbell, S. (1992) Bright light treatment of behavioural and sleep disturbances in patients with Alzheimer's disease. *Am. J. Psychiatr.* 149: 1028-1032.

Sawchenko, P.E. and Swanson, L.W. (1982) Immunohistochemical identification of neurons in the paraventricular nucleus of the hypothalamus that project to the medulla or to the spinal cord in the rat. *J. Comp. Neurol.* 205: 260-272.

Scharrer, E. and Scharrer, B. (1940) Secretory cells within the hypothalamus. The hypothalamus and central levels of autonomic function. *Research Publications Assoc. Nervous Mental Disease* 20: 170-194.

Scharrer, E. (1933) Die Erklärung der scheinbar pathologischen Zellbilder im Nucleus supraopticus und Nucleus paraventricularis. *Zs. f. Ges. Neurologie Psychiat.* 145: 462-470.

Scherbaum, W.A. (1992) Autoimmune hypothalamic diabetes insipidus ("autoimmune hypothalamitis") *In:* D.F. Swaab, M.A. Hofman, M. Mirmiran, R. Ravid and F.W. Van Leeuwen (eds.): *Progress in Brain Research, Vol. 93, The Human Hypothalamus in Health and Disease*, Elsevier, Amsterdam, pp. 283-293.

Schmale, H., Bahnsen, U. and Richter, D. (1993) Structure and expression of the vasopressin precursor gene in central diabetes insipidus. *Ann. N.Y. Acad. Sci.* 689: 74-82.

Schwartz, W.J., Bosis, N.A. and Hedley-Whyte, E.T. (1986) A discrete lesion of ventral hypothalamus and optic chiasm that disturbed the daily temperature rhythm. *J. Neurol.* 233: 1-4.

Stenzel-Poore, M.P., Heinrichs, S.C., Rivest, S., Koob, G.F. and Vale, W.W. (1994) Overproduction of corticotropin-releasing factor in transgenic mouse: a genetic model of anxiogenic behaviour. *J. Neurosci.* 14: 2579-2584.

Swaab, D.F. (1995) Alterations in the hypothalamic paraventricular nucleus and its oxytocin neurons (putative satiety cells) in Prader-Willi syndrome: A study of five cases. *J. Clin. Endocr. Metab.* 80: 573-579.

Swaab, D.F. (1996) Neurobiology and Neuropathology of the Human Hypothalamus. *In:* F.E. Bloom (ed.): *Handbook of Chemical Anatomy, Primate Volumes; in press.*

Swaab, D.F. and Hofman, M.A. (1995) Sexual differentiation of the human hypothalamus in relation to gender and sexual orientation. *Trends Neurosci.* 18: 264-270.

Swaab, D.F., Fliers, E. and Partiman, T.S. (1985) The suprachiasmatic nucleus of the human brain in relation to sex, age and senile dementia. *Brain Res.* 342: 37-44.

Swaab, D.F., Grundke-Iqbal, I., Iqbal, K., Kremer, H.P.H., Ravid, R. and Van de Nes, J.A.P. (1992) Tau and ubiquitin in the human hypothalamus in aging and Alzheimer's disease. *Brain Res.* 590: 239-249.

Swanson, L.W., Sawchenko, P.E., Rivier, J. and Vale, W.W. (1983) Organisation of ovine corticotropin releasing factor immunoreactive cells and fibers in the rat brain: an immunohistochemical study. *Neuroendocrinology* 37: 165-186.

Touitou, Y. (1995) Effects of aging on endocrine and neuroendocrine rhythms in humans. *Horm. Res.* 43: 12-19.

Trick, G.L., Barris, M.C. and Bickler-Bluth, M. (1989) Abnormal pattern electroretinograms in patients with senile dementia of the Alzheimer type. *Ann. Neurol.* 26: 226-231.

Van de Nes, J.A.P., Kamphorst, W., Ravid, R. and Swaab, D.F. (1993) The distribution of Alz-50 immunoreactivity in the hypothalamus and adjoining areas of Alzheimer's disease patients. *Brain* 116: 103-115.

Van der Woude, P.F., Goudsmit, E., Wierda, M., Purba, J.S., Hofman, M.A., Bogte, H. and Swaab, D.F. (1995) No vasopressin cell loss in the human paraventricular and supraoptic nucleus during aging and in Alzheimer's disease. *Neurobiol. Aging* 16: 11-18.

Van Someren, E.J.W., Hagebeuk, E.E.O., Lijzenga, C., Scheltens, P., De Rooij, S.E.J.A., Jonker, C., Pot, A.-M., Mirmiran, M. and Swaab, D.F. (1995) Circadian rest-activity rhythm disturbances in Alzheimer's disease. *Biol. Psych.; in press.*

Veith, R.C., Lewis, N., Langohr, J.I., Murburg, M.M., Ashleigh, E.A., Castillo, S., Peskind, E.R., Pascualy, M., Bissette, G., Nemeroff, C.B. and Raskind, M.A. (1993) Effect of desipramine on cerebrospinal fluid concentrations of corticotropin-releasing factor in human subjects. *Psychiatry Res.* 46: 1-8.

Wharton, R.H. and Bresman, M.J. (1989) Neonatal respiratory depression and delay in diagnosis in Prader-Willi syndrome. *Dev. Med. Child Neurol.* 31: 231-236.

Whitnall, M.H., Kiss, A. and Aguilera, G. (1993) Contrasted effects of central alpha-1 adrenoceptor activation on stress-responsive and stress non-responsive subpopulations of corticotropin-releasing hormone neurosecretory cells in the rat. *Neuroendocrinology* 58: 42-48.

Wierda, M., Goudsmit, E., Van der Woude, P.F., Purba, J.S., Hofman, M.A., Bogte, H. and Swaab, D.F. (1991) Oxytocin cell number in the human paraventricular nucleus remains constant with aging and in Alzheimer's disease. *Neurobiol. Aging* 12: 511-516.

Witting, W., Kwa, I.H., Eikelenboom, P., Mirmiran, M. and Swaab, D.F. (1990) Alterations in the circadian rest-activity rhythm in aging and Alzheimer's disease. *Biol. Psychiatry* 27: 563-572.

Zhou, J.N., Hofman, M.A. and Swaab, D.F. (1995) VIP neurons in the human SCN in relation to sex, age, and Alzheimer's disease. *Neurobiol. Aging* 16: 571-576.

Impact of somatostatin receptor scintigraphy in differential diagnosis of meningeoma

K. H. Bohuslavizki, W. Braunsdorf[1], W. Brenner, A. Behnke[1], S. Tinnemeyer, H.-H. Hugo[1], N. Jahn, H. Wolf, C. Sippel, M. Clausen, H. M. Mehdorn[1] and E. Henze

Christian-Albrechts-University of Kiel, Clinics of Nuclear Medicine, Arnold-Heller-Straße 9, D-24105 Kiel
[1] *Neurosurgery, Weimarer Straße 8, D-24106 Kiel, Germany*

Summary. The aim of this study was to evaluate somatostatin receptor scintigraphy (SRS) in patients with suspected meningeoma. Prior to surgery 59 patients were investigated up to 24 h following an injection of 200 MBq In-[111]-octreotide. Tracer uptake was tested for the presence of meningeoma. Histology revealed 43 meningeoma and 16 other tumours. True positive SRS was obtained in 36 patients. Tumour volume was > 10 ml in 23 patients. False negative SRS was seen in 7 patients. All of them had a tumour volume of < 10 ml. While magnetic resonance imaging (MRI) alone was decisive in 39 of 48 patients, MRI could only limit differential diagnosis to meningeoma versus neurinoma in the remaining 9 patients. Out of these, positive SRS confirmed meningeoma in 5 patients, and negative SRS excluded meningeoma in 4 patients. In conclusion, there is significant clinical benefit of functional imaging with In-111-octreotide in patients with suspected meningeoma: Large meningeoma can be excluded by scintigraphy alone, while meningeoma of any size may be confirmed in combination with specific MRI results only.

Introduction

Somatostatin receptors have been described in meningeoma both *in vivo* by scintigraphy (Maini et al., 1993, 1995; Scheidhauer et al., 1993; Hildebrandt et al., 1994) and *in vitro* in cell culture studies (Reubi et al., 1986, 1991, 1992, 1993; Koper et al., 1992). This does not hold true for neurinoma (Reubi et al., 1986, 1987). Therefore, somatostatin receptor scintigraphy (SRS) was suggested by various authors for differential diagnosis of neurinoma versus meningeoma (Scheidhauer et al., 1993; Maini et al., 1995). However, we observed negative scintigrams in some patients with histologically proven meningeoma. In consequence, the exclusion of meningeoma by a lack of tracer uptake seems questionable.

Therefore, the aim of this study was to reassess the clinical impact of SRS in patients with suspected meningeoma.

Methods

59 patients suspected of meningeoma were included in this study (18 male and 41 female). Their median age was 59 years, ranging from 26 to 83 years. All patients underwent surgery and subsequent histological evaluation. Tumour volumes were calculated from magnetic resonance imaging (MRI) images under assumption of a rotational ellipsoid, and ranged from 0.3 to 112.8 ml.

Surgical specimens were fixed in 4 % formaldehyde and embedded in paraffin for histo-pathological examination. Sections of 4 µm thickness were stained both with hematoxylin-eosin and elastica van Gieson.

MRI was performed either on a 1.0 T or a 1.5 T machine acquiring both T1- and T2-weighted spin-echo sequences. Gadolinium was administered for contrast enhancement. After an intravenous injection of 200 MBq In-[111]-octreotide digital whole-body acquisitions in anterior and posterior projection were obtained at 10 min, 1, 4, and 24 h. Single photon emission computed tomography (SPECT) was performed at 4 and 24 h.

Quantitation of regional uptake was performed by comparing the activity in the region of interest (target region) with the background activity (ROI-technique). Relative percent tumour uptake was measured by relating the activity within this area to whole body activity after correction for background activity. All quantitative data were calculated as geometric mean of anterior and posterior projection. Results are given as mean ± standard deviation. Two-tailed Student's t-test for unpaired data was used to evaluate statistical differences, with $p < 0.05$ being considered statistically significant (Sachs, 1984).

With respect to histological diagnosis patients were divided into true positive (SRS positive when histology revealed menigeoma), true negative (SRS negative when histology revealed absence of menigeoma), false positive (SRS positive when histology revealed absence of menigeoma), and false negative (SRS negative when histology revealed meningeoma).

Results

There was no correlation between detailed histological analysis, scintigraphic result and anatomical location. Therefore, pooled data are given only. Mean uptake versus time of all patients is shown in Fig. 1. Statistical evaluation of our 59 patients is given in Table I.

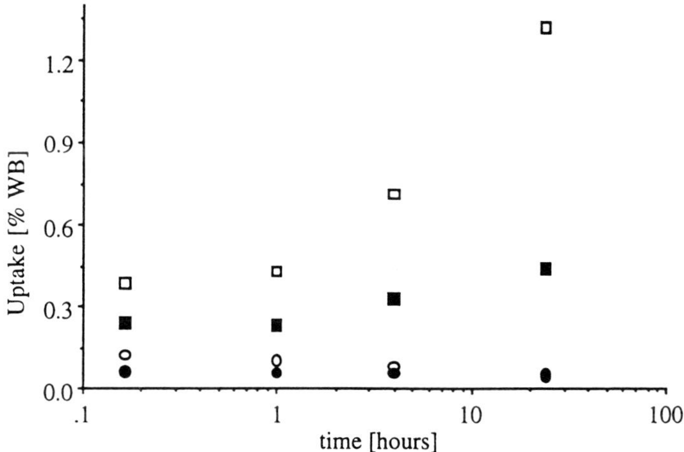

Fig. 1. Geometric mean of uptake in percent of whole body activity derived from anterior and posterior projection of somatostatin receptor scintigraphy in 59 patients with suspected meningeoma versus logarithm of time post injection in hours. Filled squares: 36 true positive; open squares: 4 false positive; open circles: 7 false negative; filled circles: 12 true negative. For standard deviation see Table I.

No additional benefit was yielded by SRS in 4 patients with false positive results. In 36 patients true positive somatostatin receptor scintigrams could be shown. An increasing uptake of In-[111]-octreotide was observed (Fig. 1, filled squares, Table I). In 5 out of these 36 patients significant information could be added by SRS as compared to MRI. One of these patients is shown in Fig. 2 in which MRI does not yield sufficient information whether meningeoma or neurinoma is the underlying process. However, SRS clearly demonstrates marked tracer uptake in planar projections as well as in transverse SPECT slices. Meningotheliomatous meningeoma was proven histologically.

Table I. Uptake of somatostatin receptor scintigraphy in percent of whole body activity (Uptake [% WB]) with respect to histological evaluation at different times post injection of In-111-octreotide.

Time p.i.	Uptake [% WB]			
	True positive	False positive	False negative	True negative
	n=36	n=4	n=7	n=12
10 min	0.24 ± 0.24	0.38 ± 0.44	0.12 ± 0.14	0.06 ± 0.03
1 h	0.23 ± 0.17	0.43 ± 0.47	0.10 ± 0.08	0.06 ± 0.06
4 h	0.33 ± 0.26	0.71 ± 0.90	0.08 ± 0.04	0.06 ± 0.04
24 h	0.44 ± 0.49	1.32 ± 1.89	0.05 ± 0.04	0.05 ± 0.07

Data represent mean ± standard deviation.

Negative SRS could be shown correctly in 12 patients (Fig. 1, filled circles). In 4 of them SRS could add significant information as compared to results of MRI. In these patients meningeoma could be excluded by SRS in 4 patients with histologically proven neurinoma in 3 cases and ependymoma in 1 case.

In 7 patients with histologically proven meningeoma SRS yielded false negative results. Consequently, uptake values did not change with time (Fig. 1, open circles). While there was neither correlation with location nor with histological type of the meningeoma, tumour volume was less than 10 ml in all patients.

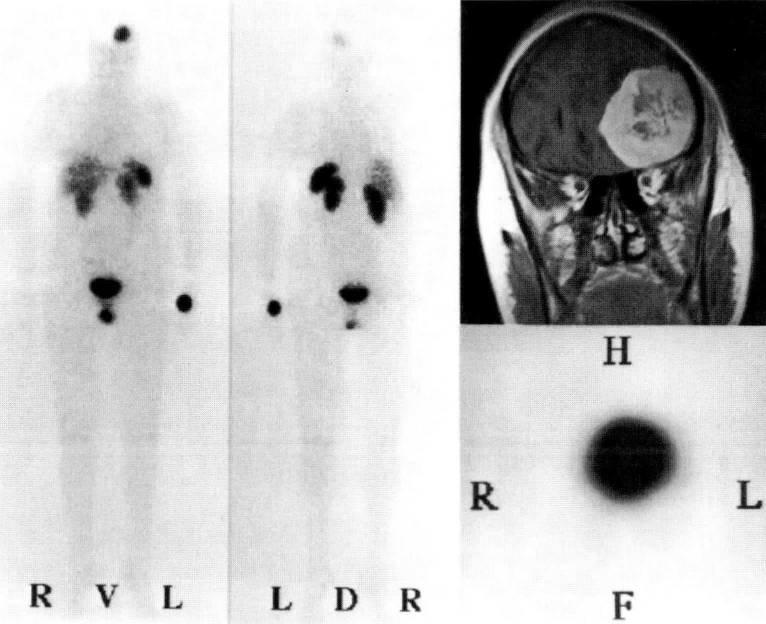

Fig. 2. Example of a true positive patient (f/51). Planar somatostatin receptor scintigraphy 4 h p.i. in anterior (R V L) and posterior (L D R) projection; coronal Gadolinium enhanced T1-weighted MRI slice (right upper) and corresponding coronal SPECT slice (right lower). Note, cystic necrotic areas inside the meningeoma, complete compression of frontal horn of the left lateral ventricle, and displacement of midline structures to the right side.

Discussion

While meningeoma and neurinoma are tumours which have similar predelection sites, surgical treatment may require different strategies due to their differing biological behaviour. Therefore, pre-operative discrimination is desired by the neurosurgeon. Usually, meningeoma and neurinoma can be discriminated sucessfully by MRI (Bydder et al., 1985; Koper et al., 1992; McConachie et al., 1994), but not in all patients.

Somatostatin receptor expression in meningeoma could be documented in nearly all cell cultures (Reubi et al., 1986, 1991, 1992, 1993; Koper et al., 1992) and *in vivo* by scintigraphic imaging using In-[111]-octreotide (Maini et al., 1993; Scheidhauer et al., 1993; Hildebrandt et al., 1994; Maini et al., 1995). On the other hand, it is known from autoradiographic studies (Reubi et al., 1986, 1987) that neurinoma do not express somatostatin receptors on their surface. Consequently, our three neurinoma exhibited true negative somatostatin receptor scintigrams in consistence with the literature (Maini et al., 1995). Functional imaging using In-[111]-octreotide was suggested to discriminate meningeoma and neurinoma (Scheidhauer et al., 1993; Maini et al., 1995).

Fig. 3. Tumour volume of patients with false negative and true positive somatostatin receptor scintigraphy. Note, all patients with false negative somatostatin receptor scintigraphy have tumour volumes of < 10 ml.

In conflict with the literature (Reubi et al., 1986, 1991, 1992, 1993; Koper et al., 1992) we found false negative SRS in 7 patients with histologically proven meningeoma. Neither histological type nor localization of the meningeoma correlated with their negative tracer uptake. Therefore, exclusion of meningeoma by a negative somatostatin receptor scintigram is not longer permitted. However, tumour volume was associated with tracer uptake. While all meningeoma of more than 10 ml in volume could be imaged positively, SRS was positive in meningeoma below 10 ml in 65 % only (Fig. 3). Thus, the known sensitivity of near 100 % (Maini et al., 1993, 1995; Scheidhauer et al., 1993; Hildebrandt et al., 1994) needs to be qualified with respect to tumour size. Our data are consistent with the literature for large meningeoma.

False positive results of SRS were observed in four patients with various diseases. Their final diagnosis could be established by MRI alone: two pituitary adenoma, an inflammatory process of the petrous bone and a glomus jugulare tumour. Tracer uptake in these tumours is known to be variable (von Werder and Faglia, 1992; Krenning et al., 1993; Scheidhauer et al., 1993). Accordingly, for these patients there is no need for SRS.

Clinical benefit of SRS in preoperative work-up of patients with suspected meningeoma has to be defined carefully. MRI is mandatory during clinical work-up. In most tumours final diagnosis can be established by MRI alone. This holds for 39 out of 48 patients in our study. However, in some cases MRI alone is not decisive and yields two possible differential diagnoses. When these two tumours under consideration have different expression of somatostatin receptors they can be discriminated by functional imaging using SRS. In our study this holds for 9 out of 48 patients. In five out of these SRS could correctly diagnose meningeoma. In 4 patients with large tumours a lack of somatostatin receptors enabled to exclude meningeoma. Histologically neurinoma and ependymoma were found, which both fail to express somatostatin receptors on their surface. In our study all patients who benefitted from SRS had a tumour volume of > 10 ml. This equals a diameter of 2.7 cm assuming a rotational ellipsoid.

It is conceivable that in a larger series patients with lower tumour volumes may benefit as well. With the preselection by MRI as mentioned above positive SRS will confirm meningeoma independent of tumour size. With small tumours negative SRS will carry no clinically useful information.

The main clinical result of this study is based on the lack of somatostatin receptors in neurinoma. It remains to be seen whether this holds true for chordoma and ganglion Gasseri tumours as well. These tumours are located at the skull base and are difficult to diagnose by MRI alone. Furthermore, scar tissue should be easily distinguished from meningeoma by SRS. This was observed in a single patient in which MRI could not discriminate between recurrent meningeoma and scar tissue. Tracer uptake in SRS correctly identified tumour recurrency and, thus, assisted in clinical decision-making.

In conclusion, functional imaging by somatostation receptor scintigraphy has significant impact in differential diagnosis of patients with suspected meningeoma. Large meningeoma can be excluded by scintigraphy alone, while meningeoma of any size may be confirmed in combination with specific MRI results only.

References

Bydder, G.M., Kingsley, P.E., Brown, J., Wiendorf, H.P. and Young, I.R. (1985) MR imaging of meningeomas including studies with and without gadolinium-DTPA. *J. Comput. Assist. Tomogr.* 9: 690–697.

Hildebrandt, G., Scheidhauer, K., Luyken, C., Schicha, H., Klug, N., Dahms, P. and Krisch, B. (1994) High sensitivity of the in vivo detection of somatostatin receptors by [111]Indium-[DTPA-octreotide]-scintigraphy in meningeoma patients. *Acta Neurochir. Wien* 126: 63–71.

Huk, W.J., Gademann, G. and Friedmann, G. (1990) *Magnetic resonance imaging of central nervous system disease.* Springer, Berlin Heidelberg New York.

Koper, J.W., Markstein, R., Kohler, C., Kwekkeboom, D.J., Avezaat, C.J.J., Lamberts, S.W.J. and Reubi, J.C. (1992) Somatostatin inhibits the activity of adenylate cyclase in cultured human meningioma cells and stimulates their growth. *J. Clin. Endocr. Metab.* 74: 543–547.

Krenning, E.P., Kwekkeboom, D.J., Bakker, W.H., Breeman, W.A.P., Kooij, P.P.M., Oie, H.Y., van Hagen, M., Postema, P.T.E., de Jong, M., Reubi, J.C., Reijs, A.E.M., Hofland, L.J., Koper, J.W. and Lamberts, S.W.J. (1993) Somatostatin receptor scintigraphy with [^{111}In-DTPA-D-Phe^1]- and [[123]Tyr^3]-octreotide: the Rotterdam experience with more than 1000 patients. *Eur. J. Nucl. Med.* 20: 716–731.

Maini, C.L., Tofani, A., Sciuto, R., Carapella, C., Cioffi, R. and Crecco, M. (1993) Scintigraphic visualization of somatostatin receptors in human meningiomas using 111-indium-DTPA-D-Phe-1-octreotide. *Nucl. Med. Commun.* 14: 505–508.

Maini, C.L., Cioffi, R.P., Tofani, A., Sciuto, R., Fontana, M., Carapella, C.M. and Crecco, M. (1995) In-111-octreotide scintigraphy in neurofibromatosis. *Eur. J. Nucl. Med.* 22: 201–206.

McConachie, N.S., Worthington, B.S., Cornford, E.J., Balsitis, M., Kerslake, R.W. and Jaspan, T. (1994) Review article: Computed tomography and magnetic resonance in the diagnosis of intraventricular cerebral masses. *Br. J. Rad.* 67: 223–243.

Reubi, J.C., Krenning, E.P., Lamberts, S.W.J. and Kvols, L. (1993) *In vitro* detection of somatostatin receptors in human tumours. *Digestion* 54 (Suppl. 1): 76–83.

Reubi, J.C., Kvols, L., Krenning, E.P. and Lamberts, S.W.J. (1991) *In vitro* and *in vivo* detection of somatostatin receptors in human malignant tissue. *Acta Oncol.* 30: 463–468.

Reubi, J.C., Laissue, J., Krenning, E.P. and Lamberts, S.W.J. (1992) Somatostatin receptors in human cancer: incidence, characteristics, functional correlates and clinical implications. *J. Steroid. Biochem. Molec. Biol.* 43: 27–35.

Reubi, J.C., Lang, W., Maurer, R., Koper, J.W. and Lamberts, S.W.J. (1987) Distribution and biochemical characterization of somatostatin receptors in tumours of the human central nervous system. *Cancer Res.* 47: 5758–5764.

Reubi, J.C., Maurer, R., Klijn, J.G.M., Stefanko, S.Z., Foekens, J.A., Blaauw, G., Blankenstein, M.A. and Lamberts, S.W.J. (1986) High incidence of somatostatin receptors in human meningiomas: biochemical characterization. *J. Clin. Endocrinol. Metab.* 63: 433–438.

Sachs, L. (1984) *Applied Statistics. A Handbook of Techniques.* Springer, Berlin Heidelberg New York.

Scheidhauer, K., Hildebrand, G., Luyken, C., Schomäcker, K., Klug, N. and Schicha, H. (1993) Somatostatin receptor scintigraphy in brain tumours and pituitary tumours: first experiences. *Horm. Metab. Res.* 27: 59–62.

von Werder, K. and Faglia, G. (1992) Potential indications for octreotide in endocrinology. *Metabolism* 9 (Suppl. 2): 91–98.

The Peptidergic Neuron
B. Krisch and R. Mentlein (eds)
© 1996 Birkhäuser Verlag Basel/Switzerland

Uptake of I-125 radiolabelled dynorphin in glioma cell cultures

H. Wolf, S. Tinnemeyer, A. Brandt, W. Brenner, C. Stauch, K. H. Bohuslavizki,
M. Schramm, M. Clausen and E. Henze

*Christian-Albrecht-University of Kiel, Clinic of Nuclear Medicine, Arnold-Heller-Str. 9, D-24105 Kiel,
Germany*

Summary. Dynorphin A (1-10) was radioiodinated directly via electrophilic substitution of the reactive tyrosine. The uptake of this labelled dynorphin in glioma cell cultures was measured. Tracer uptake was compared with the non-specific tumour tracer Tl-201. Tl-201 showed a time independent uptake over the whole incubation time of 4 hours. In contrast, increasing accumulation of radioactivity was found with I-125 labelled dynorphin. The increasing uptake over time may be evidence for a specific binding process with a possible potential for receptor imaging in man.

Introduction

Radiolabelled, biologically active peptides have a potential as imaging agents in nuclear medicine. Vasoactive intestinal peptide (Virgolini et al., 1994), somatostatin (Lamberts et al., 1990; Breeman et al. 1993) and atrial natriuretic peptide (Wolf et al., 1993; Lambert et al., 1994) have already been used for *in vivo* studies. Tumour or receptor imaging with radiolabelled biologically active peptides has a great perspective on future application in medicine (Fischman et al., 1993; Thakur, 1995). Dynorphins comprise a highly potent family of endogenous opioid peptides. Pro-dynorphin and its active derivatives are found in gastrointestinal tract, pituitary posterior lobe and brain. The potential for opioid receptor imaging might be of clinical interest. The purpose of this pilot study is both, to radioiodinate a dynorphin fragment and to compare the cellular uptake of this labelled peptide with the uptake of Tl–201 used as a reference tracer in glioma cell cultures.

Material and methods

Radioiodination. Dynorphin A (1-10) (Dyn, YGGFLRRIRP) (Sigma Chemie, Deisenhofen, Germany) was radioiodinated by the Iodogen method (Fraker et al., 1978): 10 µg Iodogen (1,3,4,6-Tetrachloro-3α-6α-diphenylglycoluril, Pierce, Oud-Beijerland, The Netherlands) was plated onto the bottom of a test tube and 10 µg of Dyn, mixed with 100 µl phosphate buffer (pH 6.8) and I-125-iodide-solution (Amersham Buchler, Braunschweig, Germany) were added. After a reaction time of 60 min, the radioiodinated peptide was separated by high pressure liquid chromatography (HPLC). I-125-Dyn was eluted isocratically on a reversed-phase RP C-18 column (Nucleosil 100-7C18, Macherey-Nagel, Düren, Germany) with a mobile phase consisting of 20 mmol/l aqueous solution of sodium dihydrogen phosphate/acetonitrile (3:2, v/v) at a flow rate of 1 ml/min.

Additionally, separation and quality control was performed using Sep-pak C-18 reversed-phase extraction cartridges (Millipore, Eschborn, Germany). The radiochemical purity of labelled Dyn was higher than 98 % and the radioiodinated peptide was stable over 24 h.

Cell culture and uptake experiments. The human glioma cells (U-118 MG) were maintained in L-15 medium (Boehringer Mannheim, Germany) supplemented with 50 ml fetal calf serum (Gibco, Eggenstein, Germany), 2.5 ml L-glutamine (200 mmol/l, Biochrom, Berlin, Germany) and 1 ml gentamicin (0.05 mg/ml) per 500 ml.

The cellular uptake of the labelled peptide was determined in 64 culture tubes. After addition of the I-125-labelled drug and Tl-201 – as a reference – in a quantity of 20 kBq/ml medium for each test, the cells were incubated from 20 min up to 4 h at 37 °C. Uptake was stopped by removing the medium. Subsequently, cells were washed 3 times with 10 ml cold saline solution and harvested with trypsin-EDTA (Biochrom, Berlin, Germany). Cellular I-125-Dyn and Tl-201 accumulation was measured in a gamma counter. Uptake was expressed as per cent of the activity applied normalized to one million cells. Results are given as mean ± standard deviation.

Results

The Iodogen method resulted in a high radiochemical yield of 74 ± 6 % of the whole activity added. The *ortho*-position to the hydroxyl in the phenolic ring of the tyrosine has by far the highest reactivity for electrophilic substitution in the peptide. After radiolabelling a single product, the monoiodinated dynorphin with tyrosine-bound iodine, is observed by HPLC analysis. The structure is given in Fig. 1. Instead of the HPLC-technique, separation is also possible by use of a simple Sep-Pak-C-18 purification.

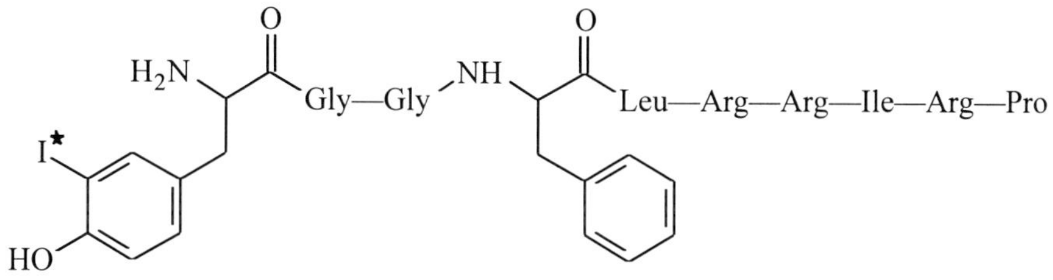

Fig. 1. Structure of radioiodinated dynorphin A (1-10).

The cellular uptake of the two tracer over the whole incubation period is shown in Fig. 2. The non-specific tracer Tl-201 reached a time independent value of 0.49 ± 0.09 % already after 20 min. In contrast, the accumulation of I-125 labelled dynorphin increased over the whole incubation period. After 90 min the cellular uptake exceeded that of Tl-201. At the end of the test, after 4 h, monoiodinated dynorphin reached an uptake of 0.86 ± 0.12 % beeing nearly twice as high as that of Tl-201.

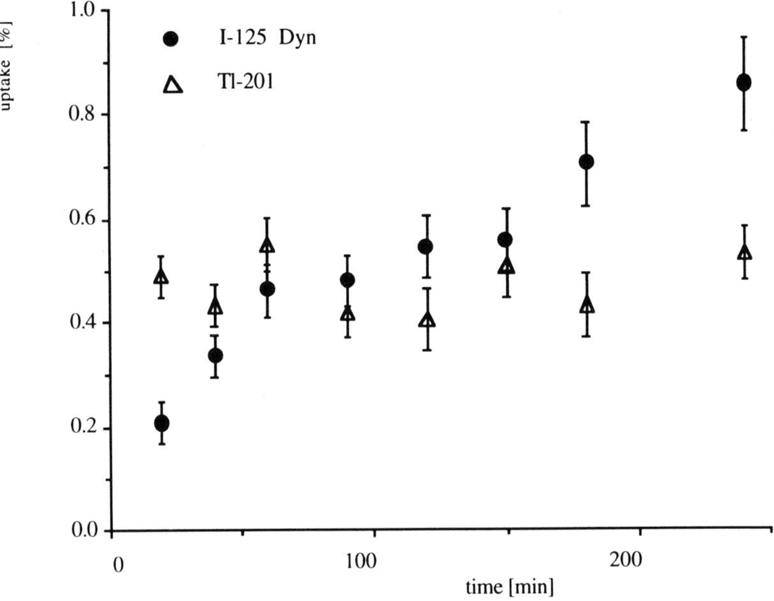

Fig. 2. Uptake of I-125-dynorphin and Tl-201 in glioma cells over time, symbols represent mean ± SD (n = 4 per incubation interval).

Discussion

If the peptide contains only one tyrosine, Sep-Pak purification of the labelling mixture has a similar efficiency as HPLC. The Sep-Pak separation technique provides two advantages: the enormously lower cost and the lower consumption of organic solvent over HPLC. Labelled dynorphin is now available in laboratories not equipped with expensive separation systems for further *in vitro* and *in vivo* studies. Our most interesting finding concerned the twice as high accumulation of iodinated dynorphin as compared to the well known tumour tracer Tl-201 after an incubation time of 4 h. Studying the uptake mechanism in detail, the existence of a receptor or a transporter has to be looked for. It is not known whether iodinated dynorphin is able to pass the blood-brain barrier. However, other oligopeptides, e.g. octreotide, cannot cross this barrier due to their hydrophilicity. Thus, *in vivo* scintigraphy studies of cerebral opioid receptors with labelled dynorphin might be rendered difficult.

Conclusion

In summary, the Iodogen method is best suited for the labelling of dynorphin, because it gives high radiochemical yields and the radioiodination of dynorphin leads to a single product. For separation of the labelled peptide from unbound iodine both, HPLC and Sep-Pak techniques can be used. In the *in vitro* study I-125 labelled dynorphin shows a higher uptake than the non-specific tracer Tl-201 in glioma cells. Furthermore, the continuous increase in uptake over time may be evidence for a specific binding process.

References

Breeman, W.A., Hofland, L.J., Bakker, W.H., Pluijm, M. van der, Koetsveld, P.M. van, Jong, M. de, Setyono-Han, B., Kwekkeboom, D.J., Visser, T.J., Lamberts, S.W.J. and Krenning, E.P. (1993) Radioiodinated somatostatin analogue RC-160: preparation, biological activity, in vivo application in rats and comparison with [I-123-Tyr-3]octreotide. *Eur. J. Nucl. Med.* 20: 1089-1094.

Fischman, A.J., Babich, J.W. and Strauss, H.W. (1993) A ticket to ride: peptide radiopharmaceuticals. *J. Nucl. Med.* 34: 2253-2263.

Fracker, P.J. and Speck, J.C. (1978) Protein and cell membrane iodination with a sparingly soluble chloramide 1,3,4,6-tetrachloro-3a-6a-diphenylglycoluril. *Biochem. Biophys. Res. Commun.* 80: 849-857.

Lambert, R., Willenbrock R., Tremblay J., Bavaria, G., Langlois, Y., Hogan, K., Tartaglia, D., Flanagan R.J. and Hamet, P. (1994) Receptor imaging with atrial natriuretic peptide, part 1: high specific activity iodine-123-atrial natriuretic peptide. *J. Nucl. Med.* 35: 628-637.

Lamberts, S.W.J., Bakker, W.H., Reubi, H.C. and Krenning, E.P. (1990) Somatostatin receptor scintigraphy in the localization of endocrine tumours. *N. Eng. J. Med.* 323: 1246-1249.

Thakur, M.L. (1995) Radiolabelled peptides: now and the future. *Nucl. Med. Comm.* 16: 724-732.

Virgolini, I., Raderer M., Kurtaran A. Angelberger, P., Banyai, S., Yang, Q., Li, S., Banyai, M., Pidlich, J., Niederle, B., Scheithauer, W. and Valent, P. (1994) Vasoactive intestinal peptide-receptor imaging for the localization of intestinal adenocarcinomas and endocrine tumours. *N. Eng. J. Med.* 331: 1116-1121.

Wolf, H., Marschall, F., Scheffold, N., Clausen, M., Schramm, M. and Henze, E. (1993) Iodine-123 labelling of atrial natriuretic peptide and its analogues: initial results. *Eur. J. Nucl. Med.* 20: 297-301.

Parathyroid hormone-related protein (PTHrP) - a paracrine factor in astrocytes and an autocrine factor in astrocytomas

A. Turzynski, G. Struckhoff[1], D. Colangelo, S. Guidotto, A. Bunge and M. Dietel

Institut für Pathologie der Charité, Humboldt-Universität, Schumannstraße 20/21, D-10117 Berlin, Germany
[1]*Anatomisches Institut der Christian-Albrechts-Universität, Olshausenstraße 40-60, D-24098 Kiel, Germany*

Summary. Parathyroid hormone-related protein (PTHrP), that has been identified as the main causative factor for the humoral hypercalcemia of malignancy, is nearly ubiquitiously expressed in tumours and normal tissues of various histogenesis. In normal tissues as well as in malignant conditions an auto- or paracrine function as growth and differentiation factor has been demonstrated. In cultured astrocytes of the rat brain we found an expression of the PTH/PTHrP receptor. Since normal astrocytes *in situ* and *in vitro* fail to express PTHrP by themselves, they presumably represent the physiological target for meningeal PTHrP via a paracrine mechanism. In normal astrocytes PTHrP induces an activation of adenyl cyclase accompanied by glial stellation, an effect possibly involved in the formation of the glial limiting membrane. Surprisingly, in the majority of the astrocytomas (grade II - grade IV, WHO-classification) PTHrP immunoreactivity can be detected. To test the biological significance of this observation we simultaneously performed the reverse transcription polymerase chain reaction for PTHrP and PTH/PTHrP receptor mRNA in three astrocytoma cell lines. In all three astrocytomas investigated the specific amplification products were detected, thus, indicating a possible autocrine function of PTHrP. In the monolayer proliferation assay the application of a monoclonal PTHrP-antibody against the receptor-binding N-terminus inhibited the proliferation of two astrocytoma cell lines, especially when they were seeded at low cell densities. Accordingly, in the clonogenic assay both cell lines showed a marked reduction in their ability to form clones. The data indicate a functional shift of PTHrP from a paracrine to an autocrine mode, occurring during the development of the astrocytomas. The simultaneous expression of PTHrP and its receptor and the effect on the proliferation *in vitro* substantiate its role as a growth factor in astrocytomas.

Introduction

Parathyroid hormone-related protein has been identified as the main causative factor for the humoral hypercalcemia of malignancy (Suva et al., 1987), the most frequent and life-threatening paraneoplastic syndrome. Due to its N-terminal sequence homology with parathyroid hormone (PTH) it binds with similar affinity to the recently cloned PTH/PTHrP-receptor (Jüppner et al., 1991; Abou-Samra et al., 1992). In contrast to PTH, PTHrP is widely distributed and can be detected - at least in certain developmental phases - in virtually every organ system. In peripheral tissues the effects of PTHrP are heterogeneous and are elicited in endo-, para- and autocrine modes (Turzynski et al., 1994, for review).

Currently, only little is known on its functions in the central nervous system. In brain homogenates PTHrP mRNA and the production of biologically active PTHrP immunoreactivity have been described (Weir et al., 1990). By immunohistochemistry this protein was only detected in the meninges and the choroid plexus (Kitazawa et al., 1991). Our present study is aimed to clarify its occurence and biological significance in normal brain and in astrocytomas.

Material and methods

Immunohistochemistry for PTHrP was performed in archival paraffin-embedded material of 34 astrocytomas using a polyclonal antibody against midregional PTHrP (anti-PTHLP, AB-2, Dianova, Hamburg) at 1 mg/ml and the avidin-biotin-peroxidase complex method for detection (Hsu et al., 1981). Controls with 1 mg/ml nonimmune rabbit IgG were included in every case.

Cell cultures of astrocytes and meningeal cells were performed as described previously (Struckhoff and Turzynski, 1995). The astrocytoma cell line 53/91 was established from an astrocytoma grade IV and maintained in Leibovitz L15 medium with 10% fetal calf serum. The astrocytoma cell lines GAMG and U 138-MG were received from the German collection of microorganisms and cell cultures (Braunschweig, Germany).

Isolation of RNA from cell cultures and subsequent reverse transcription-polymerase chain reaction (RT-PCR) were essentially performed according to our investigation in astrocytes and meningeal cells (Struckhoff and Turzynski, 1995). Due to the different species the following primer pairs were used: (i) human PTHrP 5'CT GAA ATC AGA GCT ACC TC (upstream) and 5'AGC TCC AGC GAC GTT GTG GA (downstream); (ii) human PTH/PTHrP receptor 5'AGT CTG AGG AGG ACA AGG AG (upstream) and 5'TTG ATG

TCG CTC ACA CAG TT (downstream). The annealing temperatures were 45°C and 56°C, respectively.

For proliferation assays of the astrocytoma cell lines, cells were seeded into microtiter plates (500 - 2000 cells/well) and treated with a monoclonal antibody against the N-terminus of PTHrP (0.2 - 10 mg/ml, clone 1D5, Quartett, Berlin, Germany) for four days. Proliferation was measured by the tetrazolium XTT-method according to the supplier's instruction (Boehringer, Mannheim, Germany).

For clonogenic assays each 250 cells were seeded into 25 cm^2 flasks. The cells were either untreated or treated with the N-terminal antibody (10 mg/ml). After 7 to 10 days of culture without medium change the cells were stained with toluidine blue and cell clones with more than 20 cells were counted.

DNA content was measured fluorometrically by intercalation of ethidiumbromide according to Karsten and Wollenberger (1976).

Results

In the central nervous system of the rat immunoreactivity against PTHrP is most conspicuous in the meninges. In accordance to Kitazawa et al. (1991), who investigated human brain, glial cells were not stained. In contrast, human astrocytomas (grade II - IV) showed a strong granular cytoplasmatic staining in 90 % of 34 investigated cases.

These findings *in situ* were paralleled by the immunocytochemical findings in cultured cells. Whereas no reaction product was found in cultured astrocytes, meningeal cells and the cultured glioma cell lines (GAMG, U 138-MG and 53/91) were strongly immunopositive (Fig. 1).

The unexpected expression of PTHrP in astrocytomas was confirmed on the mRNA level. By RT-PCR using a sequence-specific primer pair spanning a fragment of 306 bp mRNA for PTHrP was detected in all cell lines tested.

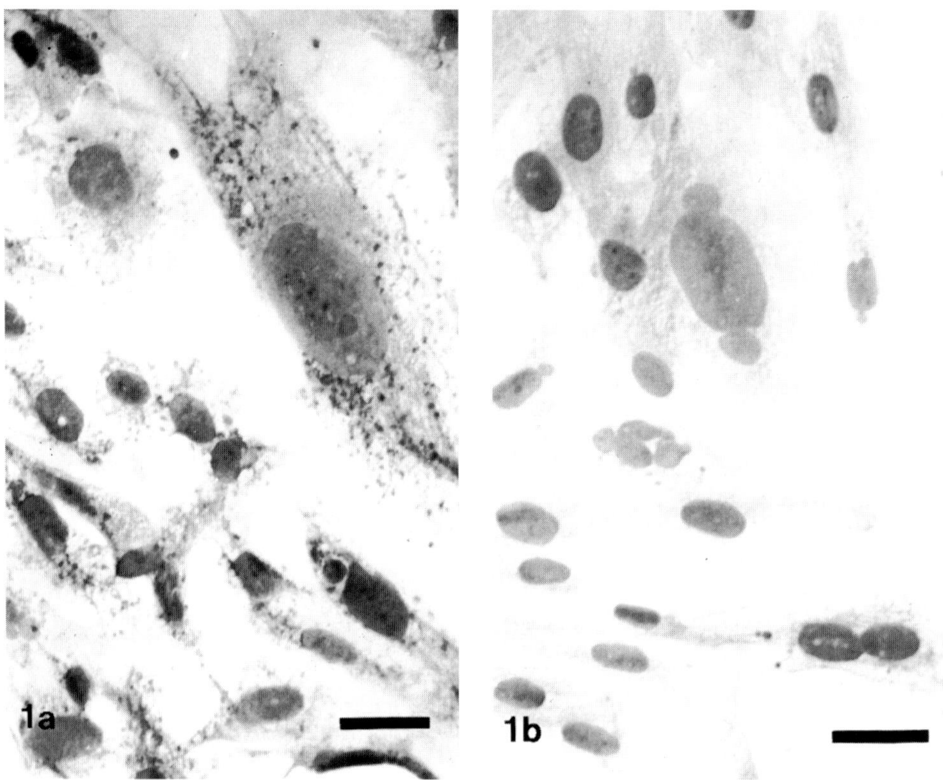

Fig. 1. PTHrP immunopositivity in the astrocytoma cell line U 138-MG demonstrated with a polyclonal antibody directed against midregional PTHrP. A granular cytoplasmatic staining is observed **(a)** which is not seen on controls with nonimmune rabbit IgG **(b)**. Nuclear counterstain with hematoxylin. Bars = 10 mm.

To identify possible targets of PTHrP under physiological and pathological conditions, all cultured cells were subjected to RT-PCR for the PTH/PTHrP receptor. For the rat cells a primer pair spanning a fragment of 566 bp was used whereas for the human astrocytomas the amplification product contained 247 bp. An amplification product was observed only with mRNA of astrocytes and astrocytomas, but not with mRNA of meningeal cells. The appropriate specificity controls were run by (i) RT-PCR for aldolase indicating integrity of the RNA preparations, (ii) omission of the reverse transcription excluding residual contamination of genomic DNA, and (iii) restriction analysis. Thus, astrocytes as well as astrocytomas express the mRNA for the PTH/PTHrP receptor.

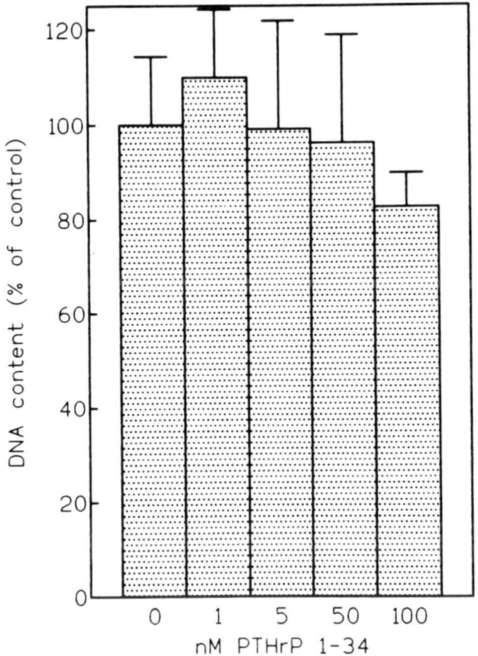

Fig. 2. Inhibition of astrocytic growth by PTHrP. 40 000 astrocytes were grown for 24 hours with the indicated concentrations of PTHrP. DNA content was measured fluorometrically by intercalation of ethidiumbromide. A slight, but significant inhibition is observed wih 100 nM PTHrP ($p<0.05$). Means and standard deviation of 4 experiments.

In astrocytes and astrocytoma cell lines the biological significance of this finding was tested. In astrocytes the treatment with PTHrP 1-34 elicited an 80-fold dose-dependent increase of intracellular cAMP levels. Concurrently astrocytes, which normally appear as epitheloid cells, develop cytoplasmatic processes within 15 minutes after the treatment. This differentiation is accompanied by a slight but significant growth inhibition. After 24 h of PTHrP treatment (100 nM) the DNA content is reduced to 80 % of untreated controls (Fig. 2).

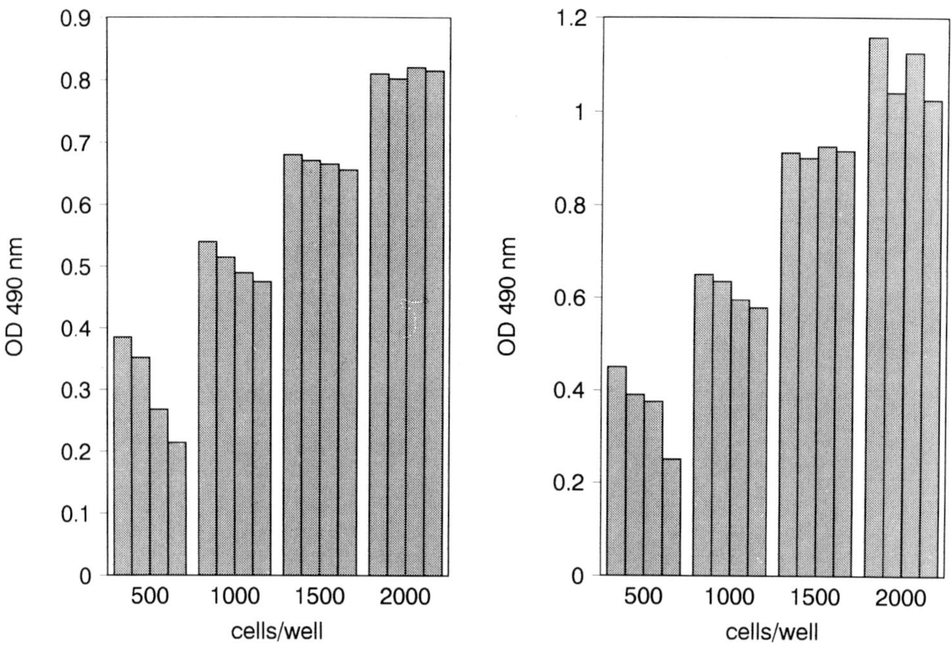

Fig. 3. Proliferation assay for two astrocytoma cell lines (**a**: GAMG, **b**: 53/91). Cells were seeded in different cell densities and cultured for four days in the presence of different concentrations of a neutralizing PTHrP antibody (from left to right within the groups 0; 0.4; 2; 10 mg/ml). A growth inhibition as measured by XTT is observed in both cell lines at a density of 500 cells/well (n=6). OD, optical density.

In the astrocytomas, that produce PTHrP, we investigated the effects after withdrawal of this factor. For this purpose a neutralizing monoclonal antibody against the receptor-binding domain of PTHrP was used. In monolayer proliferation assays the application of the anti-PTHrP antibody resulted in a 50% growth reduction after 4 days of culture (Fig. 3). The growth inhibiting effect was dependent on the cell density. It was pronounced at low cell densities (500 cells/well in a microtiter plate) and was completely abolished, when the cells were seeded at higher cell densities (2000 cells/well) in the beginning of the experiment. This growth effect is primarily due to the enhancement of the colony formation of the astrocytoma cells by PTHrP. In the GAMG and the 53/91 cell lines the neutralization of PTHrP reduced the colony forming ability to about 50% (Fig. 4). This effect can be demonstrated also with antisense oligonucleotides against PTHrP (Turzynski et al., unpublished).

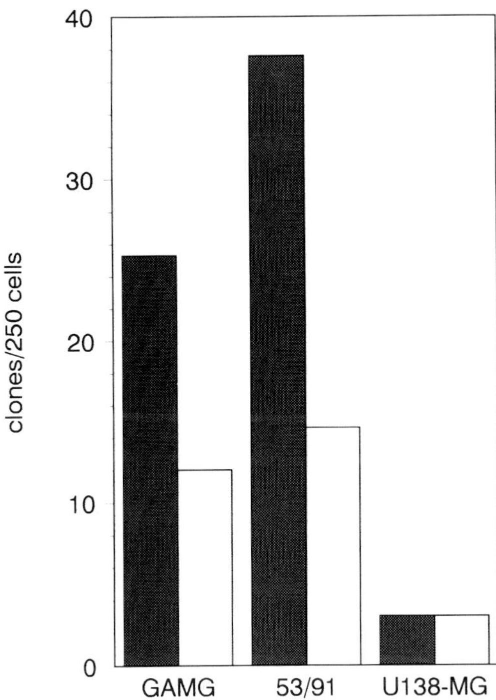

Fig. 4. Clonogenic assay in astrocytoma cell lines seeded at a density of 250 cells/25cm² flasks and cultured for one week. Dark bars: Controls; light bars: 10 mg/ml anti-PTHrP within the medium. Cell aggregates of more than 20 cells were counted as a clone.

Discussion

PTHrP, that has been identified as the main causative factor for hypercalcemia in the context of cancer disease, is widely accepted to function primarily as an auto-/paracrine factor that modulates cell functions, growth and differentiation. In the central nervous system PTHrP is expressed in the choroid plexus and meningeal cells. In accordance to previous studies (Hashimoto et al., 1994; Struckhoff and Turzynski, 1995) the PTH/PTHrP receptor can be demonstrated on astrocytes. Since the receptor mRNA is not expressed in meningeal cells, it can be assumed, that astrocytes are the primary targets for meningeal PTHrP. The development of astrocytic processes induced by PTHrP *in vitro*, might be a correlate for the formation of glial endfeet *in vivo*. Presumably PTHrP contributes to the formation of the glial limiting membrane via paracrine meningo-astrocytic loop.

By immunohistochemistry in astrocytomas a frequent expression of PTHrP is described here for the first time. Immunopositivity is cytoplasmatic and vesicular. Accordingly, PTHrP immunopositivity and PTHrP-mRNA was also found in three astrocytoma cell lines. The data presented here indicate that PTHrP is a functional factor in astrocytomas, that acts via an autocrine mode and promotes especially the colony formation. During tumour progression malignant cells infiltrate into the normal tissue adjacent to the tumour and loose the contact to the tumour mass. In this new microenviroment autocrine tumour promoting factors might initiate the settlement and proliferation of the tumour cells. Among a panel of other growth and differentiation factors PTHrP is a new factor, that gains importance in this tumour entity. However, it is worth noting, that PTHrP inhibits the proliferation of normal astrocytes but promotes colony formation in astrocytomas. The mechanism, which enables astrocytoma cells to produce PTHrP and which induces a functional shift of PTHrP from a paracrine factor in astrocytes to an autocrine factor in astrocytomas, remains to be clarified.

References

Abou-Samra, A.B, Jüppner, H., Force, T., Freeman, M.W., Kong, X.F., Schipani, E., Urena, P., Richards, J., Bonventre, J.P., Potts, J.T., Kronenberg, H.M. and Segre, G.V. (1992) Expression cloning of a common receptor for parathyroid hormone and parathyroid hormone-related peptide from rat osteoblast-like cells: A single receptor stimulates intracellular accumulation of both cAMP and inositol triphosphates and increases intracellular free calcium. *Proc. Natl. Acad. Sci. USA* 89: 2732-2736.

Hashimoto, H., Aino, H., Ogawa, N., Nagata, S. and Baba, A. (1994) Identification and characterization of parathyroid hormone/parathyroid hormone-related peptide receptor in cultured astrocytes. *Biochem. Biophys. Res. Commun.* 200: 1042-1048.

Hsu, S.-M., Raine, L. and Fanger, H. (1981) Use of avidin-biotin-peroxidase complex (ABC) in immunoperoxidase techniques: A comparison between ABC and unlabeled antibody (PAP) procedures. *J. Histochem. Cytochem.* 29: 577-580.

Jüppner, H., Abou-Samra, A. B., Freeman, M., Kong, X.F., Schipani, E., Richards, J., Kolakowski L.F., Hock, J., Potts, J.T., Kronenberg, H.M. and Segre, G.V. (1991) A G-protein-linked receptor for parathyroid hormone and parathyroid hormone-related peptide. *Science* 254: 1024-1026.

Karsten, U. and Wollenberger, A. (1976) Improvements in the ethidium bromide method for direct fluorometric estimation of DNA and RNA in cell and tissue homogenates. *Anal. Biochem.* 77: 471-477.

Kitazawa, S., Fukase, M., Kitazawa, R., Takenaka, A., Gotoh, A., Fujita, T. and Maeda, S. (1991) Immunologic evaluation of parathyroid hormone-related protein in human lung cancer and normal tissue with newly developed monoclonal antibody. *Cancer* 67: 984-989.

Mangin, M., Webb, A.C., Dreyer, B.E., Posillico, J.T., Ikeda, K., Weir, E.C., Stewart, A.F., Bander, N.H., Milstone, L., Barton, D.E., Francke, U. and Broadus, A.E. (1988) Identification of a cDNA encoding a parathyroid hormone- like peptide from a human tumour associated with humoral hypercalcemia of malignancy. *Proc. Natl. Acad. Sci. USA* 85: 597-601.

Struckhoff, G and Turzynski, A. (1995) Demonstration of parathyroid hormone-related protein in meninges and its receptor in astrocytes: evidence for a paracrine meningo-astrocytic loop. *Brain Res.* 676: 1-9.

Suva, L.J., Winslow, G.A., Wettenhall, R.E.H., Hammonds, R.G., Moseley, J.M., Diefenbach-Jagger, H., Rodda, C.P., Kemp, B.E., Rodriguez, H., Chen, E.Y., Hudson, P.J., Martin, T.J. and Wood, W.I. (1989) A parathyroid hormone-related protein implicated in malignant hypercalcemia: Cloning and expression. *Science* 237: 893-896.

Turzynski, A., Baumgart, S., Bauch, B. and Dietel, M. (1994) Morphological characteristics of tumours with humoral hypercalcemia of malignancy: Functional morphology of PTHrP. *In:* F. Raue (ed.): *Recent Results in Cancer Research:* Hypercalcemia of malignancy Vol. 137, Springer, Heidelberg, pp. 76-97.

Weir, E.C, Brines M.L., Ikeda, K., Burtis, W.J., Broadus, A.E. and Robbins, R.J. (1990) Parathyroid hormone-related peptide gene is expressed in the mammalian central nervous system. *Proc. Natl. Acad. Sci. USA* 87: 108-112.

AUTHOR INDEX

SUBJECT INDEX

A

ACTH, see adrenocorticotropin
adenylyl cyclase 126ff
adipokinetic hormones (AKH) 195, 203f
adrenocorticotropin (ACTH) 21, 33f, 48, 50, 52, 279, 310, 312ff
aging 305, 315ff, 323f
Alzheimer's disease 315ff, 323
aminopeptidase N 88f, 105, 109, 112, 115
aminopeptidase W 87f, 92
amygdala 28, 267f, 270ff
angiotensin, angiotensin II 88f, 97f, 113, 151ff
  -receptor 151, 153f,
anterior pituitary 34, 89, 106, 122
anticancer agents 121
astrocytes 109f, 112ff, 141ff, 343ff, 349
astrocytoma 343ff
AT$_1$ receptor subtype 151f
axonal transport 4

B

BARK (beta-adrenergic receptorkinase) 128
basal hypothalamus 259f, 262
behaviour 172, 267ff, 272, 274, 285f, 288, 306, 333
binding sites 111, 123, 141f, 145f, 149f, 157ff, 292
bradykinin 93, 109, 113, 115, 157ff, 225f, 233
Brattleboro rat 270
bullfrog 251ff, 255

C

calcium
  -channel 126ff
  -concentrations, intracellular 245
  -homeostasis 223
callatostatin 185-192
Calliphora vomitoria 185-192
cAMP, see cyclic AMP
CD10 309
cerebrospinal fluid 97, 288
cGMP, see cyclic GMP
cholecystokinin 30, 93, 185, 291f
choroid plexus 28, 31, 92f, 97, 344, 349
chromaffin cell 59, 224, 227ff, 301
circadian and circannual rhythms 316f
clonidine 239, 241f
cnidarians 39ff

CNP, see natriuretic peptides
cockroach 73ff, 165, 167
coexistence 151ff
cold shock 282
convertases 21, 23ff, 26f, 30ff, 35ff
core vesicles 9, 55, 57ff, 68, 81f, 84f
corpus cardiacum 10, 73f, 195
corticotropin-releasing hormone (CRH) 283, 311, 322ff
cross talk 223
cyclic AMP 126f, 129, 174, 199, 223, 226, 239-249, 347
cyclic GMP 171-174, 299ff

D

defective mutants 56
delta opioid receptor 307
deltorphin I 307
desensitization 121, 124, 128, 130
diabetes insipidus 270, 315f, 320ff
dipeptidyl aminopeptidase 39, 42
dorsal root ganglion 157-161
Drosophila 56, 61, 68, 167ff, 178, 209, 243
dynorphin 22, 308

E

earthworm 177, 180f
eclosion hormone 168-174
ectoenzyme 88ff, 92f, 103f, 112
Eisenia foetida 177ff, 181f
endopeptidase -24.11 305, 307, 309
endopeptidase 24.15 115
enkephalin 22, 90, 93, 95, 112f, 305f, 307, 309f
enkephalinase 87, 90, 309
enterochromaffin cells 13, 15ff
epinephrine 267, 269ff, 274
evolutionary pathway 211ff
excitability 165, 170f, 174, 232ff
exocytosis 5, 7, 9f, 15, 55, 59, 69, 76ff, 84ff, 181f, 202, 235

F

food intake 320, 324
functional imaging 333, 335

G

gene duplication 211, 215
gene expression 31, 36, 65, 168, 211, 279, 282ff

356

Prader-Willi-syndrome 315f, 320f
precursor processing enzymes 39
preoptic nucleus 3
processing 21f, 24, 27f, 32, 34-36, 39-46, 47ff,
    52ff, 94, 104, 203, 235, 267f, 271, 274, 291
prohormone theory 22
proopiomelanocortin 22, 27, 32ff, 47-53
proprotein convertase 21, 24
PVN, see paraventricular nucleus

R

radioiodination 337ff
rainbow trout 279f, 284
receptor 47, 51ff, 56, 103, 106, 115, 121-161,
    191, 223, 225f, 228, 231ff, 239f, 243-248,
    268, 271-274, 300, 305, 307ff, 320, 329, 331-
    335, 340, 343f, 346, 348f
    -subtypes 121, 123, 137, 141f, 144f, 147, 191
recombinant DNA 168
regulated secretory pathway 47ff, 51ff, 68
ryanodine receptor 225, 232f

S

salmon 63, 211f, 214ff
scanning electron microscopy 73f, 76ff
seasonal and circadian fluctuations 315
secretory granules 7, 9f, 16, 48, 50, 177, 181,
    195- 203
Segi's cap 16
serotonin 15
SNARE hypothesis 57
sodium 124, 135, 138, 226f, 245ff
    -channel 227, 245ff
    -regulation 124, 138
somatostatin 109, 112f, 121-149, 285, 329,
    331ff
    -receptor 121-149, 329, 333, 335
    -receptor scintigraphy 329, 331ff
SON, see supraoptic nucleus
sorting receptor 47, 51f
sorting signal 47-53
SSTR 3 135
stress 267, 270, 272, 274, 279f, 282f
substance P 93, 95ff, 109, 113, 205ff, 306
substantia nigra 29, 93, 98
suprachiasmatic nucleus 29, 315f
supraoptic nucleus (SON) 4, 29, 151, 153, 315,
    319, 320, 322
synaptic vesicles 5, 9, 10, 15, 55-61, 68, 81
synaptotagmin 57ff, 68
syntaxin 55-60, 67ff

T

Tenebrio molitor 205
thiouracil 285, 287f
transmembrane signalling 223f, 233
transmitter 7, 9, 57, 68, 85, 130
TRH 88f, 103ff
    -degrading enzyme 89, 103ff
    -receptor 103, 106

V

VAMP 56-61, 68
varicosity 81, 84
vasopressin 32, 113, 154, 211ff, 216f, 267-274,
    285, 315, 317, 320-324
vasotocin 211f, 214-217, 251-255
vasotocin gene 211, 214ff, 251f, 255
vesicle
    -docking 55, 57f, 61
    -fusion 58
    -trafficking 55ff, 61, 68f

Somesthesis and the Neurobiology of the Somatosensory Cortex

Edited by
O. Franzén, *Biomedical Center, Uppsala University, Sweden*
R. Johansson, *Umeå University, Sweden*
L. Terenius, *Karolinska Institute, Stockholm, Sweden*

This volume is a compilation of current research on somatosensation and its underlying mechanisms written by international experts from a broad range of disciplines. It is divided into six sections:

- **structural basis of information processing and neocortical neurotransmitters**
- **psychophysics of somatosensation**
- **cortical representation of somatosensation**
- **sensory-motor interface**
- **neuronal population behavior,**
- **cortical neurocomputation and modelling.**

It highlights not only important new findings but also novel methods and technologies applied to major unresolved issues in the field of neuroscience. The number of methods for investigating the neural mechanisms of somotosensory perception has grown substantially in the last decade. The book encompasses levels of inquiry from ionic channels, single unit recordings of neural activity, and functional brain imaging of the coordinated activity of large neuronal ensembles to human psychophysics of controlled somatic stimulation.

This work is of great value for researchers and students interested in the dynamic neuronal mechanisms involved in the complex processes of sensory perception and provides a picture of our present understanding of the neural representation of the external world relayed through the somatosensory system.

1996. 448 pages. Hardcover
ISBN 3-7643-5322-8

Birkhäuser Verlag • Basel • Boston • Berlin

D. Siemen, *Universität Regensburg, Germany*
J. Hescheler, *Freie Universität Berlin, Germany (Eds)*

Nonselective Cation Channels

Pharmacology, Physiology and Biophysics

In 1981, Neher and Sakmann published a pioneering paper describing the patch clamp method for measuring the currents flowing through individual ion channels. Apart from channels which are selective for a single species of ion, numerous "nonselective channels" have been found since then, which are activated by various agonists and differ in selectivity and amino acid sequence. Today it is widely acknowledged that these channels belong to the most important components of the cell membrane and fulfil a wide spectrum of different functions.

Nonselective Cation Channels is the first book to report on the immense variety and diversity of nonspecific ion channels, ranging from the nicotinic acetylcholine receptor to the gap junction channel. Written by recognized international experts, its coverage also includes the role of nonselective cation channels in cardiac and smooth muscle cell functions, and their importance in human platelets and endothelial cells. Sections on receptor-activated and mechanically sensitive cation channels are also included.

Primarily intended for scientists in basic biomedical and biological research, including physiology/electrophysiology, neurophysiology, pharmacology, toxicology, biochemistry, biophysics, neurobiology, cell biology, molecular biology and zoology, this book will also be of interest to clinicians as well as graduate and postgraduate students.

1993. 304 pages. Hardcover.
ISBN 3-7643-2888-6 (EXS 66)

Birkhäuser Verlag • Basel • Boston • Berlin

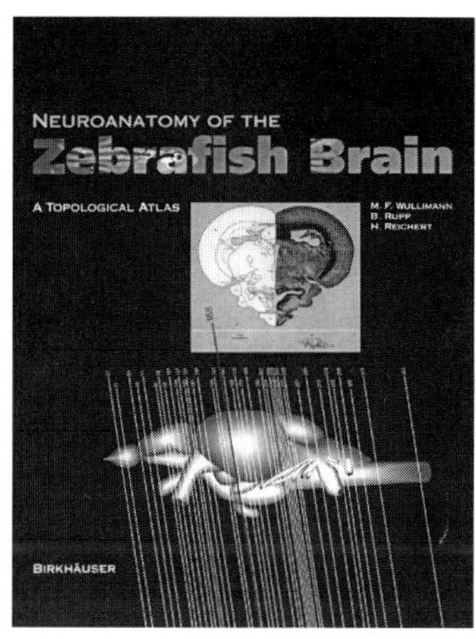

Neuroanatomy of the Zebrafish Brain

A Topological Atlas

Edited by
M.F. Wullimann, *University of Bremen, Germany*
B. Rupp, H. Reichert, *University of Basel, Switzerland*

1996. 160 pages. Hardcover
ISBN 3-7643-5120-9

The zebrafish, *Danio (Brachydanio) rerio*, is rapidly becoming one of the major model systems for the study of nervous system development and function at the molecular, cellular and systems level. This is due in large part to the extensive and ongoing maturation of sophisticated genetic manipulation and mutant generation techniques. Although considerable knowledge concerning the embryonic development of the central nervous system in this animal has been gained in recent years, there is a conspicuous lack of information concerning the structural organization of the adult brain. The present work therefore aims at providing a neuroanatomical overview of the zebrafish brain and identifies the major subdivisions, nuclei and tracts within the adult brain, based mainly on a comparison with the well-known brain of the closely related goldfish, *Carassius auratus*.

In the course of this investigation, more than 30 complete series of 12μm Bodian stained cross, sagittal and horizontal sections were produced and analyzed. Here an extensive selection of sections out of the three best series is presented. Each figure combines photographs and schematic drawings to allow rapid orientation and identification of specific structures. The discussion provides a source of information about the major subdivisions and eventual fiber connections within the zebrafish brain, as well as remarks about their functional significance.